Jürgen Gailer, Raymond J. Turner (Eds.)
Environmental and Biochemical Toxicology

Also of interest

Environmental Toxicology
Edited by Luis M. Botana, 2018
ISBN 978-3-11-044203-8, e-ISBN (PDF) 978-3-11-044204-5,
e-ISBN (EPUB) 978-3-11-043363-0

Hazardous Substances.
Risks and Regulations
Thomas Schupp, 2020
ISBN 978-3-11-061805-1, e-ISBN (PDF) 978-3-11-061895-2,
e-ISBN (EPUB) 978-3-11-061979-9

Plastics: The Environmental Issue
Oliver Türk, 2022
ISBN 978-3-11-064139-4, e-ISBN (PDF) 978-3-11-064143-1,
e-ISBN (EPUB) 978-3-11-064154-7

BioChar.
Applications for Bioremediation of Contaminated Systems
Edited by Riti Thapar Kapoor, Maulin P. Shah, 2022
ISBN 978-3-11-073858-2, e-ISBN (PDF) 978-3-11-073400-3,
e-ISBN (EPUB) 978-3-11-073406-5

Emerging Contaminants.
Remediation Technologies
Edited by Jeyaseelan Aravind, Murugesan Kamaraj, 2022
ISBN 978-3-11-075158-1, e-ISBN (PDF) 978-3-11-075172-7,
e-ISBN (EPUB) 978-3-11-075178-9

Environmental and Biochemical Toxicology

Concepts, Case Studies and Challenges

Edited by
Jürgen Gailer, Raymond J. Turner

DE GRUYTER

Authors
Jürgen Gailer
Department of Chemistry,
University of Calgary,
Calgary,
Alberta, Canada
jgailer@ucalgary.ca

Raymond J. Turner
Department of Biological Sciences,
University of Calgary,
Calgary,
Alberta, Canada
turnerr@ucalgary.ca

ISBN 978-3-11-062624-7
e-ISBN (PDF) 978-3-11-062628-5
e-ISBN (EPUB) 978-3-11-062634-6

Library of Congress Control Number: 2022931475

Bibliographic information published by the Deutsche Nationalbibliothek
The Deutsche Nationalbibliothek lists this publication in the Deutsche Nationalbibliografie;
detailed bibliographic data are available on the Internet at http://dnb.dnb.de.

© 2022 Walter de Gruyter GmbH, Berlin/Boston
Cover image: Gettyimages/Hramovnick
Typesetting: Integra Software Services Pvt. Ltd.
Printing and binding: CPI books GmbH, Leck

www.degruyter.com

We dedicate this textbook to the memory of Dr. Robie W. Macdonald

Dr. Robie Macdonald sharing a conversation with the then PhD student Alex Hare during an Arctic expedition aboard the Canadian Icebreaker *CCGS Amundsen* in 2005 (photo by Dr. Zou Zou Kuzyk).

The authors would like to dedicate this book to the memory of Dr. Robie Wilton Macdonald (1947–2022). Robie co-authored a chapter of this textbook and sadly passed away during the final production.

Robie was one of the intellectual giants in environmental, marine and Arctic sciences. With a PhD in physical chemistry (1972, Dalhousie University), Robie brought the rigour of chemical thermodynamics and kinetics to the study of biogeochemical pathways in complex and dynamic aquatic systems. By doing so, Robie laid the foundation for our contemporary understanding of the cycling of freshwater, organic carbon and chemical contaminants, and how each of these may be affected by climate variability and change. His study regions covered the Eastern Pacific and the Arctic Ocean, and rivers and lakes in western and northern Canada. His ground-breaking work in the cycling of organochlorine and mercury contaminants in the Arctic Ocean played a crucial role in the identification of their effects on marine ecosystems and Inuit and other Indigenous Peoples, and in the negotiation of international treaties such as the Stockholm and Minamata Conventions.

During his half-century career, Robie authored or co-authored over 280 papers in peer-reviewed journals, numerous reports and book chapters. He was a recipient of many prestigious awards and recognitions, including the Polar Medal, the Northern Science Award and Centenary Medal of the Canadian Polar Commission, the Royal Canadian Geographical Society Gold Medal, the Miroslaw Romanowski Medal of the Royal Society of Canada, the Head of the Public Service Award for Excellence in Policy, and the Presidents Prize of the Canadian Meteorological Society. Robie was a Fellow of the Chemical Institute of Canada, the Royal Society of Canada, the

https://doi.org/10.1515/9783110626285-202

American Geophysical Union, the Royal Canadian Geographical Society and the Explorers Club. Robie was appointed an Officer of the Order of Canada in 2019.

While spending most of his career with Fisheries and Oceans Canada, Robie managed to collaborate with, mentor and promote an extraordinary number of university-based researchers and graduate students. One of the authors of this book had the privilege to work with Robie for over two decades, and will forever miss him for his keen curiosity, subtle sense of humour, infectious story-telling, sharp wit and wisdom, and unwavering guidance and friendship.

When asked to contribute to this book, Robie said he usually avoided doing book chapters because he felt like they tend to become "lost souls". After learning it was to serve as a textbook, he gladly obliged and went full steam despite poor health. It is the authors' sincere hope that instead of becoming a lost soul itself, this book would inspire souls because we know Robie would have approved of that. (Text by Feiyue Wang, University of Manitoba).

Preface

The idea to compile this book is closely linked to a new course that was offered by the University of Calgary for the first time in the winter term of 2016. This course, in turn, came into existence because of critical conversations between JG and RT about the way that many societies around the world deal with toxins that are inadvertently or deliberately released into the global environment. To this end, mankind has effectively become a global biogeochemical force (e.g. mankind has already been changing the chemical composition of the earth's atmosphere), which prompted a Nobel laureate to refer to the age that we currently live in as the 'anthropocene' (*anthropos* = human; *cene* = geological epoch). This perspective or realization challenges the mantra that "the solution to pollution is dilution" and prompted us to develop a new interdisciplinary senior-level course which was initially entitled 'Biochemical Toxicology' and eventually changed to 'Concepts in Biochemical Toxicology'. The overall goal of this course was and is to focus on the diverse biomolecular mechanisms by which natural toxins that are produced by organisms, as well as organic and inorganic pollutants, adversely affect cell function at the single-cell level (microbes), at the multicellular level (animals and humans) as well as the organism–environment interface (entire ecosystems). Owing to the required expertise at the interface between chemistry and biology, the delivery of this interdisciplinary new course involved faculty from the Department of Chemistry and from Biological Sciences. The book draws from the lessons that we have learned by organizing and delivering this course which involved focusing on chemical events that unfold immediately after a toxin is released into the environment to capture the student's attention early and to provide important context to topics that we covered. Thus, this book is not intended to be a thorough introduction to biochemical toxicology since there are a few recent textbooks that have this flavour. Rather, this book should be of interest for students who are curious to uncover the (bio)chemical basis and mechanisms that are at the heart of the societal problems that are intricately linked to the economy and the associated improper management and disposition of organic and inorganic toxins. Since the first target of any toxin release is the immediate area that we live in, any toxin release has the potential to damage the very ecosystem that supports us, which will then propagate up the food chain. We should therefore not be surprised to learn that pollution is the largest environmental cause of disease and premature death in the world today. Although it is estimated that 9 million premature deaths are attributed to pollution-related diseases in 2015, it is what we don't know in terms of how chemical pollution adversely affects human health (Landrigan, Fuller et al. 2017) in the form of 'ticking time-bombs' that is particularly worrisome. We have to recognize that we are just now embarking on the journey to unravel the complex mechanisms by which pollutants first result in subtle changes to our local ecological habitat, which may eventually result in adverse human health effects. Given the need to find new ways

https://doi.org/10.1515/9783110626285-203

to experimentally address this problem, this textbook serves as a starting point for senior students to further explore this challenging frontier.

Landrigan, P. J., et al. (2018). 'The Lancet commission on pollution and health.' The Lancet. 391, 462–512.

Contents

Overview and author contact information

1. Introduction and textbook scope

Jürgen Gailer*
Department of Chemistry, University of Calgary, 2500 University Drive NW, Calgary, AB, Canada,
e-mail: jgailer@ucalgary.ca
Raymond J. Turner*
Department of Biological Sciences, University of Calgary, Calgary, Alberta, Canada,
e-mail: turnerr@ucalgary.ca

This chapter provides the reader with an overview of the scope of this textbook as well as the multifactorial issues that currently limit our understanding of the biomolecular mechanisms that are inherently associated with the chronic exposure of microbes, fish and humans to multiple pollutants over their lifetime. It is important to recognize that this complex problem is intricately associated with the nexus that exists between drinking water, food and resource extraction/utilization and has a local, regional and global dimension as 9 million people died in 2015 of environmental pollution-related causes. Tackling the pollutant exposure-adverse effects problem from a regulatory point of view is hampered on the one hand by temporal dynamic changes of pollutant concentrations in the environment (i.e. in drinking water, food and air) and by the complexity of biological organisms on the other. Thus making progress in addressing this multifactorial problem requires a clear focus on what organism and which pollutants one should focus on, to identify which exposure pathways are most relevant and how to unravel the biochemical mechanisms that unfold within the organism(s) involved. The reader will be made aware that important issues that need to be resolved pertain to environmental monitoring/sampling, the accurate quantification of pollutants (e.g. nanoparticles are exceedingly difficult to quantify in complex environmental matrices) and the related challenge of choosing an appropriate model biological system to study the effect of a certain class of pollutants.

2. Overview of biochemical toxicology principles

Andrii Lekhan and Raymond J. Turner*
Department of Biological Sciences, University of Calgary, Calgary, Alberta, Canada,
e-mail: turnerr@ucalgary.ca

This chapter will provide an overview of the biochemistry in the context of targets of toxins. It is presented as a primer for students from diverse multidisciplinary audience. The goal will be to remind students of some central cell biology and system biology. It will be used to make students familiar with what the biochemical targets of toxins are and what fundamentally makes a toxin a toxin. There are sections on: lipids and lipid bilayer biochemistry and how compounds affect its integrity; proteins and a reminder to enzymology inhibition kinetics; and nucleic acids and sites on DNA where toxins could interact. Additionally, this chapter will cover some basic

https://doi.org/10.1515/9783110626285-205

toxicology principles including toxin kill curves and terminology, Lipinski's rules, etc. As for toxic metals, there will be a short section pointing out speciation (reminder of Pourbaix diagram) and the relationship of metal atom characteristics (polarizability, electronegativity, reduction potential, etc.), including hard-soft acid-base theory. The goal of this chapter is to provide brief background reminders to these topics so the student is primed for the details that will be dealt with in subsequent chapters. References are chosen to point the student towards specific readings to acquire more information on each topic.

3. Bacterial response to toxins

Raymond J. Turner[*]
Department of Biological Sciences, University of Calgary, Calgary, Alberta, Canada,
e-mail: turnerr@ucalgary.ca

This chapter will explore microbes response to (i) organic toxins (ii) toxic metal ions and (iii) antibiotics. A brief commentary of microbial communities (Biofilms and Microbiomes), aquatic periphyton and food chain system ecology will be included. Microbes are often the first to be exposed to pollutants in the environment, whether through their presence in soil, aquatic or marine systems. The effect of different groups of common organic toxins on bacteria will be discussed. The concept of bioremediation will be introduced where the immense metabolic potential of bacteria that make up >95% of the species on the planet can help degrade various natural and xenobiotic organic pollutants. Microbes have evolved with metal ions in their environment and have biochemical processes in place to manage their stress at natural levels. However, high concentration pulse challenges through anthropogenic activities such as exposure to mine leachates and tailings releases leave bacteria challenged to metal ion stressors. Additionally, metals as metal salts, alloys or nanomaterials are increasingly being used as antimicrobials in infection control for humans and in agriculture. An overview of the biochemical mechanisms in bacteria of metal ion toxicity, tolerance and resistance is included. Finally, to a bacterium cell antibiotics/antiseptics/disinfectants are toxins. Thus, a brief overview of the cellular targets of antibiotics will be described as well as mechanisms of resistance and the antimicrobial resistance problem.

4. Toxic metal(loid) species at the blood–organ interface

Maryam Doroudian and Jürgen Gailer[*]
Department of Chemistry, University of Calgary, 2500 University Drive NW,
Calgary, AB, Canada, e-mail: jgailer@ucalgary.ca

Environmental pollution globally caused an estimated 9 million deaths in 2015. Related to this, past poisoning epidemics have revealed that the chronic exposure of human populations to exceedingly small daily doses of arsenic, cadmium and mercury species can – over time – severely affect human health. Today, several potentially toxic metals and metalloids have been accurately quantified in the bloodstream of the average population, but what these numbers mean from a public health point of view remains unclear. Considering that the biomolecular origin of many neurodegenerative diseases (e. g. Parkinson's disease) remains unknown, the question arises as to whether these seemingly unrelated facts may be connected. To answer this fundamental question will require – in our opinion – a much better understanding of the bioinorganic chemistry of potentially toxic metal/metalloid species that unfolds in the bloodstream. Conceptually, these processes determine which and how much of a toxic metal(loid) species and/or metabolite will impinge on toxicological target organs. While one goal from this overall research effort is to gain much-needed insight into the toxicology of metal(loid)s in the bloodstream, the discovery of blood-based detoxification mechanisms for toxic metal(loid) species can also serve as a starting point for developing potential treatments to ameliorate their adverse health effects in affected populations. Since the global emission of potentially toxic metals and metalloid compounds will increase due to the fast-growing technological advancements, expanding the knowledge of their biochemistry at the blood–organ interface is critical before more stringent measures can be implemented to reduce their emission into the global environment to mitigate the magnitude that environmental pollution has on human health.

5. Structural and chemical aspects of the molecular toxicology of heavy metals and metalloids

Graham N. George[*],[a,b,c] Ben Huntsman,[a] Olena Ponomarenko,[a] Emérita Mendoza Rengifo,[a] Monica Y. Weng,[a] Julien, J. H. Cotelesage,[a] Natalia V. Dolgova[e] and Ingrid J. Pickering[a,b,c]

a. Molecular and Environmental Sciences Group, Department of Geological Sciences, University of Saskatchewan, 114 Science Place, Saskatoon, Saskatchewan S7N 5E2, Canada
b. Toxicology Centre, University of Saskatchewan, Saskatoon, Saskatchewan S7N 5B3, Canada
c. Department of Chemistry, University of Saskatchewan, Saskatoon, Saskatchewan S7N 5C9, Canada
d. Calibr – California Institute for Biomedical Research, Scripps Research, La Jolla, California 92037, USA, e-mail: g.george@usask.ca

The chemical compounds of heavy metals and metalloids can be among the most toxic species to which humans are commonly exposed. The toxic properties of some of these compounds have been known since antiquity. The medieval physician Paracelsus (1493–1541) is acknowledged as the father of modern toxicology, and amongst his famous quotation can be translated as 'only dose makes the poison'. We now understand that Paracelsus was only partly correct; it is not only the dose, but it is also the molecular form in which the metal or metalloid is presented that makes the poison. Indeed, the molecular form, or chemical speciation, of an element controls whether it is toxic, benign or beneficial. This chapter will review the scientific toolbox that the modern molecular toxicologist can employ, ranging from computational chemistry and quantum mechanical approaches to understanding structure and chemistry to advanced *in situ* probes such synchrotron X-ray spectroscopy and imaging. Finally, a number of specific examples illustrating how these methods give toxicological insight will be presented.

6. Using isotopic abundances to follow anthropogenic emissions: an example of sulfur, oxygen and boron isotopes in a Canadian watershed

A. L. Norman[*], J. Xie, C. Kruschel and M. Wieser[*]
Department of Physics and Astronomy, University of Calgary, 2500 University Drive NW, Calgary, AB, T2N 1N4, Canada, e-mail: alnorman@ucalgary.ca; e-mail: mwieser@ucalgary.ca

Sulfur and boron isotope abundance data were applied as natural tracers to determine the origin of contamination in a study of the Castle River watershed in southern Alberta. The combination of sulfur and boron are sensitive indicators since their isotope composition is characteristic of local industrial emissions which are distinct from the background. Sulfate and boron concentrations as well as the $\delta^{34}S$ values in the Castle River decreased downstream and are consistent with Castle River water that contains contributions from long-term atmospheric sulfur deposition from nearby industrial emissions closer to the headwaters. As distance from headwaters and watershed areas both increase, air pollutant deposition is diluted by naturally occurring sulfate and/or deposition from vehicle and/or volcanic emissions. Boron isotopic compositions showed similar behaviour with an increase in a terrestrial, rather than atmospheric signature downstream. The strong relationship between boron and sulfate isotopes during baseflow conditions in fall and winter months suggests industrial emissions impacts in the upper headwaters region are diluted as additional water feeds into the Castle River downstream.

7. The role of metalloids (As, Sb) in airborne particulate matter related to air pollution

Daniel Sánchez-Rodas[*], Ana M. Sánchez de la Campa and María Millán Martínez
Center for Research in Sustainable Chemistry-CIQSO, University of Huelva, Spain, e-mail: rodas@uhu.es

Air pollution affects human health worldwide. In addition to gaseous contaminants, particulate matter (PM) is also of especial concern as particles are inhaled, reaching the lungs, alveoli and even the blood stream, depending on their size. Metalloids of great toxic concern such as arsenic (As), but also antimony (Sb) to a minor extent, can be found in PM mainly due to anthropogenic activities. Legislation has been established in some countries (e.g. European Union, China) for arsenic concentrations in air. Sampling of particles with diameter equal or lower than 10 (PM10) or 2.5 microns (PM2.5) is accomplished by pumping air through filters, followed by a chemical treatment (e.g. acid digestion) for the determination of As and Sb by spectroscopy and/

or mass spectrometry. Speciation analysis of As and Sb present in PM can be accomplished by selective extraction of the individual species followed by separation techniques (e.g. liquid chromatography) prior to their detection. A case study is provided about the determination of As in an urban area located near an important As emission source, namely a copper smelter, indicating how monitoring over a long period of time (>15 years) can provide valuable information about the evolution of air quality in relation to industrial activity and the implementation of emission abatement techniques. Also, a case study is presented in relation to Sb in PM, indicating how the speciation analysis of the different oxidations states of Sb can be employed to identify different emission sources, such as traffic or the metallurgy industry.

8. Toxic trace metals in the environment, a study of water pollution

Eve Kroukamp[*] and Victor Wepener
PerkinElmer Inc., Canada; North-West University, South Africa,
e-mail: Eve.Kroukamp@PerkinElmer.com; e-mail: evekroukamp@gmail.com

The concentrations of elements in natural waters can vary greatly depending upon the underlying geology and land surrounding the catchment area. Due to increasing pollution and land disturbance from anthropogenic activities, vast quantities of potentially harmful elements can enter natural waters. Once in the water system, these metal(loid)s are considered mobile and interact with the physical and chemical parameters of the receiving environment, undergoing a number of transformations which make them more or less available for biological uptake. This chapter will explore (i) the sources of metals in the environment; (ii) physicochemical interactions of metals with the receiving environment and how these affect the mobility, bioavailability and ultimate fate of these contaminants; (iii) the ecological impact of metals in the environment with case studies of the effects of metal pollution on single species and whole ecosystems; and (iv) strategies towards treating industrial waste, the mitigation of the entry of these pollutants into the environment and the remediation of impacted environments.

9. Toxicity of nanomaterials

Raymond J. Turner[*]
Department of Biological Sciences, University of Calgary, Calgary, Alberta, Canada,
e-mail: turnerr@ucalgary.ca

Nanoscience and nanotechnology are now impacting essentially every industry and modern technologies. In the industrialized world, nanomaterials cannot be avoided anymore. Nanomaterials have unique and exceptional properties compared to their equivalent macro-sized materials. This chapter introduces nanomaterials, their types and uses. The reader will become exposed to the physicochemical parameters that influence nanomaterial's special properties. Synthesis methods will be introduced so the reader will realize that not only the nanomaterial may be hazardous, but the synthesis can also be harmful. We also consider the lifecycle of use of nanomaterials from synthesis to the manufacturing of nanomaterial enhanced products, to their disposal and decomposition. The parameters and properties influencing their potential toxicity are discussed followed by highlighting specific nanomaterials that are of concern. Finally, the chapter provides an overview of analytical methods that can be employed to monitor nanomaterials in various environmental niches. The field of nanomaterials and nanoscience is now quite extensive, yet in this chapter it is aimed to expose the reader to the fundamental and salient points to be considered around their use and toxicity to biological systems.

10. Toxicology of trace metals in the environment: a current perspective

Som Niyogi[a]*, Kamran Shekh[b] and Solomon Amuno[c]
a. Department of Biology and Toxicology Centre, University of Saskatchewan, Saskatoon, SK, Canada, e-mail: som.niyogi@usask.ca
b. Yordas Group, Hamilton, ON, Canada
c. School of Environment and Sustainability, University of Saskatchewan, Saskatoon, SK, Canada

Trace metals are persistent environmental contaminants, and their ubiquity poses major threats to ecological and human health. While the toxicity of trace metals often varies among organisms, non-essential metals are generally more toxic than essential trace metals. Nevertheless, some essential trace metals (e.g. Cu, Se) can also cause adverse health effects in organisms when their concentration exceeds physiological thresholds. To evaluate the adverse consequences of metal pollution in the environment, a holistic understanding of various interconnected factors that strongly influence the toxicity of metals is required. With that in mind, this chapter discusses: (i) metal bioavailability in the environment; (ii) epithelial transport processes involved in the uptake and accumulation of metals via different exposure routes; and (iii) modes of toxic action of metals – key molecular and cellular events that lead to the adverse effects at the whole organism level. The chapter also discusses the usefulness of biochemical and histopathological biomarkers in assessing the population levels effects of metals in both aquatic and terrestrial ecosystems using a case study that investigated the impacts of industrial metal pollution in the Canadian sub-arctic. Finally, this chapter provides a review of the current regulatory approaches which are employed for the environmental risk assessment of metals in North America and the European Union.

11. Socioeconomic, political, and legal ramifications of environmental and biochemical toxicology: the complicated story of mercury

Feiyue Wang[a*] and Robie W. Macdonald[b]
a. Centre for Earth Observation Science, University of Manitoba, Winnipeg, MB, Canada
e-mail: feiyue.wang@umanitoba.ca
b. Institute of Ocean Sciences, Department of Fisheries and Oceans, Sydney, BC, Canada

Mercury (Hg) is among the first elements discovered and used by humans. Our relationship with Hg over the past four millennia has been an exceptional story evolving from an assumption of mythological properties that benefit medicine to widespread applications in industry. During the last century, we have become aware of the dangers posed by the casual treatment of Hg as witnessed by local catastrophes, like Minamata disease, and most recently that the widespread release of this 'element without borders' poses a global threat. In particular, methylmercury (MeHg), which is formed in the receiving environment and is especially prone to enter aquatic foodwebs, is a potent developmental neurotoxin to humans. Recognition that long-range transport of Hg, which may lead to severe adverse biological effects in distant locations, led to a legally binding global treaty, the Minamata Convention, which entered into force in 2017. Definitive scientific evidence indicating the need to reduce global Hg emissions came from the least expected region: the Arctic. In this chapter, we follow the development of the Hg story to examine socioeconomic, political, and legal ramifications of environmental and biochemical toxicology. We show how a fugitive element like Hg can lead to a global problem threatening food safety and Indigenous ways of living in Arctic communities, and how a solution can only be found by coordinated global political action not just on Hg but on climate change.

12. Suggested problems, case studies, assignments

Jürgen Gailer[*]
Department of Chemistry, University of Calgary, Calgary, 2500 University Drive NW, AB, Canada, e-mail: jgailer@ucalgary.ca
Raymond J. Turner[*]
Department of Biological Sciences, University of Calgary, Calgary, Alberta, Canada, e-mail: turnerr@ucalgary.ca

Jürgen Gailer and Raymond J. Turner

Chapter 1
Introduction and textbook scope

1.1 Foreword

Here we present the scope of this textbook and important commentary about important issues that are associated with pollutants and the associated toxicological problems. This chapter will provide the learner with insights into the issues and challenges around various toxin types and should prepare the reader towards thinking differently about the toxicological problems of the past, those of the present and those of the future.

1.2 Scope

This book is intended to be of interest for senior students and junior scientists or those who want to gain a deeper insight into the many ways by which human activities adversely affect human health. Our own curiosity into this topic prompted us to initiate a new course at the University of Calgary in 2016 which was eventually entitled 'Concepts in Biochemical Toxicology'. Thus, this book is also intended to accompany this course in the future and may serve as a blueprint to initiate courses with a similar flavour at other institutions, especially since its focus is – at least in our opinion – timely and certainly relevant on our path to transition to a more eco-friendly economy in the years to come.

When we started to contemplate to put a textbook together that addresses these issues, we soon agreed that our overall approach should not aspire to assemble yet another comprehensive biochemical toxicology textbook, of which there are quite a few (see list in suggested readings). Rather, our approach was to put toxicology at the forefront of the numerous issues that are associated with the highly complex environment–human interface and to explicitly point out that classical toxicology approaches are no longer sufficient to address the problems we face today and the future. Instead, we decided to dissect the pollution-related toxicological issues into its relevant multiple facets and to identify the political-sociological-economic system as the main driver of the many detrimental changes that currently unfold in

Jürgen Gailer, Department of Chemistry, 2500 University Drive NW, University of Calgary, Calgary, AB, Canada, e-mail: jgailer@ucalgary.ca
Raymond J. Turner, Department of Biological Sciences, University of Calgary, Calgary, Alberta, Canada, e-mail: turnerr@ucalgary.ca

https://doi.org/10.1515/9783110626285-001

our environment (climate change and ongoing pollution due to our consumer society and population growth in many parts of the world). To define the overall scope of this textbook more clearly, its content is intended to address the following conceptual learning objectives which are intertwined throughout the chapters:

- Defining and understanding toxic pollution problems and their biochemical targets.
- Considering the toxicity to different ecosystems and lifeforms.
- Exploring the tools that are available to monitor and study toxins both in the environment and within the affected lifeforms.
- Focus on metal toxins which cannot be degraded and therefore constitute an inherently more serious problem than organic pollutants.
- Philosophic considerations
  - Will the human species survive the Anthropocene?
  - Influence of the political-sociological-economic world.

Below the reader will find a brief description of some important issues that one needs to be aware of before embarking on the individual book chapters that follow. While it is not absolutely necessary to read this section before delving into the individual chapters themselves, the following sections are intended to serve as a primer which should help the reader to better understand the content of the chapters that follow.

1.3 Philosophy of the problem

The earth's crust, which may be regarded as the 'agar media plate' on which life evolved, contains not only 'benign' chemical elements that are essential to a large variety of life forms (e.g., iron, copper, zinc) but also chemical elements that are inherently toxic (e.g., arsenic, cadmium, mercury). Ever since we realized that all of these chemical elements represent useful building blocks for a variety of consumer goods (cell phones, computers, TVs, etc.), we began to extract them at an unprecedented scale from the earth's crust. The associated energy consumption (e.g., coal combustion) and the improper disposal of consumer goods that contain these elements into the environment have inadvertently infused additional amounts of chemical species into their respective 'global biogeochemical element cycles' (i.e., the cycling of chemical elements through all environmental compartments). Accordingly, all organisms, including humans, today are exposed to gradually increasing concentrations of certain chemical compounds simply via the diet, the drinking water and the inhalation of polluted air. In addition, we currently produce >100,000 chemicals on an industrial scale, and since many of these enter the environment through leaking tanks, chemical explosions, their improper disposal in landfills or their deliberate release into the environment, certain human populations are exposed to these environmental chemicals. Owing to the resulting chronic exposure of a variety of organisms to a plethora of environmental chemicals, a variable fraction of them cross lipid bilayer membranes in the lungs, the gastrointestinal tract and/or the skin and enter the bloodstream, for which we know the concentration of a considerable number of environmental chemicals [1]. We hardly know anything, however, about the concentrations of these bioavailable pollutants in human organs (the liver, the brain, etc.) of the average population. Unfortunately, the low concentrations of environmental pollutants that are present in human tissues make it essentially impossible to extract health relevant information since classical toxicology approaches – which traditionally only assess the acute toxicity of a chemical in any given model organism within a 24 h period – are no longer useful. Related to this, the staggering number of polluted sites that should be urgently cleaned up is often met with lacklustre political will to do so, insurmountable bureaucratic/jurisdictional hurdles and/or – perhaps most frequently – the lack of funding. These challenges are exemplified by the overall slow clean-up rate of numerous superfund sites in the USA [2]. Although a few success stories of superfund sites that were cleaned up are widely popularized, political-sociological challenges often delay the expedient clean-up of the majority of sites, which unnecessarily results in costs that are associated with environmental contamination and often adverse human health effects. Clearly, we need to aspire to learn from our past mistakes and strive towards a 'can-do' attitude in terms of reducing pollutant release (from both point and non-point sources) and to clearly recognize the urgency of cleaning up all existing superfund sites.

Avoiding the generation of new superfund sites is – in our opinion – also not receiving the importance that it should. This is inherently achievable as numerous examples exist where well-intended economic business decisions resulted in environmental and economic disasters. We will highlight some examples, which include polychlorinated biphenyls, mercury, lead and polyfluorinated hydrocarbons to illustrate this point from the US Environmental Protection Agency (EPA) top 10 priority list.

Based on past pollution case studies, we intend to instil in students the notion to holistically think about the problem. For example, we tend to focus our attention on where the damage occurs in an organism (organ) or specific cellular process (key enzyme inhibited or biochemical process) without realizing that the underlying fundamental chemical processes may unfold over months or years to result in the observable adverse health effects in a variety of potential organisms. In this context, a better understanding of the environment–human interface ultimately rests on the recognition that environmental pollutants can be absorbed into the systemic blood circulation from the skin, the respiratory system and/or the gastrointestinal tract throughout life! These simple facts are often overlooked by researchers, physicians and regulatory agencies.

1.4 Regulatory and toxicological problems

A large part of the environmental toxicology–related problems that we face today are directly related to the fact that we (a) know exceedingly little about the human health effects of the vast majority of the >100,000 chemicals that inevitably enter the environment and (b) that an ineffective regulatory framework – the Toxic Substance Control Act (TSCA) in the United States – has hampered progress in terms of reducing chronic human exposure to potentially toxic environmental pollutants. It is not difficult to infer from these facts that the manufacturing industry, which does provide jobs, has the required muscle to stall the necessary evaluation of the adverse effects of environmental chemicals on human health [3].

Yet ignoring the existence of the problem has never been a useful strategy. In fact, in our industrialized society we have seen an exponential increase of citizens acquiring allergies due to the ingestion of food, which is in certain regions increasingly laced with environmental chemicals, such as pesticides, fungicides and pesticides.

Simultaneous exposure of humans to multiple pollutants represents another key challenge that cannot be effectively addressed using classical toxicology approaches. While simple feeding studies have provided insight into whether the toxic effect of two chemicals is additive, synergistic or antagonistic, the detection of a cocktail of environmental chemicals in the human bloodstream (Centers for Disease Control (CDC) [1] represents a fundamental bottleneck in terms of gaining insight into the exposure-adverse health effects problem. One of our societies' major waste streams, wastewater treat plant (WWTP) effluent, adversely affects the ecosystem downstream owing to the fact that it contains a plethora of chemicals from personal care products, metals (e.g., Cd, Pb, nanoparticles), detergents, chelating agents and numerous pharmaceutically active compounds. While we may think that each of these chemicals 'minds their own business', it has been surmised that the potentially toxic metal Cd, which is present in many WWTP effluents at low ppb concentrations, forms complexes with the widely used chelating agent ethylenediamine tetra acetic acid (EDTA). Since EDTA is present in most WWTP effluents at low μM concentrations, the formation of Cd-EDTA complexes is potentially problematic as these complexes have shown to be bioavailable to the plant roots of food crops. Therefore, the irrigation of food crops with water that contains WWTP effluent, which is a widespread practice around the world and is likely going to be exacerbated by climate change, may inadvertently introduce this and conceivably other toxic metals into the food chain and compromise food safety [4].

1.4.1 The environment–organism interface

In the context of identifying pollutant sources and exposure pathways one must clearly differentiate between occupational exposure (exposure to chemicals in an occupational setting; welding, soldering, grinding and the associated release of metals either inform of particles or vapours) and environmental exposure (pollutants that are present in air, water, food and to a lesser extent soil). In this context, it is becoming increasingly evident that we also have to consider other pollutant sources, such as textiles, construction materials (sick house syndrome), cosmetics, household cleaning products and personal care products as well as their pharmaceutical ingredients. Potential toxicants are often ingested with contaminated food, which may contain chemicals that were introduced during processing and/or during extensive storage (e.g., plasticizers leaching from packaging into the food). In this context, the consumption of milk and fruit juice, for example, is often overlooked as a vehicle of toxicant delivery. Strangely, while pathogen delivery through food sources is recognized as a common health problem, it is still not fully appreciated that food is rapidly becoming our major source of exposure to environmental chemicals. One must consider all possible exposure pathways of a toxin: oral ingestion, inhalation, dermal exposure and deliberate injection (i.e., chemotherapy).

1.5 Toxicological problems and environmental monitoring

The outcome of the industrial revolution in combination with exponential population growth has led to the establishment of toxin priority lists in different parts of the world. A rather large number of new compounds with unique properties (including nanomaterials), however, are added to the arsenal of chemicals that are already produced at an industrial scale every year, for many of which their environmental fate and their acute and chronic toxicological outcomes are entirely unknown. The nanoscience revolution over the past 20 years has led to nanomaterials which are used today in almost every facet of our lives while the assessment of their toxicological effects in the receiving environments is only now beginning to be systematically studied. Furthermore, environmental consciousness has led to interests of developing easily degradable materials (e.g., bioplastics), yet we do not understand fully the fate of the resulting compounds in terms of their microbial degradation. We are slowly starting to appreciate that the complete lifecycle of chemicals in the environment should be an integral part of the development process.

1.5.1 Sampling and quantification of emerging pollutants

A variety of challenges are associated with sampling in the environment, at contaminated sites and contaminated animals. This includes temporal changes of a pollutant concentration (µM, ppb or ppb range) at any given environmental site since the latter may rapidly change on an hourly, daily, monthly and/or annual scale. Changes of a pollutant concentration at a certain environmental site may also be driven by climate change–related global events. Additionally, sensing unique compounds (e.g., nanomaterials) in real time in environmental compartments is still in its infancy.

The analysis of collected samples is often difficult since one may not know what pollutants one should be looking for. A physician who is trying to evaluate the illness of an individual in their community may not even be considering that a toxicant may be critically involved. Additionally, a toxicant may be exasperating a common ailment. This problem hints at the general difficulty of which analytical approaches should be employed in parallel to medical diagnostic tests. While some strategies exist about how to deal with toxicants that society has been dealing with for decades, it is exceedingly difficult to analyse environmental samples considering that every year a plethora of new compounds are released into the environment for which we have data neither on their adverse effect on the environment nor on their chronic toxicity in human populations, including children. The fact that state-of-the-art analytical techniques can detect exceedingly small concentrations of pollutants in environmental samples further emphasizes the need to refine sampling methods and the interpretation of the obtained results.

1.5.2 Choosing model systems

The analysis of certain organisms that inhabit a certain ecosystem (e.g., periphyton or amphibians in rivers, plants, herbivores) can provide information about the health of the food chain. The identification of the right sentinel organism, however, in this context is critical. Furthermore, new ideas such as the development of affordable biosensors may be of considerable practical use to detect a pollution event.

A number of model organisms including protozoans, insects, worms, fish and even various tissue cultures from various organisms and simple haemolytic assays are well established in toxicological monitoring studies. One problem that needs to be addressed is the fact that the prevalence rate of certain diseases is increasing dramatically (type 2 diabetes, autism, etc.), yet the mechanisms that may link exposure to disease are missing. With regard to age (children and the elderly are often more susceptible to a toxicological insult), sex, culture and customs, how can we appropriately conduct studies with a model organism to distinguish between 'normal' aging and chronic low-level poisoning? Can we identify model organisms to monitor toxicants for disease states?

1.6 Mechanisms of action

The number of established mechanisms of action of environmental pollutants is small compared to the total number of mechanisms of action of all toxicants that are 'out' there. The application of novel tools and conceptual approaches to study the effect of pollutants in appropriate model systems is urgently needed to provide fundamentally new insight into hitherto unknown mechanisms of action.

1.6.1 Systems biology

Some book chapters touch upon how modern 'omics' tools, such as genomics, proteomics, metabolomics, transcriptomics, metallomics, lipidomics and other emerging 'omics' approaches may be applied to study mechanisms of toxicity. The application of these powerful tools will allow us to grasp the full complexity of biological systems to visualize toxicological effects at the systems level. If the chronic toxicity of metals, for example, is associated with the induction of an essential element deficiency, the application of metallomics tools is crucial in terms of uncovering the underlying mechanism of action. In a similar fashion the low-level toxicant concentration in biological tissues may influence multiple biochemical pathways in subtle ways, which may eventually result in a disease state. The application of next generation 'omics' approaches have recently put us in the position to delineate how epigenetic influences of toxicants will affect future generations. The proper application of this unique toolset, however, forces us to take into consideration the immense diversity of ecosystems/population (genetics, diet, lifestyle) and the fact that little appreciation has been given to understand how mixtures of anthropogenic compounds influence different environments and populations.

1.7 Philosophical considerations

1.7.1 Will the human species survive the Anthropocene?

With the onset of the industrial revolution man has inflicted dramatic changes onto our only habitat, planet earth. Changes in land use and the unprecedented extraction of elemental building blocks from the earth's crust have resulted in man perturbing global biogeochemical cycles of numerous chemical elements on a local/regional and a global change, which has already resulted in changes of the very chemical composition of the air, the water and the food that we consume. In this context, students need to appreciate that pollutants can be released either from point sources (e.g., living in the vicinity of a coal-fired power plant) or from non-point sources (e.g., application of agricultural fertilizers) and that correspondingly, man needs to acknowledge that the release of pollutants into the environment ultimately rests on the behaviour **of every human being**. Since all life forms have been exposed to 'natural' toxins (e.g., toxic metals/metalloids that are present in the earth's crust) throughout evolution over 4.5 billion years, detoxification mechanisms have evolved which protect the organisms from adverse health effects. Every detoxification mechanism, however, has an inherent capacity, and if that capacity is exceeded it is not that difficult to envision that disease prevalence rates may begin to increase as is actually the case [5]. Recalcitrant pollutants, such as radioisotopes, metals and persistent organic pollutants deserve particular attention. Several chapters will expose students to case studies and associated group discussions and debates in class.

1.7.2 Influence of the political-sociological-economic world

All environment-health issues that we currently face as a species are the result of the cumulative political-sociological-economic decisions that we have made over the past ~ 200 years. We rely on political leaders and regulatory bodies to encourage environmental stewardship (e.g., by providing rewards) and discourage adverse unsustainable behaviour (e.g., by delivering penalties). Owing to the fact that the economy invariably drives the ideologies that politicians tout to become elected, it is not surprising that ecosystem health is frequently undermined by economic interests. The average citizen is usually unaware of this undesirable situation until blatant adverse health effects become impossible to ignore.

Faced with this conundrum the critical question is: how can government regulations make industry comply and recognize the toxicological burden of their products once they are released into the environment to pave the way for novel products that are inherently more environmentally and toxicologically safe?

We feel that it is important to solidify in the minds of students as the next generation of leaders that pollution is inherently mankind's legacy of its lifestyle. Whenever the economy is booming, resources are mined, transported and consumed the result is too often more pollution that adversely affects human health and the quality of life. Thus, we are both the problem and hold in our hands the inevitable solution to resolve this conundrum.

Prompted by the recognition that the way in which mankind has been conducting itself on planet earth is becoming increasingly unsustainable. The international political community has put forth Sustainable Development Goals (SDGs), which were ratified by the United Nations in 2015. The corresponding document takes into account that the biosphere and human behavioural processes are fundamentally and inextricably interconnected and interdependent. The SDGs are exceptionally diverse and outline a strategy to address the mounting pollution-related problems that we face [6]. Our ultimate goal should therefore be to manage aquatic and terrestrial systems sustainably to provide safe and abundant drinking water, food and clean air for all lifeforms on the planet. In the words of Stephen Jay Gould: 'Through no fault of our own . . . we have become the stewards of life's continuity on earth. We have not asked for that role, but we cannot abjure it. We may not be suited to it, but here we are.'

Suggested readings

Timbrell JA. Introduction to Toxicology (3rd edition). CRC Press, 2001.
Timbrell JA. Principles of Biochemical Toxicology (4th edition). CRC Press, 2009.
Smart RC, Hodgson E. (editors). Molecular and Biochemical Toxicology (5th edition). Wiley, 2018.
Vanbden Heuvel JP, Greenlee WF, Perdew GH, Mattes WB (editors). Comprehensive Toxicology. Vol 14 Cellular and Molecular Toxicology. Elsevier, 2002.
Gupta PK. Fundamentals of Toxicology: Essential Concepts and Applications. Academic Press. 2016.
Carson R. Silent Spring, 1962.
Steffen W, Crutzen PJ, McNeill JR. The Anthropocene: Are humans now overwhelming the great forces of nature? Ambio. 2007, 36, 614–621

Bibliography

[1] Fourth national report on human exposure to environmental chemicals, 2009, Centers for Disease Control and Prevention, Atlanta, GA, USA, http://www.cdc.gov/exposurereport/pdf/FourthReport.pdf, accessed December 10, 2021.
[2] P. Voosen Wasteland, National Geographic 2014, December, 128–145.
[3] W.E. Wagner and S.C. Gold, Legal obstacles to toxic chemical research. Science, 375, 2022, 138–141.
[4] S. Khan, Q. Cao, Y.M. Zheng, Y.Z. Huang, Y.G. Zhu. Health risks of heavy metals in contaminated soils and food crops irrigated with wastewater in Beijing China. Environ Pollut. 2008, 152, 686–692.
[5] J. Gailer, Probing the bioinorganic chemistry of toxic metals in the mammalian bloodstream to advance human health. J. Inorg. Biochem. 108, 2012, 128–132.
[6] O.A. Ogunseiten, J.M. Schoenung, J.D.M. Saphores, A.A. Shapiro. The electronics revolution: from E-wonderland to E-wasteland. Science 326, 2009, 670–671.

Andrii Lekhan and Raymond J. Turner

Chapter 2
Overview of biochemical toxicology principles

2.1 Introduction

The study of the biological response to a toxin and how an organism's biochemistry is affected continues a trajectory of increasing importance in our highly industrialized world. This chapter is divided into two sections where we first provide a brief overview of biochemical principles that are related to toxicology in order to get the reader in the context of how toxins "poison" biochemical processes and what are the key biology targets of a toxin, and thus we will cover:
– Fundamental biological processes
– Biomolecular targets and interactions

With this background in place, the next step will be to provide an overview of fundamental chemistry related to toxicology. The reader will become familiar with:
– Principles of toxicokinetics and the parameters used in toxicity prediction:
    – Absorption
    – Distribution
    – Metabolism
    – Excretion
– Chemical principles behind toxicodynamics:
    – Nucleophilicity
    – Hard-Soft Acid-Base theory and ligand binding
    – Metal ion properties and speciation
– Approaches for toxicity assessment

The goal here will not be to provide a contextual introductory description of these processes, yet to provide enough background that will enable the reader to grasp the issues, methods, mechanisms, and challenges described in subsequent chapters in this textbook on specific topics.

Andrii Lekhan, Raymond J. Turner, Biological Sciences, University of Calgary, Calgary, Alberta, Canada, e-mail: turnerr@ucalgary.ca

https://doi.org/10.1515/9783110626285-002

2.1.1 What are toxins and where do they come from?

Before we get into the biology, biochemistry, and chemistry of toxins, we have to address this question: Where does a toxin come from in the first place? What is overlooked in many texts is the origin of the toxins of concern, as it is important to understand why toxins exist in our environment. Otherwise, there is little chance to fully understand or improve the biological response to them.

A toxin is a natural or artificial compound that leads to unfavourable changes in the biological system upon exposure to it. As such a toxin causes a toxic effect. Many toxins are naturally produced and are part of biological warfare between species defined through ecosystem-driven evolution. But natural toxins, although significant and interesting, are of minor concern under anthropogenically managed planet. Anthropogenic activity yields artificial compounds, which are not otherwise present in nature. These compounds are grouped together as xenobiotics. For the most part, often when one thinks of something as a toxin, they are thinking of "pollutants". There are many threats to mankind highlighted by the popular press, social media, experts, and politicians including but not limited to economic stability, infectious diseases, chronic diseases, social unrest, food security, climate change, etc. Yet, pollution is rarely mentioned or thought of in these contexts. Jobs and a strong economy tend to be the driver of most country's political conversations in the world. Their political and social priorities become obvious with policies that reduce funding to environmental protection agencies in order to free up regulations holding back the industrial machine. However, the Global Commission on Pollution and Health has suggested that pollution is responsible for at least 9 million deaths per year, which is 16% of total premature deaths in the world [1], higher than any other cause of death, although this tends to be difficult to define. If a patient dies of a heart attack or stroke, this is what is officially listed as the cause of death. However, what led these biosystems to fail? Chronic exposure to toxins leading to biosystem failure is difficult to diagnose. This is in part due to the fact that humans consider acute challenges to their health more consciously than chronic challenges. Related to this, the world metre clock (https://www.worldometers.info) estimations suggest close to 1 million tons of toxic chemicals are released into the environment per month in 2021. Thus, humans are constantly bombarded with toxins, challenging the biosystems and leading to cumulative damage.

Toxins may be natural compounds yet are toxic at higher concentrations. The basic principle of toxicology is that "the dose makes the poison"; an idea originally expressed by Paracelsus. "All things are poison, and nothing is without poison; the dosage alone makes it so a thing is not a poison." In this respect even water can be considered a toxin, considering that drowning is essentially being poisoned by water, the water load decreases oxygen absorption (water has fewer dissolved O_2 molecules than air) and thus decreased O_2 levels in our cells for complex IV of the electron transport chain to accept electrons leading to loss of ATP production.

If we look at the US Environmental Protection Agency's (EPA) priority list we see a mixture of natural existing organics, xenobiotics, and metal/metalloid ions. Metals as toxins fall into two categories: those that are essential trace elements for life (Fe, Cu, Co, Ni, Zn, etc.) and those that have no positive activity in biology. The second category is often referred to as heavy metals (Hg, Pb, Bi, to lesser dense Cu, Ag, Cd, Sn Tl); this is not an accurate chemical term but is considered a useful definition by mineralogists (density more than 5 g/cm^3). This term was adopted by toxicologists and physiologists as the heavy metals tend to be quite toxic elements. But this list leaves off others toxic elements such as Cr, Zn, Co, Rh, As, Sb, Se, and Te.

It is seen worldwide that pollution is a highly neglected risk factor to continued life on the planet and yet it has not received the attention it deserves. Two thirds of human exposure to pollutants is seen in our air, both indoor and outdoor, and the remaining is water and soil. Surprisingly, even though pollution can lead to a large loss to potential gross domestic product (GDP), it is rarely mentioned in economic conversations [2]. In most governments, the environment, health, and economy are discussed and managed separately, preventing concerted effective strategies to mitigate underlying issues.

Where is pollution coming from? The Covid-19 pandemic brought to light the difference there can be when anthropogenic activities are reduced – clear air, less noise, clean water, decrease in offensive odours, less crowding, etc. [3]. It illustrates that the main source of environmental pollution is indeed human activity. A strong economy fundamentally means that people have money and spend it on *consumer goods*. Consumer goods production, however, requires mining for materials, fuel consumption in manufacturing and transportation, and subsequent disposal of outdated or damaged products. Thus, the newest gadgets or clothing styles indirectly and directly lead to a considerable pollution generation. Therefore, it is no one else but the consumer who is responsible for the pollution – either directly or indirectly. It also means that humans produce, on purpose, toxins that are a major threat for their health and life. Part of the problem is the lack of clear connection between pollutant(s) and public health, as there is rarely a single source of pollution and establishment of direct causes and effects is difficult, particularly for chronic challenges to biological systems. Additionally, air and water pollution can move across political jurisdictions, which produces additional barriers for adequate response, regulations, and policies.

2.2 General biology/biochemistry – the biomolecular targets of toxins

If we consider toxin exposure, it is somewhat obvious that before the toxin actually causes detrimental changes to its final target, it should first reach it. Depending on the lifeform under the exposure, there will be a considerable amount of deliberation on how the toxin gets to the cell and affects the whole biological system. Once a toxin reaches a cell, there is a path through the cell biology to the final biochemical target and the process by which the toxin deprives that target of its functionality. Such topics will be discussed in specific examples through this textbook.

Biochemical targets of toxins fall into the categories that group together biomolecules with similar properties. Depending on the biochemical target group, toxin–target interactions may also vary. These include interactions with the amino acids of proteins, partitioning in with the fatty acids of lipids in membranes and lipid bodies, interactions with the bases and backbone of nucleic acids, binding to carbohydrate monomers and polysaccharides. Toxins may also interact with the smaller biomolecules of metabolites and cofactors. All of these interactions can lead to dysfunction of the biological system(s) they are involved with.

The purpose of this section is to remind the reader of the biomolecules and biochemical processes of a cell. The reader is directed to any introductory biochemistry textbook for further descriptions and information on any of the topics presented here.

2.2.1 Biology central dogma

Once a toxin is inside the cell significant lethal damage to the cell can occur. Let us remind ourselves of what is required biochemically to define a cell to be alive:

- Replication: copying the genetic material (DNA) to distribute to daughter cells upon cell division.
- Transcription: Converting genetic code from DNA into RNA, selection of a gene and messaging the gene to be made.
- Translation: turning the message into polypeptides by the ribosome.
- Protein folding and targeting: taking the polypeptide and making sure it folds correctly into a functional structure, assembles, and interacts with other proteins for functionality and targets to the correct place in the cell for its function. This is a recent consideration for the central dogma that was previously overlooked, but it is no less important than the conventional elements.

DNA ⟶ RNA ⟶ PROTEIN ⟶ FUNCTIONAL PROTEIN

DNA POLYMERASE   RNA POLYMERASE   RIBOSOME   CHAPERONE

Scheme 2.1: Information flow and processes of the central dogma steps of biology.

The final components required for an effective biological system are energy, mass flow, and mass exchange (what biochemists refer to as central metabolism). Energy metabolism is the oxidation of a reduced molecule to obtain free energy (ΔG) or reducing equivalents to drive the electron transport chain (ETC). The end result is the energy currency of ATP units. Mass flow is the movement of carbon, nitrogen, phosphorus, and other trace elements from one molecular source into another by anabolism or catabolism to make cellular building blocks or energy storage molecules. Mass exchange is the movement of biomolecules between cells. Any toxin that affects one or more of these steps will typically kill a cell, as these are the minimum requirements for life.

There is an increasing trend in usage of omics technologies, where each of the central dogma stages can be studied to understand the response of the biological system to a toxin challenge. Genomics sequencing methods tell us what genes are present and of any mutations. Transcriptomics by gene arrays or RNA-sequencing tells what genes are being expressed. Proteomics tells us what transcripts actually made it to become a protein. Interactomics can give us information if all subunits in a complex are present and if the proteins are properly targeted. Finally, metabolomics gives us a picture of the physiological state of the organism. Omics methodologies are of a great use in toxicology, as one can extract mechanistic insights by exploring biological systems under toxin challenge and comparing them to pristine systems.

2.2.2 Cell membrane

Most biomolecular targets of toxins are inside the cell, although the membrane and subsequent components can be targets of toxins in their own right. Regardless of the type of toxin, all see the surface of the cell membrane as their first contact. The membrane is a mixture of lipids and proteins. The proteins on the membrane may be associated as peripheral proteins on either of the membrane surfaces or integral membrane proteins passing completely through the membrane. Often the extracellular side of the membrane is decorated with carbohydrates as well. The result is several zones or phase transitions as one passes from the bulk external environment through the carbohydrate extensions which will be polar and potentially charged, then protein surface to the lipid head groups to the lipid core. The fate of a given toxin and its efficacy will depend on how it is able to deal with this barrier.

2.2.2.1 Lipids

The membrane is a bilayer of lipids, with polar head groups to the aqueous sides with a lipid tail core. The lipid composition of the membrane is different between the cytoplasmic leaflet and outer leaflet, as was shown in red blood cells [4]. The lipid composition can also vary spatially on the cell surface. The membrane proteins often select different lipid types as their annular lipid layer around the protein. The dynamics of the lipids are also critical for integral membrane protein structure and function. Spatial localization of lipid types (lipid domains, lipid rafts) is also key to organizing various membrane proteins [5]. The traditional model of the cell membrane, the fluid mosaic model, states that proteins are randomly scattered through a sea of unorganized lipids. This is no longer considered fully accurate, as the density of proteins is much higher than originally thought and there is a considerable organization of the membrane lipids.

Due to the hydrophobic and amphipathic nature of many organics, the primary target of the cytoplasmic cell is the lipid membrane. Depending on the nature of the organic compound, it will either be transported into the cytoplasm through an integral membrane transporter, or it will partition into the lipid bilayer and get "stuck" there. Once in the lipid bilayer, it will affect the lipid components of the membrane through changes in head groups and acyl chains spatial organization and/or by means of interaction with acyl chains. This changes lipid fluidity, and as such will change the phase transition temperature of the membrane (liquid to crystalline). Subsequently, the membrane functionalities are altered, as it is well accepted that the dynamics of the lipids influence membrane permeability and membrane protein function. Thus, disruption of lipid-packing, lipid–lipid, and lipid–protein interactions can lead to disruption of critical membrane biochemical processes. Overall, an organic toxin partitioning into the membrane will change receptor and signalling efficacy and solute transport in and out of the cell.

2.2.2.2 Transporters

Solute movement across the membrane is primarily through integral membrane proteins referred to as transporters and channels [6]. Movement through channels tends to be concentration gradient–driven, whereas transporters use cell energy to move the solutes typically against a concentration gradient. These solute movement facilitators tend to be substrate-specific, yet solutes of similar physicochemical properties can use the same vehicle. Thus, a toxin that is similar to a solute normally imported into the cell can cheat its way in. It was previously thought that hydrophobic toxins simply diffuse across the membrane, yet now it is appreciated that most toxins enter cells via the natural solute uptake systems [7], as without a protein carrier the hydrophobic molecule would become stuck in the membrane.

With this information in hand, we can consider a type of toxicity mechanism. A toxin that uses the solute transporter can generate biochemical dysfunction through the toxin outcompeting a natural solute for the movement into the cell. A related mechanism is a toxin interacting with an uptake proteins' binding site, but unable to move through the channel and thus "plugs" the passage. Subsequently, both of these mechanisms lead to the toxin starving the cell of an essential nutrient.

In addition to solute uptake transporters, there are also solute exchangers where solutes are exchanged with each other, usually following their concentration gradient. There are also specific efflux transporters to move solutes out of the cell (metabolic wastes and synthesized carbon/energy molecules addressed for other cells in an organism). Protection of some toxins arises from such efflux transporters [8, 9].

2.2.2.3 Membrane bioenergetics

Cell membranes are the location where most of the biochemical energy for a cell is generated, in eukaryotic cells, this is the chloroplast and mitochondrial membranes, and for bacteria, it is the cytoplasmic membrane itself. Here the electron transfer chain (ETC) exists taking the reducing equivalents produced primarily from central metabolism (glycolysis and tricarboxylic acid cycle) cascading down through favourable free energy transfers resulting in establishing an electrochemical gradient (proton motive force (pmf)) across the membrane. This gradient is then harvested for the synthesis of the energy currency molecule adenosine triphosphate (ATP).

Given most cells cannot function without energy produced by the ETC, any toxin that disrupts the electron flow will starve the cell of energy. Another indirect toxicity is the production of reactive species. The most common is reactive oxygen species (ROS) but there can also be nitrogen (RNS) or sulphur species (RSS) generated (thus reactive O, N, S species or RONSs). ROS are naturally produced from the ETC from steps where there is a stoichiometric disconnection in electron exchange. For example, a 2-electron carrying molecule is giving electrons away one at a time. The membrane electron shuttle, "quinone", has evolved to deal with this problem and generates a long-lived, stable, semi-quinone species. Yet electrons can escape, reacting with oxygen to generate a cascade of oxygen radical species. Similarly, the 4-electron reduction of O_2 at the terminal oxidase is a problematic electrochemical process that spins off ROS normally. For this reason, a vast majority of life on the planet (prokaryotes) still avoids oxygen as it is quite toxic through this mechanism. Aerobic organisms have evolved ROS repair systems to address these natural levels of ROS.

Thus, for a toxin that disrupts the electron flow in the ETC, either directly by damaging a protein or protein complex or indirectly by affecting the lipid fluidity of the membrane; the typical result would be an increase in biochemical radical molecules leading to RONSs and subsequent cell damaging reactions. In fact, almost all toxin challenges to a cell will eventually lead to a RONSs pulse, because as the cell dies and biochemical processes fail, the membrane becomes compromised leading to a sputtering and failing ETC system.

2.2.2.4 Other membrane components

Beyond transporters and oxidation-reduction enzymes of the ETC, the membrane contains biosynthetic enzymes associated with the membrane. The obvious participants are those involved in fatty acid and lipid biosynthesis. But in some cell types, there can be cell wall and cytoskeleton proteins also associated, some of which may be involved in cell division or cell to cell communication. The final group are receptors, integral membrane proteins that capture molecular signals on one side of the membrane and propagate a signal internally. Any of these additional components can also be either direct or indirect targets of a toxin, leading to cell system failures.

2.2.3 Proteins and enzymes

A typical disruption of central dogma processes involves the toxin binding to a protein. This is similar to pharmaceutical drugs activity, where they bind to specific proteins/transporters or enzymes to destroy their function. In fact, one can consider a drug as a type of toxin. Toxin binding can act as an enzymatic inhibitor, allosteric regulator, inhibitor of proper folding, an inhibitor of protein–protein interactions (PPI) or plugging an ion channel or transporters' pore. Anything binding to a protein follows the ligand-binding theory of:

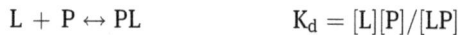

$$L + P \leftrightarrow PL \qquad\qquad K_d = [L][P]/[LP]$$

Where [L], [P], and [LP] are the concentrations of the protein, ligand (i.e., toxin) and the protein-ligand complex, respectively. The dissociation constant, K_d, represents the affinity of the protein for the ligand. Ligand binding is influenced by all the various physiological parameters of temperature, pH, ionic strength. Additionally, the nature of the interactions between the L and P will influence the energy of the interaction and involve all the normal biochemical bonds (hydrogen bonds, electrostatic, van der Waals, hydrophobic). The bonding will be with amino acid side chain functional groups as well as with the peptide backbone amine and carbonyl groups that are effective H-bond donors and acceptors, respectively.

At least 20% of genomes code for enzymes, the catalytic components of metabolism. They mediate the reaction rate of a given step appropriate for a particular cells' physiology with minimal waste of materials and energy. Enzymes are also regulated in accordance with the metabolic flux of a pathway. The reader is reminded that the key measures of activity come from the Michaelis-Menten model of enzymology [10]. There is the Vmax, which approximates the limiting rate of the reaction and Km, the Michaelis constant, that reflects the affinity of the enzyme for the substrate. To an enzymologist, a toxin that affects a specific enzyme is referred to as an inhibitor. Toxins can both inhibit enzyme function and increase or decrease their rate, or even turn on or turn off the enzyme through a process referred to as allosteric regulation. Enzyme active sites are specific to their substrate, but close mimics of substrate, product, or transition state can also bind to active sites. Thus, toxins can act as competitive inhibitors. There is also uncompetitive and non-competitive inhibition, which can occur through binding to the enzyme in regulatory or allosteric sites. Additionally, some compounds can become irreversible inhibitors if a chemical bond is generated at the site of binding. Depending on the mechanism and site of ligand binding on the enzyme, the effect can be to change either Km or Vmax or both. All types of inhibition lead to imbalance to a particular reaction or metabolic pathway and thus harm to the cells' physiology [11].

Beyond toxins affecting enzymatic activities, they can also be involved in affecting protein folding and important PPIs. Protein folding into an active conformation in cells can be spontaneous, but more often requires "foldases" and "chaperones" as well as prosthetic group "insertases". A toxin binding to such folding machinery can lead to the prevention of maturation to the functionality of many different proteins and enzymes. An additional process to folding is protein degradation, which is key to protein turnover, to remove proteins no longer needed in the cell for a given cell state. In eukaryotes, this may involve the ubiquitination pathway [12], but all cells use a proteasome or a variety of proteases. Similar to folding, compounds that interact with such protein degradation cell machinery and change rates of protein turnover can be quite toxic to a cell leading to both short-term and long-term effects to the organism.

Most proteins work in complexes with other proteins mediated by specific protein–protein interactions (PPI). Such complexes may be transitory such as regulatory reactions by enzymes (phosphorylation), or for secondary modifications (glycosylation, lipidation). Other complexes are more permanent, where several individual polypeptides fold (monomers) and come together to make a multimer or quaternary complex. Some complexes may be small with only two monomers coming together; others can be large with many different monomers assembling together to make a larger functional entity such as actin or nuclear pore complex in eukaryotes or the pili or flagella in prokaryotes. Proteins can also be found in large groups that may incorporate other biomolecules such as the ribosome, a mixture of rRNA and proteins, into a huge complex. The protein complexes in the electron transport chain or photosystems can also incorporate 5–50 different proteins. Thus, the results of PPIs disruption can be catastrophic.

The last step in the path to a functional protein formation is correct cellular targeting. This could involve transport of the protein into an organelle or across membranes. These processes are mediated by the protein chaperones and protein translocation machinery, as well as the cytoskeleton network that can act like the cells' subway system. Again, if any of these systems are inhibited by a toxin interaction, the cargo enzyme, even if properly folded, will not have biochemically relevant functionality.

2.2.4 Nucleic acids

Nucleic acids are involved in the cells' information storage (DNA), information relay (RNA) regulation, and signalling. We consider these separately, although what has been discussed above about proteins applies. The toxin can bind to nucleic acids in different ways. There can be interactions with the phosphate backbone, competing out the natural cation ligands and affecting the pitch of the helical axis. These types of binding can also affect the melting temperature and thus gene regulation, transcription, and translation rates. Toxins may partition (intercalate) between the bases, leading to base–pair mismatch or catalysing reactions leading to mutations. Additionally, larger toxins may bind differentially to the major or minor groove of the DNA superstructure and potentially cause regulatory issues and inhibit the ability of transcription factors to bind.

One can again follow ligand-binding theory for toxin molecule binding to nucleic acids. The difference is the outcome. A compound that binds to NA that leads to cell death because of inhibition of transcription, translation, or gene regulatory processes can be considered as a biotoxic compound. But toxins binding to DNA can be genotoxic. In this case, the binding of the toxin to the DNA leads to mutations during replication and as such the mutation is passed down to daughter cells. The binding interaction can also lead to double-strand breaks, which can lead to gene fragmentation, deletion, or rearrangement, all of which are typically more serious forms of mutation.

2.3 Principles of toxicology and toxicity prediction

What makes a compound toxic? In the sixteenth century, Paracelsus stated that the only thing that distinguishes a poison and a remedy is the dose. As simple as this 500-year-old wisdom may seem, there is more to the dose than a number of venom drops. Here, we will discuss the chemistry of toxins and define general toxicity prediction approaches based on fundamental principles.

2.3.1 Toxicokinetics

The common trait of toxic compounds is that they are foreign to the biological system and are either taken from the environment in one way or another or are produced from the environmentally sourced compounds during their metabolization. The uptake of the foreign compound, its movement within, transformation, and excretion from the biological system provide the toxicokinetics of the compound. Toxicokinetics describes the foreign compound behaviour in the living system with no regard to the toxic effect specifically. In the interest of simplicity, toxicokinetics can be referred to as an answer to the question "what the biological system does to the toxin?" There are four principal components of toxicokinetics: *absorption*, *distribution*, *metabolism*, and *excretion* (the abbreviation ADME is often used).

2.3.1.1 Absorption

In order for a foreign compound to affect the biological system, it first must be taken up from the environment. The ability of a toxic substance to be absorbed by the organism is called bioavailability. With regard to unicellular organisms, compound uptake is a more straightforward process, as the cell is exposed directly to the environment. In more complex organisms, however, final toxicity sites (organs, tissues, and even specific cells) are usually separated from the toxin entrance sites (respiratory and digestive tracts, skin and mucosae); thus, actual toxicity is determined by the extent of toxin availability to systemic uptake and the rate of the latter. The extent of the bioavailability (F) shows a fraction of the total amount of a foreign compound that can be found in the systemic circulation after extravascular exposure. F value ranges from 0 (no absorption) to 1 (full absorption of the compound); an F value of 0.75, for example, means that 75% of the foreign compound received extravascularly from the environment can be found in the systemic circulation (e.g., bloodstream). The extent of bioavailability is usually a complex parameter and is calculated as a sum of particular bioavailability events for each barrier within the biological system that a foreign compound should pass to reach the systemic circulation. The following is an example of the compound total extent of bioavailability calculation for oral exposure in mammals:

$$F = Fa \times Fg \times Fh$$

where Fa is the extent of the absorption through the intestinal wall, Fg is the extent of the passage through the gut wall metabolism, and Fh is the extent of passage through the hepatic metabolism.

As unintentional intravascular exposure to toxic compounds happens rarely, only molecules with particular sets of physicochemical properties make it to the systemic circulation. These properties include lipophilicity of the molecule, its size and geometry. Lipophilicity is the affinity of the molecule to the lipids of a membrane and relates to affinity to non-polar solvents versus polar ones. One of the ways to assess the lipophilicity of the molecule is to measure its partition coefficient (P) – distribution of a compound between a hydrophobic organic solvent and water after mixing and subsequent separation. The most commonly used partition coefficient is K_{ow} (octanol-water partitioning). The higher the molar ratio found in octanol, the higher is the molecule's lipophilicity, which increases the molecule's solubility in the membrane lipids.

Bioavailability prediction currently lacks a clear theoretical approach; however, empiric rules were formulated by Lipinski [13] and further modified by Veber [14] that allow relatively accurate prediction of the compound's bioavailability (Box 2.1). According to Lipinski, there is a 90% chance that compounds that possess two or more characteristics listed on the left column of the table will have poor bioavailability.

After the experimental study of 1,100 drug candidates, Veber suggested a slightly different scheme for bioavailability prediction. If the compound possesses all the characteristics from the right column of the table, it is most likely to have good bioavailability.
The rules initial intent was to define if a compound would be pharmacologically active,
and druggable space of molecular development of new compounds is still confined
within Lipinski's rules. However, the rules apply well to predicting if a compound
could or would be toxic, i.e., bioavailable, yet with negative rather than positive effects.

Lipinski's rules	Veber's rules
Molecular weight > 500 D	Rotatable bonds count < 10
Hydrogen bond donors count > 5	Polar surface area < 140A
Hydrogen bond acceptors count > 10	Hydrogen donor and acceptor count < 12
logP > 5	

Box 2.1: Lipinski's and Veber's approaches for bioavailability prediction.

2.3.1.2 Distribution

Depending on the site of entrance of the toxin, a foreign compound first reaches peripheral blood flow in case of skin uptake, pulmonary blood flow in case of lung uptake, the portal vein in case of gastral intestinal tract (GIT) uptake. Further distribution
of a compound depends on its physicochemical properties. Hydrophilic molecules tend
to remain in the systemic circulation, while lipophilic molecules tend to accumulate in
the fat-rich tissues. With minor differences influenced by the compound's chemical nature, the profile of its plasma concentration follows a predictable trend (Figure 2.1).

A characteristic of a compounds' prevalence in the biological system is referred to
as its half-life ($T_{1/2}$), the time where the concentration is decreased by half. Compounds
with greater half-lives persist longer in the organism, thus can lead to cumulative
chronic effects. If the dosing happens more often than the half-life of the compound,
there will be greater accumulation in the organism. Half-life, besides being directly
measurable, is a parameter that can be calculated theoretically from primary toxicity
parameters: volume of distribution (V_d) and clearance (CL). V_d represents the volume
of body fluids into which the compound is distributed (V_d = Dose/Concentration in
plasma). Highly lipophilic compounds tend to be accumulated in fat-rich tissues and
are often hardly detectable in blood, making the volume of distribution for the toxin
very large. CL is a characteristic of a compounds' elimination efficiency; it is interpreted
as the rate of toxin removal from a defined volume of plasma. Given that volume

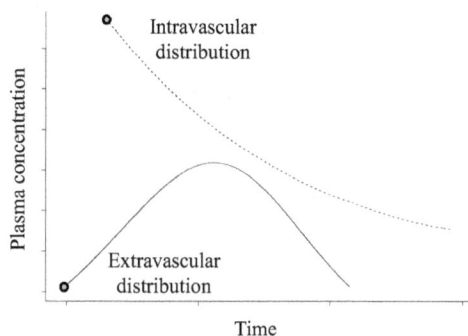

Figure 2.1: Typical toxin's plasma concentration profiles for intra- and extravascular distribution.

of distribution and clearance are known, the half-life of the compound can be calculated with the relationship:

$$T_{1/2} = (0.693 \times V_d)/CL$$

Half-life strongly depends on the molecules' properties, but even compounds with similar physicochemical properties can have different half-lives depending on slight structural differences [15] (Box 2.2). Branched-chain organic compounds, for example, tend to have longer half-lives than linear alkanes with the same carbon atom number. Within the branched compounds, ones with quaternary carbon atoms are more stable. Longer chains are degraded quicker than shorter ones. Polar groups and double carbon bonds decrease the half-life of the compound. While double-bonded nitrogen and halogen substituents increase the half-life. Positioning of the halogen substituent plays a role as well as para-positioning increases the half-life of the compound compared to meta- and ortho-positioning. Besides halogen substituents, other substitutions tend to decrease the half-life of the foreign compound. The half-lives of polyaromatic systems increase with the number of rings.

An important aspect of toxin distribution is its plasma protein binding, which makes it harder for the toxin to enter a toxicity site due to association with the larger molecule, but at the same time may prevent the toxin excretion. Some toxins can bind to albumin [16], while lipophilic molecules, like dichlorodiphenyltrichloroethane (DDT), can also bind to plasma lipoproteins [17]. Some toxicity sites are less readily accessible for a toxin, such as the brain, which is separated from the system blood flow by the blood–brain barrier.

Parameter	Example of a compound with	
	Longer half-life	Shorter half-life
Chain length		
Branching		
Polar groups presence		
Functional group count		
Hydrolysable groups		
C-C double bonds		
N-N double bonds		
Aromatic rings count		
Halogenated substituents count		
Halogenated substituents localization		
Quaternary carbon atoms		

Box 2.2: Relation of structures to their residence half-life.

2.3.1.3 Metabolism

After the foreign compound is absorbed into the organism, it can be metabolized and undergo biotransformation. Outcomes of toxic compound metabolism are the modifying of the physicochemical properties of that compound towards:
- *Change* of half-life (decrease in system exposure time)
- *Change* in absorption and distribution
- *Change* in biological activity.

Foreign compound's metabolites tend to be more hydrophilic, compared to the initial molecule and thus less likely to be deposited in fat tissue and more likely to be excreted with urine. Notable reactions of metabolism that decrease the solubility of a compound are almost exclusively represented by acetylation and methylation. A common misconception about the metabolism of foreign compounds is that metabolization leads to the decrease of the compounds' toxicity, which is not always the case.

The process of biotransformation can be divided into two distinct phases – phase 1 and phase 2 (Scheme 2.2). However, if the foreign compound possesses key functional groups, it can skip phase 1 and undergo phase 2 reactions right away. Phase 1 reactions add a polar group to the foreign hydrophobic molecule, which increases the compound hydrophilicity and allows Phase 2 reactions to take place. Phase 2 reactions involve conjugation of the hydrophilized toxin with a bulk molecule that facilitates faster excretion by means of increased solubility and potentially reducing the compound's toxicity.

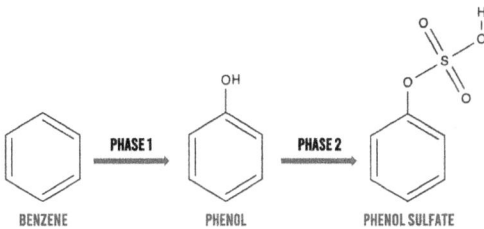

Scheme 2.2: Example of metabolic transformation – benzene and phenol [18]. Hydroxyl group is added during phase 1 that increases the hydrophilicity of the molecule and further conjugation with a sulphate group during phase 2. Notably, phenol as an external toxin already possesses a hydroxyl group and thus undergoes conjugation right away.

Let's look at these phases in more detail. Phase 1 reactions of foreign compounds' metabolism include the following:
- Oxidation of aromatic (e.g., benzene) and aliphatic (e.g., vinyl chloride) hydroxylation, oxidative dealkylation of N-, O- and S-linked alkyl groups.
- Reduction of azo- and nitro groups (more common), reduction of aldehydes, ketones, epoxides, and double bonds (less common).

- Hydrolysis of esters and amides.
- Hydration of epoxides.
- Both reductive and oxidative dehalogenation to cleave the halogen atom from the foreign molecule.

Phase 1 reactions can either cause a decrease or increase in compound toxicity and half-life. An example of possible biological activation of the toxin after phase 1 metabolism is acetaminophen oxidation in mammals (Scheme 2.3). The initial molecule (paracetamol) is used as a minor analgesic, and common therapeutic doses are safe. However, if the dose is significantly high, the acetaminophen metabolism pathway shifts from conjugation to oxidation of the substrate that yields hepatotoxic compound N-acetyl-p-benzoquinone imine [19].

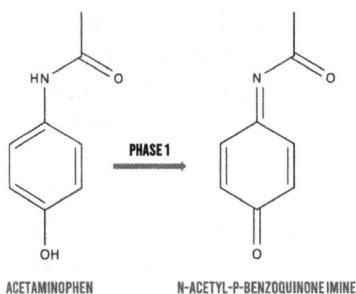

ACETAMINOPHEN   N-ACETYL-P-BENZOQUINONE IMINE

Scheme 2.3: Acetaminophen conversion to the more potent hepatotoxic N-acetyl-p-benzoquinone-imine.

Phase 2 reactions are also known as conjugation reactions. These reactions take place with molecules that possess electrophilic sites (either initially, or as a result of a previous phase 1 reaction(s)) and involve the addition of a "bulk" molecule to the target compound. Phase 2 renders foreign compounds either more hydrophilic (often) or less hydrophilic (rarely). Reactions that lead to the increase of molecule hydrophilicity and thus increased clearance of the compound are the following:

- Sulphation of hydroxyl group.
- Glucuronidation of hydroxyl, amino, carboxy and thiol groups.
- Glutathione (GSH) conjugation. Sulfhydryl group of the GSH acts as a nucleophile and targets the pre-existing or phase 1-yielded electrophilic site of the xenobiotic molecule.
- Amino acid conjugation of carboxylic groups.

On the other hand, phase 2 reactions, which decrease the hydrophilicity of the molecule, also exist and include:

- Acetylation of aromatic amino compounds and sulphonamides.

- Methylation of hydroxyl groups, amino groups and thiol groups and metal ions (Scheme 2.4).

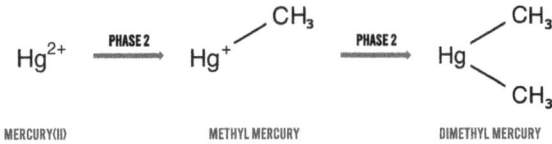

$$Hg^{2+} \xrightarrow{\text{PHASE 2}} Hg^+{-}CH_3 \xrightarrow{\text{PHASE 2}} Hg\begin{smallmatrix} CH_3 \\ CH_3 \end{smallmatrix}$$

MERCURY(II)　　　　　　　METHYL MERCURY　　　　　　　DIMETHYL MERCURY

Scheme 2.4: Conjugation of mercury with methyl groups yields neurotoxic methyl-mercury compounds [20].

2.3.1.4 Excretion

The final step towards biological system clearance of the foreign compound is ex-cretion. On the cellular level, this process is maintained by means of specific and/ or non-specific transporters that pump the compound to extracellular space (see Section 2.2.2.2). While for bacteria and other unicellular organisms this is sufficient, in more complex biological systems further elimination of the compound from sys-temic circulation is required. Mammals, as an example, tend to have several excre-tory pathways:

- *Urinary excretion*: maintained by the kidney and is relevant for smaller water-soluble molecules that can pass the glomerular pores in the kidney under nor-mal blood pressure or have specific active transporters there. Two parameters that affect the efficiency of urinary excretion are ionization of the compound and binding of the compound to plasma proteins. The more charge that is car-ried by the compound, the more water-soluble it becomes and thus there is more chance for it to be excreted with urine. If the compound is tightly bound to the plasma protein, however, it won't pass the glomerular pore and will re-main in the circulation until the plasma protein is degraded.
- *Biliary excretion*: maintained by the liver and is relevant for larger polar mole-cules of molecular weight around 300, such as GSH conjugates of xenobiotics, which are excreted to the gastrointestinal tract with the bile. An important con-sideration for biliary excreted xenobiotics is that they can be reabsorbed in GIT either as initial conjugated compounds or after being metabolized by gut micro-flora. The process of reabsorption of toxins is called enterohepatic circulation, which may increase the toxicity of the compound (if it is being metabolized) and/or prolong the exposure time. During oral exposure to the compound, en-terohepatic circulation locks a significant part of the xenobiotic in the gut-liver circuit and significantly increases the hepatic damage.

- *Lung excretion*: This is relevant for volatile xenobiotics and their metabolites, such as benzene, and is caused by the concentration gradient of the compound, directed from the bloodstream to the alveolar space.
- *Minor excretion pathways*: including excretion with breast milk, sweat, tears, semen, saliva, etc.

2.3.2 Toxicodynamics

The mechanisms by which a compound causes unfavourable effects in the biological system are altogether called toxicodynamics. Following on the conceptual simplification provided for toxicokinetics, toxicodynamics can be viewed as an answer to the question "what the toxin does to the biological system?" While the macro effects of toxicity often can be found by visual or surgical examination, the exact biochemical cause can or, at least, could be traced to the molecular level and described in terms of chemistry. Here we will briefly overview to remind the reader, but the reader is directed to standard chemistry textbooks for more extensive descriptions and examples.

2.3.2.1 Covalent binding

The strongest type of toxin–target interaction is the covalent bond, which is formed by means of electron orbitals overlap with common electron pairs. In biological systems, the most common mechanism of covalent bond formation between biomolecules and toxins is electrophilic addition or substitution. Most proteins, lipids, and nucleic acids possess nucleophilic sites – atoms, which due to their atomic environment, have an excess electron density and can serve as an electron donor. Toxins that tend to form covalent bonds with target biomolecules, on the other hand, are usually electrophiles. Electrophilic molecules have electron-deficient regions and are attracted to nucleophilic sites of targets.

In order for a compound to perform an electrophilic attack on a biomolecule, the former should meet the definition of a good nucleophile. A good electrophile is dependent on several molecular properties, known under the acronym CARIO:

- *Charge*: A more negatively charged molecule tends to demonstrate more nucleophilic properties, compared to non-charged or positively charged compounds.
- *Atom*: The nucleophilicity tends to increase with the size of the atom. In larger atoms, the lone electrons pair is located further from the nucleus and thus is more readily available for reaction.
- *Resonance*: Resonance structure is a favourable state of the electron pair in the respective molecule and thus is less likely to be involved in bond formation.

Molecules that carry resonance structures tend to be worse nucleophiles, compared to molecules without resonance.

- *Inductive effect*: The stability of the intermediates during the nucleophilic attack is dependent on the atomic environment of the nucleophilic site. The more electronegative atoms surround the nucleophilic site, the less stable intermediate will be formed during the nucleophilic substitution, thus making a weaker nucleophile.

- *Orbital effect*: the less contribution of s-orbital character to the overall outer hybridized orbital, the more nucleophilic molecules are formed.

2.3.2.2 Coordinative interactions

A coordinate bond is a polar chemical bond with one atom contributing both electrons to the common electron pair, while the other acting as an electron acceptor. The resulting compound is called a complex, and the electron-donor molecules within the complex are called ligands. Ligands that donate multiple pairs of electrons towards coordinate bonds are called polydentate ligands. The process of polydentate ligand binding to a central atom is called chelation. Chelation are stronger interactions over a monodentate ligand, while particular chelation geometries are more sterically favourable and thus are preferred over less favourable ones during the selection of the ligand.

Preferences for covalent versus the coordinate bond formation of particular compounds can in part be predicted based on principles of hard and soft acids and bases (HSAB) theory [21]. HSAB theory allows considerably accurate prediction of Lewis acids (electron acceptors) and Lewis bases (electron donors) preferable interactions. "Hard" compounds are those with a small radius, low polarizability, and high charge density, and tend to interact with other "hard" compounds in an ionic electrostatic manner. Similarly, "soft" acids and bases have large radii, high polarizability, and low charge density, prefer other "soft" compounds, and interact mostly in a covalent or coordinate manner. Acids and bases with intermediate hard/soft properties are called "borderline" (Box 2.3). HSAB theory plays a significant role in metal toxicology (see Section 2.3.3 below).

2.3.2.3 Non-covalent interactions

Non-covalent interaction is the type of interaction that does not involve a covalent bond of the toxin to the target molecule. The most abundant interactions in that regard are electrostatic and hydrophobic.

	Acids	Bases
Hard	Na⁺, K⁺, Mg²⁺, Ca²⁺, Cr³⁺, Al³⁺, Ga³⁺, Co³⁺, Fe³⁺	Alcohols, Phosphate, Amines, Carbonate, H_2O, OH⁻, NH_3, Nitrate
Borderline	Cu²⁺, Zn²⁺, Pb²⁺, Bi³⁺, Ni²⁺, Co²⁺, Fe²⁺	Imidazole, Nitrite, Aniline, Pyridine, Azides
Soft	Cu⁺, Au⁺, Ag⁺, Hg⁺, Hg²⁺, Cd²⁺	Cyanide, Thioethers, Phenyl groups, Thiols, Ethylene, H_2S, H_2^-

Box 2.3: Selected examples of hard and soft acids and bases of both metals and biochemical ligands (adapted from [22].

2.3.2.3.1 Electrostatic interactions

Electrostatic interactions include ionic, dipole-dipole, ion-dipole, and hydrogen bonds. As there is no common electron pair between participating atoms and the binding is based on positive and negative charge attraction, electrostatic interactions are weaker than covalent and coordinate and strength is proportional to the distance between charge pairs. Most natural protein–biomolecule interactions bind by means of electrostatic interactions. In this case, toxins that have a more favourable electrostatic profile can bind to the protein with higher affinity than the intended substrate, leading to inhibition and changed regulatory effects. Electrostatic interactions also play a role in the genotoxicity of positively charged foreign compounds that are electrostatically attracted to DNA sugar-phosphate backbone.

2.3.2.3.2 Hydrophobic interactions

Hydrophobic interactions are driven by entropy changes of the solvent that leads to the clustering of hydrophobic molecules. In an aqueous environment, water forms a rigid clathrate cage around the hydrophobic molecules. Full or partial destruction of that type of cage leads to a significant increase in the entropy of the system, as a highly ordered structure is being dissipated. A large positive value of ΔS makes tight packing of hydrophobic molecules thermodynamically favourable. These interactions are relevant to toxin incorporation into biological membranes or hydrophobic patches on proteins, nucleic acids, and polysaccharides. Incorporation of the foreign compound into the membrane causes the change in membrane fluidity and thus impairment of intended functionality. It further contributes to leaks and susceptibility to osmotic pressure. The continuity of the electron transport chain is lost and energy production and ROS yield due to electron outflow to unintended acceptors. Intercalation of hydrophobic compounds into stacks of nitrogenous base pairs causes genotoxic effects by preventing the normal replication and transcription of the DNA. Hydrophobic interactions also play an important role in toxin's deposition in fat tissues and thus toxin distribution.

2.3.3 General overview of metal and metalloid ion toxicity

Metal pollutants come from a variety of sources including but not limited to: ore and petroleum mining, wastewater management, waste combustion, and through industries of manufacturing and product production (pharmaceutical, electronic, textile, agriculture, forestry, energy). An evolving issue worldwide leading to increasing metal pollutant release is e-waste [23]. Metal ions are easily concentrated in the food web. We see bioaccumulation of toxic metals in seafood, dairy, meat, and fruit and vegetables. Unnatural metal exposures have clear linkages to various diseases such as cancer, type II diabetes, asthma, intestinal syndromes, neurological syndromes, and autism. Finally, some sources of metal pollutants have a dual danger due to their radioactivity (e.g., U, Pu, Po).

The exact molecular form in which a particular metal compound is present in the environment is dramatically dependent on the environment itself. Oxidation-reduction potential (ΔE^O) and acidity (pH) of the aqueous system affect the elements' speciation such as redox state (Cu(I) versus Cu(II)), coordination sphere, and counter ions. Differences in speciation may decrease the toxicity in some cases or yield toxic species from initial non-toxic substances in others. Speciation also refers to the molecular form of the metal that is seen by the organism or cell, which is often coordinated by one or more organic molecules. These all define the speciation and, subsequently, the cellular uptake pathway and potentially the biochemical targets. When it comes to the toxicity of metals, speciation plays a big role.

Overall, inorganic chemicals are classified generally based on the electronic structure of the atom, their valence orbitals, characteristic inner orbitals, crystal field theory, kinetic reactivity/lability, metal complex thermodynamic stability, hard and soft acid and base theory, Lewis acidity (metal ions), basicity (ligands) and softness index [24]. All of these schemes have been explored for relationships to metal ion toxicity. Explorations into the ability to predict the toxicity of metals by their chemical and physical properties have been underway for more than a century. Toxicity of metal ions is a combination of the intrinsic property of the element and the consequences from the way the metal atom or ion interacts with a given cell's biology and biochemistry.

2.3.3.1 Prediction of metal(loid) toxicity

In the review of Walker and colleagues [25], reader can find an overview of the concept of qualitative/quantitative cationic-activity relationships, or (Q)CAR, which links the elements' toxicity to its physicochemical properties. In the classical metal toxicology text of Lucky and Venugopal [26], the simplest explanation for a metals' toxicity was its binding capacity to the living system(s). But what drives and facilitates binding and bioaccumulation? Table 2.1 lists the key elemental properties used for predicting the toxicity of metals and Table 2.2 gives some inferences to biochemical relationship possibilities. Beyond these parameters, other properties have also been used, including serum concentrations, calmodulin and metallothionein levels and their metal binding activities, ATP binding constant, freshwater concentrations, and earth crust abundance.

Various studies in different animal systems have shown correlations of metal toxicity to the physicochemical properties of the metals [27, 28]. Metal toxicity studies are challenging, not only because of the variety of metal species present, but also because of the vast number of bioassays and outcomes, such as endpoints (IC_{50}, LD_{50}, EC_{50} – see Section 2.4). Other variables in studies are exposure duration (acute vs chronic), and the model system (organism, tissue, or cell). Thus, it is not surprising that conclusions from such studies vary widely for not only metals but also organic pollutants. Regardless, most studies find reasonable correlations of a metal's toxicity to σ_π, ΔE^0, IP, X_m, Log AN/ΔIP, log K_{OH}, and pK_{sp}. These correlations provide information about the biochemical mechanisms of metal toxicity which include affecting solute and essential ions transport, the ability to electrostatically or

Table 2.1: Physical and chemical properties of metal elements used for predicting toxicity.

Property	Symbol
Atomic Number	AN
Atomic volume	V
Atomic weight	W or *AW*
Covalent bond stability	$\Delta\beta$
Covalent index	$X^2_m \, r$
Covalent radius	*CR*
Crystal ionic radii	Å or *AR*
Density	Dd
Dipole moment	D
Electrochemical potential	ΔE^0
Electronegativity	X or X_m
Ionic charge	Z
Ionic index	Z^2/r
Ionic potential	Z/r
Ionic radius	IR
Ionization potential	IP
Melting point	MP
Molar refractivity	MR
Oxidation electrode potential	E^0
Polarizability	A
Softness parameter (Pearson)	σ_p
Softness parameter (Williams)	σ_w

Adapted from [25]

covalently bond to biomolecules, and their ability to disrupt essential oxidation-reduction reactions in a cell.

Considerably more work is required to establish standard methodologies in order to facilitate cross-comparisons in QSAR studies that will provide definitive guidelines for humans and multispecies ecosystems. As with predictions from organic compounds for the level of toxicity, even with reasonable predictions for metals described here, a full understanding of the biochemical mechanisms and subsequent biological system response is far from properly studied and fully understood.

Table 2.2: Some relationships of physicochemical properties to biochemistry.

Property	Symbol	Biochemical inference
Metal sulphide solubility	pK_{sp}	thiol biochemistry
Electrochemical oxidation potentials	ΔE^O or E^O	respiratory chains vs general reduction
Electron density	Log AR/AW	respiratory chains vs general reduction
Electronegativity	X_m	coordination to biomolecules
Polarizability	A	covalent bonds to biomolecules
Softness	σ	covalent bonds to biomolecules
Covalent parameters	$X^2_m\ r;\ \Delta\beta$	covalent bond with soft ligands effects ion channels/transporters
Ionic index	Z^2/r	ability to form ionic bonds
Ionic radius	IR	affects ion channels
Hydration radius	HR	affects ion channels /transporters
Ionization potential	IP	affects speciation state
Hydrolysis potential	log K_{OH}	affects speciation state

2.3.3.2 Metal and metalloid chemistry and toxicity potential

A question that confuses most when starting in the field is, *why is it that sometimes a metal ion is toxic and others not?* The answer lies in speciation. The metal atoms' species is the state of a metal atom in the environment and how it is presented to the cell. Differences in speciation are seen in such things as its charge, ligated/chelated state, types of ligands, redox poise, counterions, and hydration state, which will be different in the external environment, compared to that seen in the biological organism, cells or even cell compartments.

The physicochemical properties discussed above strongly influence the metal (loids) toxicity through the speciation, electron configuration and geometry of the metal coordination. Generally, speciation is influenced by temperature, pH, redox potential, type of solvent, and presence of other molecules (and their types). Electron configuration is the distribution of electrons of a metal atom in atomic or molecular orbitals. This is a key factor influencing the bonding and chelation of organic- and inorganic-geochemicals and cellular biomolecules to metal(loid) atoms. The geometry and metal coordination consider the ligand atoms that are bound to the metal atom and the orientation of the ligands.

An outcome from the electron configuration and the metal coordination is what types of biomolecules different metal(loids) will interact with. Additionally, the fluidity of these parameters allows for some metals to be more promiscuous in their binding. The Irving-Williams series [29] looks at the binding preferences of metal cations and their ability to take on different geometric arrangements. The series Mn(II) < Fe(II) < Co (II) < Ni(II) < Cu(II) > Zn(II), defines that Cu(II) has the highest affinity of interaction with a wide variety of ligands than the other metals. The ramification of this is that Cu (II) at equivalent concentration will outcompete other metals for their ligands. For this reason, cells have systems to buffer and manage the free Cu ion concentration in the cells, being sure to have them constantly bound up in specific Cu-binding proteins.

A useful tool to understand metal(loid) speciation is the Pourbaix diagram [30], a plot of E^0 to pH (Figure 2.2). The diagrams were first explored in the field of corrosion but in our context can give what metal species are possibly present in an environmental sample. Additional parameters of ionic strength and ion content may also be included. The lines on the graph represent the equilibriums between two speciation states, i.e., a line would indicate the condition where 50% of each species is present. There may also be phase state transition, such as for liquid to solid for example. This is a convenient tool to quickly identify the major species present for easily measured parameters.

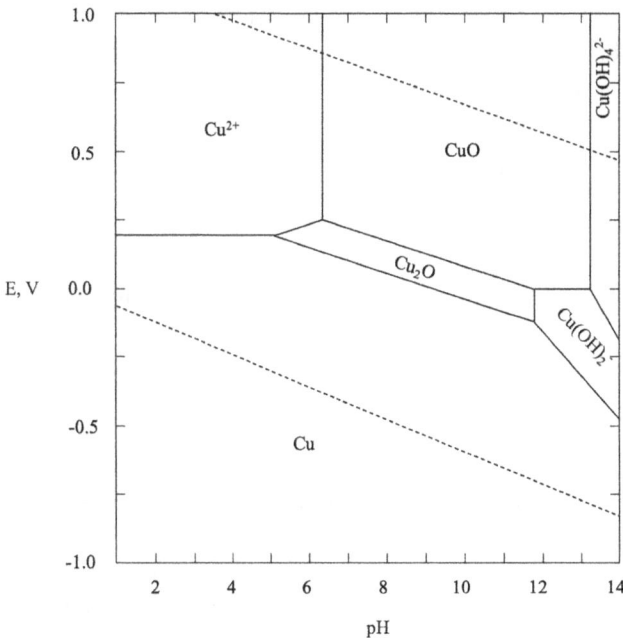

Figure 2.2: Example Pourbaix diagram. Copper in aqueous environment at 25°C.

A popular tool to model the solution state equilibria of possible chemical species of metals with known ion and organics is the Visual MINTEQ program (https://vmin teq.lwr.kth.se). This works reasonably well for the calculation of metal speciation, solubility, and sorption for simple aqueous solutions. However, it is considerably more difficult to predict speciation in biological systems due to the inherent complexity and diversity of biochemicals that could be present.

2.3.3.3 Biochemical sites of metal and metalloid ion interaction

When we think of metal toxicity biochemistry, the focus is typically on metal binding to proteins as inhibitors, toxic metals replacing essential metals in proteins, being genotoxic through binding to DNA, or catalysing Fenton-like reactions coupled with Haber-Wiess reactions that generate reactive oxygen species (ROS) [22]. A general summary of metal/metalloid ion toxicity processes to a cell is presented in Figure 2.3. Overall metal(loid) toxicity mechanisms can be generalized, but it must be remembered that each individual metal(loid) may have preferential targets. Such biochemical targets/mechanisms include:

i)   Interaction with lipids. An overlooked target of metals is the lipids of the cell membrane. It has been found that various toxic metals (Cd, Hg, Co, Ni, etc.) influence the fluidity and spacing of lipids. There even appears to be some selectivity in the lipids they interact with [31, 32]. Given the importance of the fluidity of the lipids to cell continuity and membrane protein function makes lipids an overlooked primary site of metal(loid) toxin attack. Changes in fluidity affect the functionality of integral membrane proteins such as changes in solute transporter kinetics, ion channels can get locked in closed or open conformations, and receptors can become unresponsive or locked in a given response state.

ii) Competition with essential solute carriers. Many toxic metals can outcompete essential metal ion uptake transporters and other solute carriers, thus starving the cell of key biochemical nutrients.

iii) Exchange of the catalytic metal in the active site of an enzyme. For example, Pb (II) can easily swap out Zn(II) in various proteases, or W for Mo in respiratory molybdoenzymes. Common damage is various metals replacing the Fe atom in iron-sulphur clusters, [Fe-S], in oxidoreductases. This destroys the electron flow in the enzyme as well as the released Fe which catalyses ROS production leading to oxidative damage.

iv) Exchange of a structural or regulatory metal in an enzyme. Zn(II) is common for this biochemical activity, but Ni(II) and Cu(II) can exchange the Zn atom out, leading to a loss of reversibility of the structural switch.

v)   Metal ligand binding to amino acids in proteins affecting the structure, function, and allosteric regulation. The imidazole side chain of histidine, carboxylate of aspartate and glutamate, thiol ligands from methionine or cysteine, nucleophilic hydroxyls of serine and threonine amino acids are all key metal ligands. This can be further complicated by secondary amino acid modifications such as phosphorylation which can provide additional electrostatic ligands.

vi) Catalytically damaging amino acids such as oxidation of the R-SH of cysteine. But also oxidizing R-SH of oxidation buffering molecules such as glutathione (GSH) and redoxin proteins.

vii) Genotoxic events, in part from ROS or metal-catalysed damage to the DNA bases. But also, from metals exchanging out natural cation counter ions to the sugar-phosphate backbone of nucleic acids. This can affect the ability of DNA-binding proteins, transcription factors, and polymerases to bind productively.

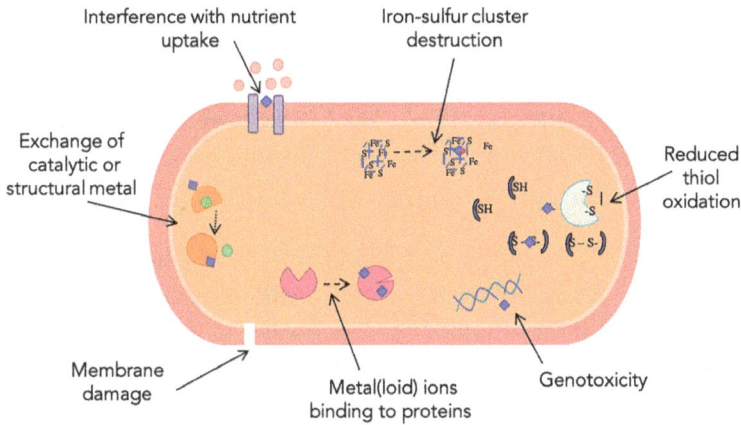

Figure 2.3: Summary of biochemical targets of metals or metalloid ions. The small blue diamonds represent the toxic metal(loid) ion. (Photo credit. Dr. Natalie Gugala. Turner Research Group).

2.3.3.4 Reactive Oxygen species (ROS)

A common observation of a cell exposed to metal(loid)s is the production of reactive oxygen species. Radical chemistry allows for attack and damage to all biomolecules from lipid peroxidation to enzyme active site inactivation and DNA damage leading to genotoxicity. However, one has to be cautious of misinterpretation of such observations, as there are a number of paths to ROS generation oxidative damage from metals [33]. The most direct route of ROS production is Fenton-type reactions.

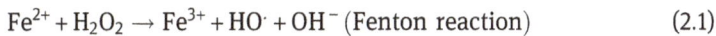

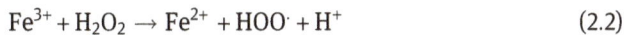

$$Fe^{2+} + H_2O_2 \rightarrow Fe^{3+} + HO^{\cdot} + OH^{-} \text{ (Fenton reaction)} \tag{2.1}$$

$$Fe^{3+} + H_2O_2 \rightarrow Fe^{2+} + HOO^{\cdot} + H^{+} \tag{2.2}$$

$$HOO^{\cdot} \leftrightarrow O_2^{\cdot-} + H^{+} \tag{2.3}$$

$$O_2^{\cdot-} + H_2O_2 \rightarrow O_2 + HO^{\cdot} + OH^{-} \text{ (Haber – Weiss reaction)} \tag{2.4}$$

But to do this, the metal must be redox-active at the correct E^0 (i.e., the potential inside the cell); thus, for the most part, only Fe and Cu are known to catalyse these reactions well in biological systems.

A second pathway to ROS is if the metal(loid) can out-compete Fe atoms in [Fe-S] clusters. If this occurs, the Fe is released and able to produce ROS. Subsequently, the produced $O_2^{\cdot -}$ can then react with other [Fe-S] centres disrupting them. This then releases more Fe to catalyse more reactions leading to a self-propagating cascade. So, in this way, a biologically redox inactive metal can lead to a considerable amount of Fenton produced ROS.

The third avenue to ROS has to do with the electron transfer chains (ETC) in bacterial cytoplasmic membranes, mitochondria matrix membrane (and to some extent, thylakoid membranes of chloroplasts) [34]. The ETC oxidoreductase complexes are known to spin off some free electrons when the system is stressed.

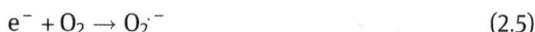

$$e^- + O_2 \rightarrow O_2^{\cdot -} \tag{2.5}$$

Additionally, the key electron carrying molecule of the ETC is quinone (Q) that carries two protons and 2 electrons between complexes I and II to complex III of the ETC. Since a single electron is transferred at a time, a semiquinone ($HQ^{\cdot}$ or $Q^{\cdot \cdot}$) is produced which has a very long half-live. When membrane damage occurs, there is an increased propensity of the Q to donate the excess electron to molecular oxygen, thereby generating ROS.

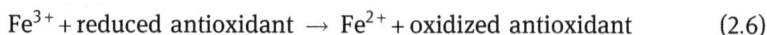

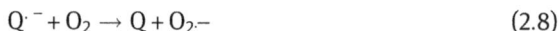

$$Fe^{3+} + \text{reduced antioxidant} \rightarrow Fe^{2+} + \text{oxidized antioxidant} \tag{2.6}$$

$$\text{Cytochrome } Fe^{3+} + QH_2 \rightarrow \text{cytochrome } Fe^{2+} + HQ^{\cdot}(\text{or } Q^{\cdot -}) + H^+ \tag{2.7}$$

$$Q^{\cdot -} + O_2 \rightarrow Q + O_2 - \tag{2.8}$$

If membrane integrity is lost due to metal ions binding up lipids or organics partitioning into the lipid and subsequently affecting fluidity and integrity, it leaves the cytochromes and quinone radicals unable to efficiently target their electrons through the ETC. Thus, they land up passing their electrons on to oxygen or water. Here the metal may never enter the cell, nor be redox-active, but will elucidate a ROS response. Similarly, we will see organic-based toxins producing a ROS response as well, all indirectly through damaging the continuity of the ETC via membrane damage.

With the oxygen radicals produced, several other biologically radicles can subsequently be produced. Radical chemistry is complicated and cascades through different atoms and molecular systems. For this reason, many are now considering grouping all reactions together as one gets carbon, sulphur, nitrogen, and oxygen radical centres.

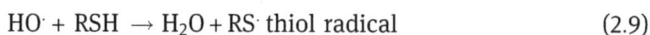

$$HO^{\cdot} + RSH \rightarrow H_2O + RS^{\cdot} \text{ thiol radical} \tag{2.9}$$

$$Q^{\cdot -} + NO \rightarrow ONOOH \rightarrow HO^{\cdot} + O_2N^{\cdot} \text{ nitrogen radical example} \tag{2.10}$$

$$HO^{\cdot} + (R)_3CH \rightarrow H_2O + (R)_3C^{\cdot} \text{ carbon – centered radical} \qquad (2.11)$$

$$(R)_3C^{\cdot} + O_2 \rightarrow (R)_3COO^{\cdot} \text{ peroxyl radical} \qquad (2.12)$$

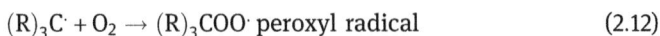

The abbreviation RONS has been suggested over ROS to include oxygen, nitrogen, and sulphur, but this still omits carbon radicals. Regardless, one should be very careful in interpreting the presence of reactive species in response to metal(loid) or organic toxin challenges and assigning a direct cause and effect.

2.4 Measuring toxicity

In many cases, it is important to establish an exact quantitative measure of toxicity as there are always:
- The conflict between beneficial and detrimental effects of various compounds, particularly drugs. Quantitative knowledge of toxicity allows selection of the frequency and dose of medication distributed to achieve maximum therapeutic and minimum side effects.
- The conflict between economic and ecological aspects of various industrial activities. Quantitative knowledge of toxicity allows the determination of safe xenobiotic contamination levels and dictates policies and regulations.

2.4.1 Toxicity terminology

A common starting point of the toxicity measurement is the amount of substance that the biological system receives by single administration – the dose. Based on the effects observed after one dose of a toxic compound received by the organism, there are several characteristic endpoint concentrations of the toxin that is often recorded (Figure 2.4): lethal dose (LD), toxic dose (TD), and effective dose (ED). Variations on these include but are not limited to lethal concentration (LC), lethal concentration and time (LC_t), lowest lethal dose (LD_{Lo}). Most toxicity data are reported with an indication of percentage of the population affected such as:
- LD_{50}, LC_{50} – the dose of a compound that causes the death of 50% of the organisms/cells in the treatments
- TD_{50} – the dose of a compound that causes an observable toxic effect in 50% of organisms/cells tested
- ED_{50} – the dose of a compound that causes observable desirable effects in 50% of treatments.

Effective dose is measured for toxic compounds that are intentionally used for medical or recreational purposes. While 50% effect numbers are the ones most widely

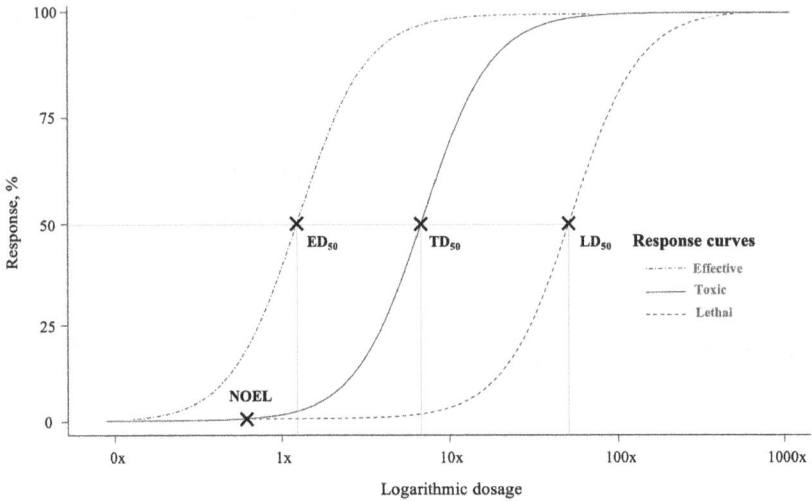

Figure 2.4: Dose–response curves. Each curve represents the trend of one of the three types of response that a group of tested systems demonstrate after identical dose distribution within the group (NOAEL; No Observed Adverse Effects Level or NOEL; No observed effect level).

reported and compared, others also can be of interest, especially with regard to toxicity studies of medical drugs. TD_1 – the dose that causes a toxic effect in 1% of treatments, or ED_{99} – the dose that causes the desired effect in 99% of treatments. These are used to determine the margin of safety of a drug, the relative difference between maximal efficiency and minimal toxicity:

$$Margin\ of\ Safety = TD1/ED99$$

Another important parameter is the therapeutic index, a value relevant to pharmacology:

$$Therapeutic\ index = LD50/ED50$$

To estimate the present risk arising from a given toxin, one can use the above parameters to derive the predicted-no-effects-concentration (PNEC) [35]. Another parameter that is required is the environmental concentration (MEC). Now a risk quotient (RQ) can be calculated as

$$RQ = MEC\ /\ PNEC$$

and

$$PNEC = LC_{50}/AF$$

AF is the assessment factor for any given environmental compartment. The assessment factor of 1,000 is applied if toxicity assessments were examined under acute condition. Chronic conditions are more difficult as one needs to factor in parameters that can influence exposure potentials, such as toxins adsorption rate and environmental flows. The RQ values are interpreted using the following scale: RQ < 1 means no significant risk, RQ of 1–10 means small adverse effects, RQ of 10–100 means potential adverse effects, and RQ > 100 means significant adverse effects.

2.4.2 Toxicity measurements to microorganisms

When considering antimicrobials, a slightly different approach is used in measuring the toxicity of a compound. The process of intermediate toxicity effects measurement is complicated in bacteria; thus, effective dose (ED) and toxic dose (TD) are either irrelevant or too hard to measure. Also, values such as LD50 are of a small practical relevance in clinical studies of antimicrobials as 50% decrease in bacteria count does not play a significant role in infection treatment, although we see this often in marketing of antiseptics. For example, seeing an advertisement of killing 99.99% of bacteria, which in reality is marginal given the high numbers of bacteria (i.e., a surface may have >10^7 cells/cm^2; thus, the antiseptic will leave 10^4 cells/cm^2 on the surface).

In a laboratory and clinical environment, the antimicrobial properties of the compounds are characterized using minimal inhibitory concentrations (MICs) and minimal biocidal concentrations (MBCs) [36]. MIC is a concentration that causes inhibition of bacterial growth by not allowing the division of cells. MIC does not account for the killing of bacteria and thus is a measure of bacteriostatic properties. MBC, on the other hand, is a measure of bactericidal properties and is determined as a concentration of the antimicrobial that causes the death of bacteria.

MIC and MBC are measured on free-swimming (planktonic) bacteria, although it is now recognized that planktonic bacteria are not the prevalent form in nature and that naturally occurring bacterial communities are found as biofilms – attached to the surface (see Chapter 3). In a biofilm, usually multispecies, bacteria form a microcosm, which is different from the actual environment – often protecting the bacterial community from physical and chemical influence from outside. This additional protection makes MIC and MBC values irrelevant with regard to biofilm treatment. Rising interest in biofilm research has driven the development of standard techniques for biofilm susceptibility measurement [37]. Minimal biofilm inhibition concentration (MBIC) of antimicrobials is the concentration that prevents biofilm formation on the tested surface. Minimal biofilm eradication concentration (MBEC) is the concentration required to kill the pre-existing biofilm.

More detailed approaches to toxicity measurement are the kill curve (essentially an inverted dose–response curve) and time-kill assays. Kill curve illustrates the survival rates of species under the increasing amounts of toxin. Comparison of kill curves may be of use for the identification of resistant strains of organisms, which demonstrate survivability under concentrations lethal to the wild strains. The Time-kill assay, on the other hand, allows for the study of the dynamics of antimicrobial action of a particular concentration of toxic compounds over time as it shows the percentage survival during the time course of toxin treatment (Figure 2.5). Time-kill assays allow the identification of tolerant species variants that, although susceptible, demonstrate a decreased response compared to wild strains of the organism.

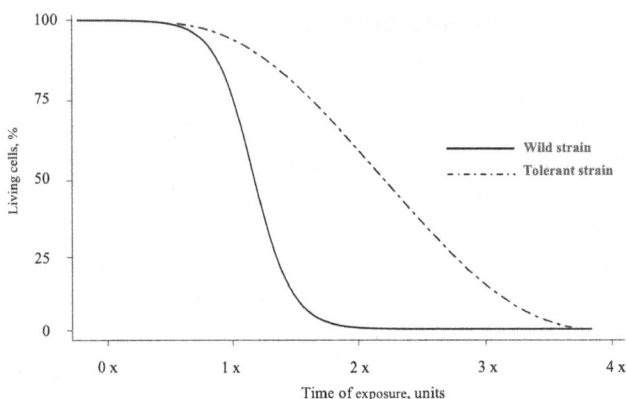

Figure 2.5: Toxin dose curves exploring effects of time-dependent exposure of wild and tolerant strains of microorganism to a toxin.

2.4.3 General considerations of measurements

Both the time of exposure and the duration is very important, but unfortunately these are often overlooked. The parameter of concentration is relatively easy to deal with, yet time is more difficult. Time of exposure can represent the stage of the life cycle of the organism or the duration of the exposure. Another time parameter is the time after exposure has ended where the toxic effect is being evaluated. Most studies take a healthy population of an organism grown typically to early or mid-life and challenged with a particular concentration and then the population is evaluated at short intervals afterwards of convenience, typically around the time, some symptom is noted. This approach ignores the difference between acute, chronic, or sustained effects. Unless specifically being investigated, this approach also misses the effects of repeated exposure and thus accumulative effects. Toxicology is also challenged by ethical considerations around what model system to use. Early toxicology used model systems of rats, mice, rabbits, dogs, and chimpanzees. Nowadays most studies are done on lower life form model systems such as worms, insects, and fish and for mammals, cells, tissue cultures, and organoids are used.

Environmental pollution is a subject of controversies entrenched in political policies and regulations. One of the main reference points during policy introduction is the level of exposure, under which no toxic effect is observed in a healthy individual – No observed adverse effects level (NOAEL). Taking NOAEL and other toxicity parameters of particular compounds (LD, TD) into account, the acceptable daily intake (ADI) may be estimated. ADI is the total amount of substance that can be continuously taken up by the organism without measurable detrimental effects on its health. An example regulation parameter, that sets daily exposure limits in industrial settings, is the threshold limit value (TLV) often used in the USA.

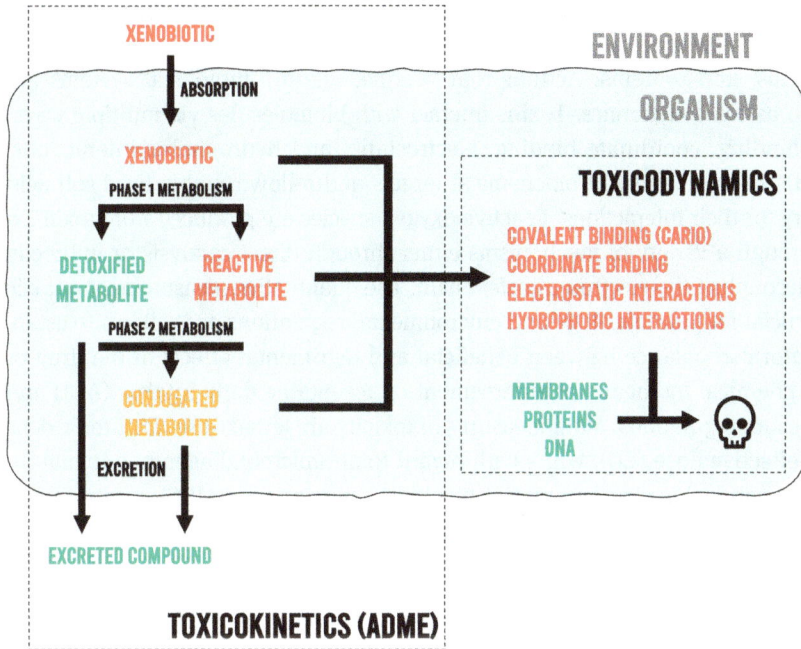

Figure 2.6: Summary of toxicodynamics and toxicokinetics.

2.5 Summary

Compounds that cause unfavourable effects or changes in the biological system are called toxins and the effects themselves are called toxicity. Natural and artificial toxins exist, and the latter is referred to as xenobiotics. No matter the nature of the compound, the principles of behaviour in the biological system are common and depend on physicochemical characteristics of the toxin. Toxicokinetics and toxicodynamics are summarized in Figure 2.6. Toxin's movement and changes within the organism are described by toxicokinetics, which can simply be referred to as "actions that biological system does to a toxin". Toxicodynamics has four main components, grouped under ADME memo: absorption, distribution, metabolism, and excretion. Important part of the absorption is bioavailability that can be predicted following Lipinski's and Veber's rules. Distribution of the toxin depends on the lipophilicity of the compound and complexity of its structure. The main goal of metabolism is to decrease the compound's plasma half-life by means of increasing hydrophilicity, which, however, may increase the compound's toxicity. Excretion is maintained by specific protein transporters on the cellular level and particular organs on systemic level: kidney for small hydrophilic molecules, liver for larger or/and more lipophilic molecules and lung for volatile compounds. Minor parts of the toxin may be excreted with other body fluids.

Many toxins damage the processes of biology's central dogma. Most of the cell machinery and structural elements are subject to toxicity, including biomembranes, nucleic acids, and proteins. Actions that "toxin does to a biological system" are referred to as toxicodynamics. Toxins interact with biomolecules via multiple ways: covalent binding, coordinate binding, electrostatic, and hydrophobic interactions. Metal(loid)s have a variety of biochemical targets and follow roughly hard-soft acid base theory for their interactions. Reactive oxygen species are produced from toxin exposure through a variety of mechanisms either through direct catalysis or indirectly through decoupling the electron transfer chain. The quantitative measurement of toxic effect is crucial for pharmacology and environmental regulations as it allows to establish the informed balance between beneficial and detrimental effects of the drug or polluting chemical by means of enforcement of acceptable daily intakes (ADI) and other exposure regulations. Main measures of toxicity are lethal dose (LD), toxic dose (TD) and effective dose (ED), while with regard to antimicrobial agents, minimal inhibitory concentration (MIC), minimal biocidal concentration (MBC), and minimal biofilm eradication concentration (MBEC) are used.

Recommended reading

For an animated view of solute transport and toxin delivery see video https://www.youtube.com/watch?v=s23vNwLE-Jw; accessed August 9, 2021.

Alberts B, Johnson A, Lewis J, Raff M, Roberts K, Walter P. Molecular Biology of the Cell. New York, Garland Science. 2002.

Nelson DL, Cox MM (editors) Leininger Principles of Biochemistry 7th Edition W.H. Freeman and Company. 2017. *Suggested Introductory Biochemistry Textbook*

Ottoboni MA, Frank P. The Dose Makes the Poison; A Plain Language Guide to Toxicology. 3rd edition. John Wiley & Sons Inc. 2016

Eaton DL, Gallagher EP, Vandivort TC. General Overview of Toxicology. In: Comprehensive Toxicology. 3rd edition, (McQueen DA editor) Elsevier Ltd. 2018

Timbrell J. Introduction to Toxicology (3rd edition). CRC Press, 2001.

Brunton LL, Knollmann BC, Hilal-Dandan R. Goodman & Gilman's: The Pharmacological Basis of Therapeutics, 13th edition. New York, N.Y: McGraw-Hill Education LLC. 2018.

Guengerich FP. Life and times in Biochemical toxicology. *Int J Toxicol*, 2005, *24*, 5–21.

Bibliography

[1] GBD 2015 Risk Factors Collaborators. Global, regional, and national comparative risk assessment of 79 behavioural, environmental and occupational, and metabolic risks or clusters of risks, 1990–2015: A systematic analysis for the Global Burden of Disease. Lancet. 2016, 388, 1659–1724.

[2] Myllyvirta L. Quantifying the Economic Cost of Air Pollution from Fossil Fuels. Center for Research on Energy and Clean Air [Online]. Available: https://energyandcleanair.org/wp/wp-content/uploads/2020/02/Cost-of-fossil-fuels-briefing.pdf [Accessed 15 September 2021]

[3] Rume T, Islam SMD. Environmental effects of COVID-19 pandemic and potential strategies of sustainability. Heliyon 2020, 6, e04965.

[4] Verklei AJ, Zwaal RF, Roelofsen B, Comfurius P, Kastelijn D, van Deenen LL. The asymmetric distribution of phospholipids in the human red cell membrane. A combined study using phospholipases and freeze-etch electron microscopy. Biochim. Biophys Acta 1973, 323, 178–193.

[5] Bogdanov M, Mileykovskaya E, Dowhan W. Lipids in the assembly of membrane proteins and organization of protein supercomplexes: Implications for lipid-linked disorders. Subcell Biochem. 2008, 49, 197–239.

[6] Stillwell W. Membrane Transport. An Introduction to Biological Membranes. 2016, 423–451.

[7] Jindal S, Yang L, Day PJ, Kell DB. Involvement of multiple influx and efflux transporters in the accumulation of cationic fluorescent dyes by Escherichia coli. BMC Microbiol 2019, 19, 195.

[8] Du D, Wang-Kan X, Neuberger A, van Veen HW, Pos KM, Piddock LJV, Luisi BF. Multidrug efflux pumps: Structure, function and regulation. Nat Rev Microbiol. 2018, 16, 523–539.

[9] Liu X. ABC Family Transporters. Adv Exp Med Biol. 2019, 1141, 13–100.

[10] Cornish-Bowden A. One hundred years of Michaelis–Menten kinetics. Perspect Sci. 2015, 4, 3–9.

[11] Holdgate G, Meek T, Grimley R. Mechanistic enzymology in drug discovery: A fresh perspective. Nat Rev Drug Discov 2018, 17, 115–132.

[12] Lecker SH, Goldberg AL, Mitch WE. Protein Degradation by the Ubiquitin–Proteasome Pathway in Normal and Disease States. Jasn. 2006, 17, 1807–1819.

[13] Lipinski CA, Lombardo F, Dominy BW, Feeney PJ. Experimental and computational approaches to estimate solubility and permeability in drug discovery and development settings. Adv Drug Deliv Rev. 1997, 23, 3–25.

[14] Veber DF, Johnson SR, Cheng HY, Smith BR, Ward W, Kopple KD. Molecular properties that influence the oral bioavailability of drug candidates. J Med Chem. 2002, 45, 2615–2623.

[15] Mackay D, Shiu W-Y, Ma K-C, Lee SC. Handbook of Physical-Chemical Properties and Environmental Fate for Organic Chemicals. CRC Press: 2nd Edition, 2006.

[16] Vanholder R, De Smet R, Glorieux G. et al., European Uremic Toxin Work Group (EUTox). Review on uremic toxins: Classification, concentration, and interindividual variability. Kidney Int. 2003, 63, 1934–1943.

[17] Plack PA, Skinner ER, Rogie A, Mitchell AI. Distribution of DDT between the lipoproteins of trout serum. Comp Biochem Physiol C Comp Pharmacol. 1979, 62C, 119–125.

[18] Snyder R, Hedli CC. An overview of benzene metabolism. Environ Health Perspect. 1996, 104, 1165–1171.

[19] Sidney D, Nelson MD. Molecular Mechanism of Hepatotoxicity Caused by Acetaminophen. Semin Liver Dis. 1990, 10, 267–278.

[20] Broussard LA, Hammett-Stabler CA, Winecker RE, Ropero-Miller JD. The Toxicology of Mercury. Lab Med. 2002, 33, 614–625.

[21] Pearson RG. Hard and soft acids and bases. J Am Chem Soc. 1963, 85, 3533–3539.

[22] Lemire JA, Harrison JJ, Turner RJ. Antimicrobial activity of metals: Mechanisms, molecular targets and applications. Nat Rev Microbiol. 2013, 11, 371–384.

[23] Ogunseitan OA, Schoenung JM, Saphores J-DM, Shapiro AA. The Electronics Revolution: From E-Wonderland to E-Wasteland. Science. 2009, 326, 670–671.

[24] Walker JD, Hickey JP, QSARs for Metals – fact or Fiction? Metal Ions in Biology and Medicine. John Libbey Eurotext, Montrouge France, 2000, 401–405.

[25] Walker JD, Enache M, Dearden JC. Quantitative cationic-activity relationships for predicting toxicity of metals. Environ Toxicol Chem. 2003, 8, 1916–1935.

[26] Luckey TD, Venugopal B. Metal Toxicity in Mammals, Vol 1 Physiologic and Chemical Basis for Metal Toxicity. Plenum, New York, NY, USA, 1977.

[27] Khangarot BS, Ray PK. Investigation of correlation between physicochemical properties of metals and their toxicity to the water flea Daphnia magna Straus. Ecotoxicol Environ Saf. 1989, 18, 109–120.

[28] Xuewu H, Jianlei W, Ying L, Xingyu L, Juan Z, Xinglan C, Mingjiang Z, Daozhi M, Xiao Y, Xuezhe Z. Effects of Heavy Metals/Metalloids and Soil Properties on Microbial Communities in Farmland in the Vicinity of a Metals Smelter. Frontiers Microbiol. 2021, 12, 2347.

[29] Irving H, Williams RJP. Order of stability of metal complexes. Nature. 1948, 162, 746–747.

[30] Pourbaix M. Applications of electrochemistry in corrosion science and in practice. Corros Sci. 1974, 14, 25–82

[31] Payliss BJ, Hassanin M, Prenner EJ. The structural and functional effects of Hg(II) and Cd(II) on lipid model systems and human erythrocytes: A review. Chem Phys Lipids. 2015, 193, 36–51.

[32] Umbsaar J, Kerek E, Prenner EJ. Cobalt and nickel affect the fluidity of negatively-charged biomimetic membranes. Chem Phys Lipids. 2018, 210, 28–37.

[33] Imlay JA. Pathways of oxidative damage. Annu Rev Microbiol. 2003; 57, 395–418.

[34] Zhao R, Jiang S, Zhang L, Yu Z. Mitochondrial electron transport chain, ROS generation and uncoupling (Review). Int J Mol Med. 2019, 44, 3–15.

[35] European Union, Technical guidance document on risk assessment TGD. Tech Guid Doc Ris Assess Part II, 2003.

[36] Wiegand I, Hilpert K, Hancock R. Agar and broth dilution methods to determine the minimal inhibitory concen.ration (MIC) of antimicrobial substances. Nat Protoc. 2008, 3, 163–175.

[37] Macia MD, Rojo-Molinero E, Oliver A. Antimicrobial susceptibility testing in biofilm-growing bacteria. Clin Microbiol Infect. 2014, 20, 981–990.

Raymond J. Turner

Chapter 3
Bacterial response to toxins

3.1 Introduction

Microbes are often the first to be exposed to pollutants in the environment, whether through their presence in soil, aquatic, marine systems, or as part of the microbiome of a living organism. This chapter will provide a brief introduction to the prokaryote grouping of microorganisms and their major life form in the environment, as sessile communities referred to as a biofilm. We will then look at the response of bacteria toward organic and metal pollutants as well as antibiotics. The chapter will look primarily at general mechanisms of toxicity as well as how bacteria resist toxins.

3.2 Microbes are abundant everywhere

Life on our planet is separated into essentially two groups: eukaryotic and prokaryotic organisms. Microorganisms can be either eukaryotic such as yeast and fungi or prokaryotic which include bacteria and archaea. Eukaryotes are in fact an outbranch that evolved from Archaea. The newly established family tree of life (Figure 3.1) now also includes a new large group (in light blue shading) that has been referred to as "microbial dark matter" [1]. From this new tree, we see that the prokaryotes make up greater than 95% of the species on the planet. The diversity of the prokaryotic arms of the trees is much larger than what we see on the macroscale. Consider the branches of the eukaryotic tree limb that are close together and show mammals, fish, insects all tightly grouped together. Now let's take into account that there are up to 10^{10} prokaryotic cells in a gram of soil, of which there are 10^4 to 10^6 different species. This means we have the equivalent of a rain forest of different plants, animals, and insects in the palm of your hand. This is possible as prokaryotes range in size from 0.5 to 3 μm in size.

Additionally, we also recognize now the concept of the microbiome, the community of organisms that are associated with higher life forms. For example, the human microbiome makes up ~5% of our body mass and is critical to our ability to survive.

Acknowledgments: The author would like to acknowledge funding from the Natural Sciences and Engineering Research Council of Canada

Raymond J. Turner, Department of Biological Sciences, University of Calgary, Calgary, Alberta, Canada, e-mail: turnerr@ucalgary.ca

https://doi.org/10.1515/9783110626285-003

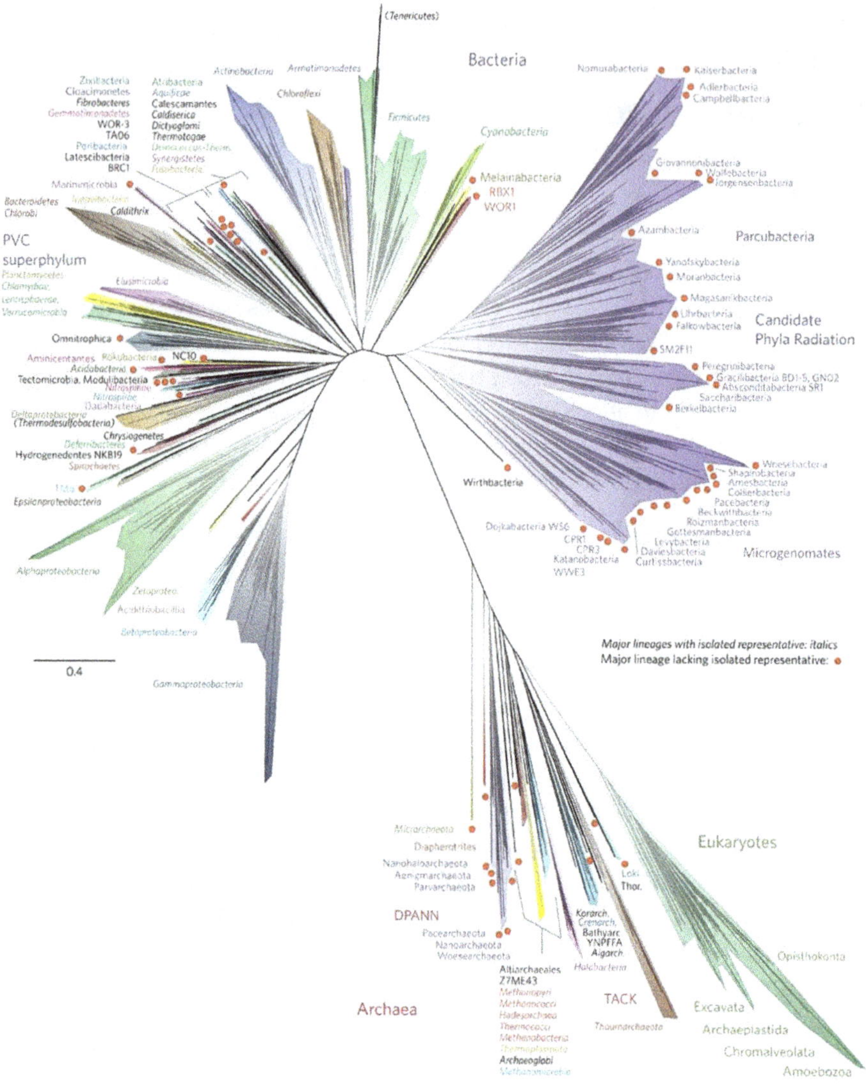

Figure 3.1: Current view of the tree of life, encompassing the total diversity represented by sequenced genomes. Red dots indicate new branches of life not formally known. The upper right blue branches represent the majority of the so-called microbial dark matter (from [1]).

Microbiomes are also found associated with plants, insects, fish, etc. Microbiomes can also be associated with abiotic surfaces such as in soils, river rocks, and drinking water distribution systems. An important consideration is that when we think of bacteria, we normally have a negative view, e.g., bacteria cause disease, yet only a very small fraction, less than 1%, of bacteria are infectious [2].

Microbes may be found as single, free-living organisms referred to as the planktonic form. They may also be assembled together in groups or communities which are often attached to surfaces; this form is referred to as a biofilm. As a surface-attached version of microbial life, we see biofilms featured in every aspect of life – they influence the environment, geochemistry, agriculture, industrial processes, and animal and human well-being. An easy visual example of a biofilm is the plaque on your teeth, which is a community of oral bacteria. Another important example is biofilms in aquatic or marine environments. These biofilms can become quite thick and complex and can become visible on various surfaces (e.g., as a green scum or film). They become a complex mixture of algae, cyanobacteria, heterotrophic microbes, and detritus (dead organic material) that is attached to submerged surfaces in most aquatic ecosystems and is referred to as the periphyton. As the periphyton matures, one will observe small eukaryotic organisms grazing and feeding on the microbial mat, which attracts higher-level organisms from snails to fish to birds to mammals. Thus, the periphyton becomes the definition of the bottom of the food chain. Because of this, one can consider the microorganisms to be *Ground Zero* of toxin exposure in the environment or to higher organisms. In this way, microbiomes of living organisms (skin, gut, eyes, ears, airway) are first to be exposed to any toxin.

Beyond the importance of various microbiomes in nature, we have to consider how prokaryotes contribute to the planets' biogeochemistry. They participate in the global carbon, nitrogen, sulfur, and phosphorus cycles. In fact, bacteria are crucial to water distribution and planet climate as their metabolism produces volatile compounds that can form aerosols that serve as nucleation sites for water molecules in the atmosphere to seed cloud formation. Microbes, particularly prokaryotes and archaea occupy every environmental niche on the planet, many of which are toxic to eukaryotes. Many microbes are able to survive in anoxic as well as oxic environments, at extremes of pH (e.g., pH 1.5–10), temperature (as low as –14 and >100°C) ionic strength, redox potential, and utilize extremely varied carbon and energy sources. Given this information, we are starting to appreciate the importance of bacteria in our planets' homeostasis and for survival and states of higher life forms.

3.3 Prokaryotes and toxins

What do we consider to be toxins to microbes? Given the history of humans' fight with bacteria around infectious diseases, we have thought a lot about developing toxins to try to control them. In our fight against pathogenic bacteria, we use antiseptics and antibiotics. To control spoilage of foods, cosmetics, and pharmaceuticals, preservatives are used. In this, we also need to include personal care products such as underarm deodorant (the odor originates from our microbiome metabolic volatile byproducts) and "disinfectant sprays" to control foot odor or athletic wear odor. We also need to include our cleaning agents, household and industrial, which are toxins to bacteria. On the industrial side to control biofouling (robust biofilm formation), a variety of general biocides are applied which may also be toxic to higher life forms. Finally pesticide use in agriculture can also be toxic to microbes. This brief introduction demonstrates a lot of toxins targeting bacteria and other microbes.

Beyond such purposely applied anti-bacterial compounds, bacteria also suffer from exposure to industrial pollutants. Pollutant contamination to soil and/or aquatic environments will lead to decreased bacterial numbers as well as a dramatic decrease in microbial biodiversity (the number of different species present). Such pollutants that affect microbial growth are the same as what concerns humans and may include either organic and inorganic (metals) pollutants or mixtures thereof.

3.3.1 Biofilms and toxin exposure

Bacteria have developed a variety of mechanisms to tolerate toxins in all forms; here let us consider the general tolerance provided by a biofilm lifestyle [3]. The structure of a biofilm includes the microbial cells surrounded by an extracellular matrix (ECM, Figure 3.2). This matrix is composed of polysaccharides (EPS: exopolysaccharides), DNA (extracellular, eDNA), proteins, and lipids, and various metabolites secreted by living bacteria or released by dead cells. The prevalence of EPS has also led to the term of biofilms being a "slime" layer, as anyone will know if they have ever slipped on a rock in a stream.

The chemistry of the biofilm matrix molecules provides a number of chemical functional groups that will interact and bind or repel various toxins. The final composition and mixture of biomolecules present in the matrix will be dependent on the type and number of species present in the biofilm community [4]. Additionally, the macroenvironment will dictate their physiology. Thus, the matrix could have a variety of biochemicals providing positive, negative, polar, and hydrophobic regions or more uniformly with a single chemical class. Thus, the biofilm matrix can act as a sponge for solute molecules and subsequently concentrate them. This is in part an ecological advantage in nutrient-poor conditions, where absorption and concentration are key to obtain nourishment. However, in the case of toxins, this concentrating effect can be the demise of the cells in the biofilm. Additionally, since the toxin is concentrated in the biofilm, the toxin becomes more concentrated for exposure by organisms grazing on the periphyton and can therefore enter the food chain.

Figure 3.2 captures the idea that there are a number of mechanisms underlying the biofilm mode-of-life that are responsible for enhanced tolerance to toxin perturbations (see also reviews [5–7]). Biofilms are well-protected, socially interactive, highly diverse communities of microbes that can readily adapt to changing environmental pressures. This is a cumulative result of cohesive and protective functions achieved by a strong extracellular matrix, adaptive mechanical features, toxin molecule mass transfer in bulk volume vs biofilm, along with active genetic stress response

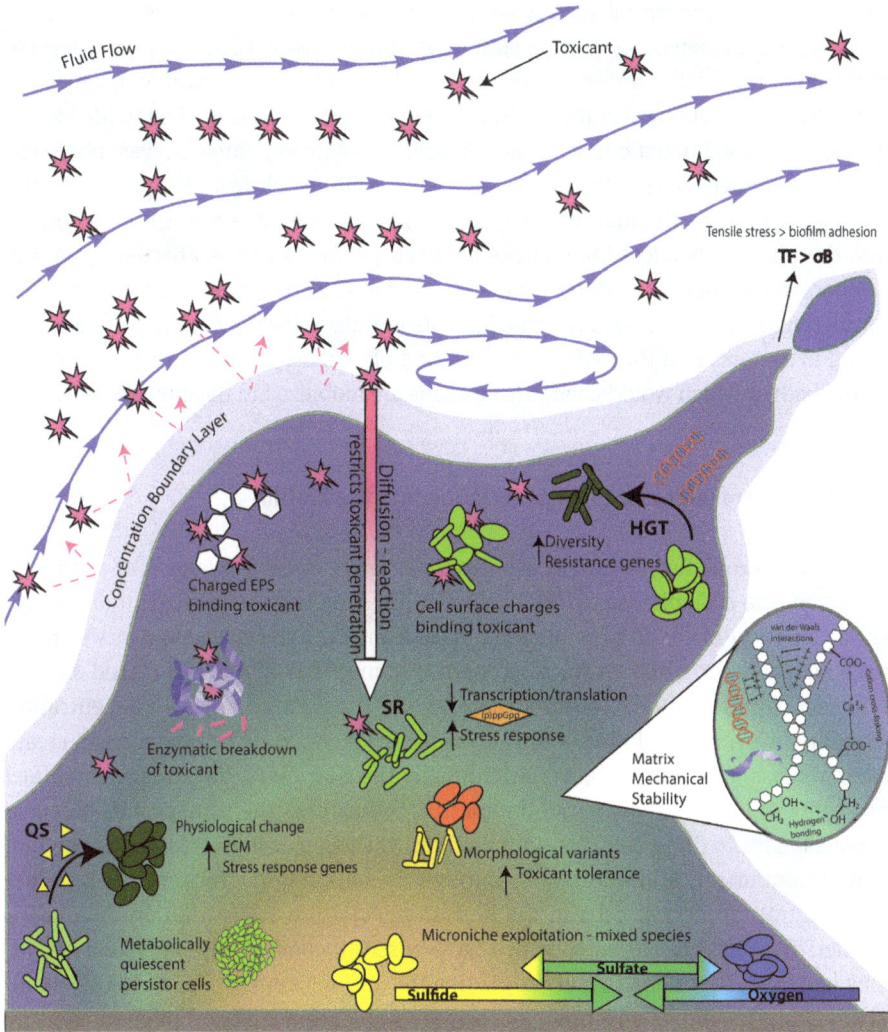

Figure 3.2: Overview model of the mechanisms responsible for toxin tolerance in biofilms. See text for details. (Credit: Dr. M. Demeter – Turner Research Group).

systems and ecological adaptive radiation and interspecies interactions. We will now briefly describe each of these contributions.

3.3.1.1 Mechanical stability and retention of biomass

Biofilms are viscoelastic which is mediated by the EPS matrix which helps to resist mechanical removal. Microbial ECM also affects the erosion process of sediments acting in part as the glue between microparticles. Biofilm viscoelasticity can be described using two physical parameters: the biofilm elastic modulus (G), which describes the reversible stretching of the biofilm under tension, and the biofilm failure strength (σ_B), which describes the overall strength of the biofilm. As a tensile force (T_F) is applied to the biofilm (through increased aquatic flow or mechanical, a biofilm will stretch reversibly to the extent determined by G, after which the biofilm will detach from the surface of the shear stress exceeds the failure strength (i.e., $T_F > \sigma_B$). The matrix integrity is derived from the physicochemical interactions between polymers and is affected by matrix polymer composition and concentration. These physical parameters are important in considering if conditions are favorable for cells to initiate biofilm formation (adhesion) or lead to dispersion of the biofilms via breaking off smaller units and fragments of the matrix, both of which would change how toxins are mobilized in the environment.

3.3.1.2 Mass transfer and toxin penetration into the biofilm

In biofilm microbiology, external mass transfer refers to the movement of solutes or gases from the bulk phase into the biofilm. It has long been recognized that one potential function of the ECM is to resist mass transfer. External mass transfer resistance may arise from the chemical properties of matrix biopolymers as well as from fluid mechanics around biofilm structures via the formation of a concentration boundary layer. Hydrophobic, and thus water-repellent, matrix polymers can create external mass transfer resistance. Additionally the flow along the surface can affect the ability of toxins to enter into the matrix. Diffusion is the transport mechanism responsible for toxin penetration into biofilms. Diffusion of most toxins through biofilms is significantly slower than it is through the bulk water, which means toxicants will penetrate to access cells in the biofilm slowly in a concentration-dependent manner. Thus deeply buried cells will be exposed to considerably lower concentrations of a toxin (a serious concern regarding the antibiotic treatment of an infection). However, extracellular enzymes can mediate a diffusion-reaction phenomenon by catalyzing reactions of toxic substrates that diffuse through the ECM at a different rate than the original molecule.

3.3.1.3 Physiological stress responses

Signaling pathways are important for the ability of biofilm organisms to withstand environmental stressors and for the cell community to work together. Two important systems related to biofilms are the cell-density dependent quorum sensing (QS) [8] and the regulatory system of cyclic-di-guanosine mono phosphate (c-di-GMP) [9]. QS is the regulation of gene expression in response to extracellular signaling molecules, produced by neighboring bacteria. QS communication circuits regulate a diverse array of physiological activities [10], and can influence biofilm development in many ways in different bacterial species. The other system is the practically ubiquitous second messenger of c-di-GMP that through a variety of sensor kinases, diguanylate cyclases, specific phosphodiesterases, and c-di-GMP binding proteins regulates EPS biosynthesis and other physiological biofilm traits. Fundamentally, c-di-GMP regulates the switch between a motile state and the formation of the sessile biofilm state.

Another important system is the starvation-activated stringent response (SR) [11], which allows bacteria to adapt and respond to nutrient deprivation and may also protect nutrient-limited and biofilm bacteria from antimicrobial stress from toxins. This system follows an intracellular regulatory process by the second messenger molecule guanosine-3′,5′-bis pyrophosphate abbreviated, (p)ppGpp. The stringent regulation seems to vary considerably from organism to organism. However, in several bacterial species, activation of the stringent response downregulates genes involved in cell replication and macromolecule production and upregulates genes involved in stress tolerance.

It is worth mentioning as well the GacSA sensor kinase, a key two-component regulatory system linked to the Rsm signal transduction pathway [12]. This system produces small regulatory RNA molecules that govern a wide variety of physiological states. This leads to locked-in physiological and morphological phenotypic variants. Additionally, GacA is a positive regulator of *lasR* and *rhlR* for QS molecule production demonstrating the interconnectivity of the systems. Under unstressed conditions, very few phenotypic variants are produced, but under stress, more variants are produced and in *gacS* mutants, there is a further increase in the frequency of variants upon exposure to various toxin stressors.

Finally, bacteria under toxin stress can shut down completely and become metabolically silent, yet not dead, and in this state bacteria are referred to as persister cells [13]. Such a survival mechanism avoids toxins attacking central biology systems and recovers after the toxin is removed. Figure 3.3 illustrates the rather complicated and integrated web of regulation of different systems that provide protection from external stressors. The biofilm lifestyle is integral to this protection, particularly a mixed-species biofilm. By no means are these systems fully understood.

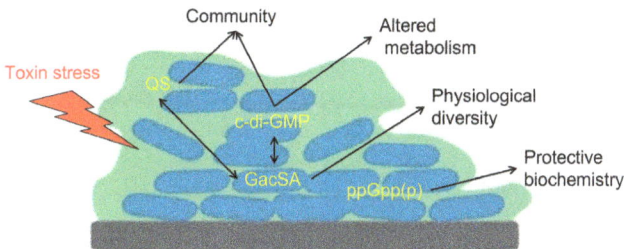

Figure 3.3: Bacterial cells (blue) in an extracellular matrix (green) attached to a surface (grey). Toxic stress leads to a number of different signaling response features that influence the biofilm, cell physiology, and protective features.

3.1.4 Microbial ecology and evolutionary processes

Biofilms are spatially structured environments with landscapes of selective pressures that drive not only ecological succession but also the evolution of microbial traits. Horizontal gene transfer (HGT) may occur in both planktonic and biofilm forms of bacterial life by natural transformation, transduction via viruses, or conjugal transfer. However, due to the high cell density and limited diffusion, HGT is considered to be as much as 1,000-fold higher in biofilms than for comparable planktonic cell populations. Toxin resistance genes and specialized metabolism capable of degrading toxins are often found on mobile genetic elements that are easily exchanged between cells by HGT processes. Micro niches within biofilms present new challenges and resources to microorganisms. Adaptive radiation may allow an organism to rapidly diversify into new forms that can occupy these niches. This may be through HGT, but also through genetic switching of phase variations of regulatory units or simply mutations, all of which are carried to subsequent daughter cells. The resulting consequence of such micro-niches and collective physiological diversity is that it can lead to increased resistance of a community to environmental perturbation. This ecological principle is known as the "insurance hypothesis" [14]. Overall, biofilms and periphytons self-generate diversity that effectively produces "insurance effects."

3.3.1.5 Interspecies interactions

Nearly all environments contain multiple coexisting species of prokaryotes and microbial eukaryotes. Emerging evidence suggests that biofilm stress resistance may be dictated by the numbers and types of microbial species in the community. QS and other metabolites are used in interspecies communication allowing the biofilm community to mount a collective defense against a toxin. An interesting thing that can happen in a community is the concept of shared communal metabolic effort. "Syntropy" is a term that describes the ability of one species of bacteria to take the metabolic waste from another species as a source of carbon, nutrients, or energy source.

3.1.6 Microbial biofilms: Summary

The biofilm lifestyle protects community cells to allow for unique physiological states and mechanisms of response to toxin exposure. There are clear trade-offs between certain mechanisms of tolerance, such as barriers to external mass transfer, and metabolic capacity, and the consequences of toxin concentrating effects. There is still much to be learned on the biochemistry of biofilms. In the context of toxicology and environmental sciences, the relationship between toxins and biofilms will need to be further explored to better understand the full effect of toxins on the periphyton, biosorption, and concentration through the trophic layers as well as health via the human microbiome.

3.3.2 Bacterial cells and toxin exposure

There are two fundamentally different prokaryotic cell structures that are referred to as the Gram-positive and gram-negative organisms based on their response to the staining procedure for microscopy. This designation relates to a significant difference in the cell wall structure (Figure 3.4) and subsequently the barriers for a toxin to enter. These different structures provide slightly different modes of defense against the toxin. For a toxin to get into the cytoplasm of the cell in order to affect the central dogma systems of biology it must pass the various membrane and cell wall barriers.

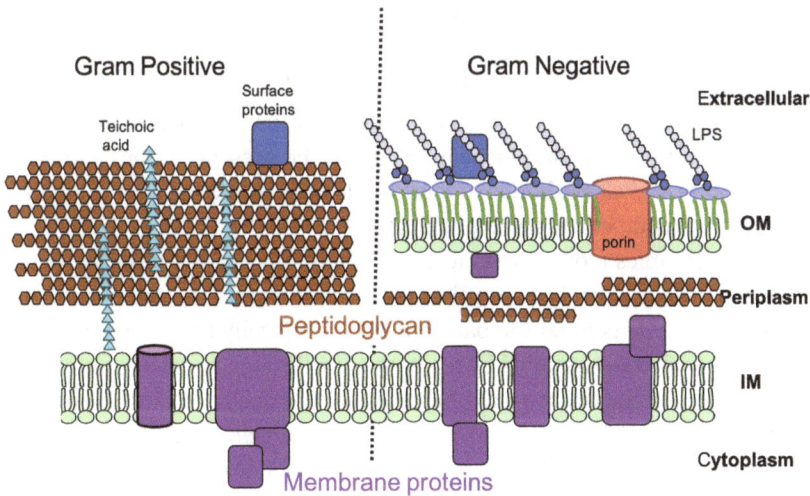

Figure 3.4: Structure of Bacterial cell barriers. Gram-negatives have an outer membrane (OM) in addition to the inner membrane (IM) also referred to as the cytoplasmic membrane. The OM contains porins and lipopolysaccharide (LPS). The gram-positive have a much thicker peptidoglycan wall structure.

3.4 Types of toxins and bacterial response

3.4.1 Toxicity of organic pollutants to bacteria

Here the effect of different types of common organic toxins (e.g., PAH, PCBs, alkanes) on bacteria will be discussed. Key cellular targets and the response of the cell to adapt to the toxin will be covered. Additionally, the concept of bioremediation will be introduced, where the immense metabolic potential of diverse bacteria can help degrade or store/immobilize various natural and xenobiotic organic pollutants.

3.4.1.1 Common traits of organic pollutants

Most hydrocarbons are toxic to cells to some degree due to their high hydrophobicity. If we overview the primary organic toxins on the EPA list, we find that various groups of organics can be quite toxic to bacteria including polyaromatic hydrocarbons (PAH) and halogenated compounds such as pentachlorophenol (PCP), polychlorinated biphenyl (PCB), and dioxin. Hydrocarbon toxicity and subsequently biodegradation by bacteria are dependent on a number of environmental factors including pH of soil/water, temperature, oxygen availability, and nutrient content [15]. Overall the toxicity of a pollutant is related to how well it can be degraded (used as a carbon source). This comes down to fundamental biochemistry of: (i) the presence of enzymes that are able to hydrolyze and cleave bonds in the organic toxin structure; (ii) the kinetics of the degradation vs competitive damaging processes.

Amphipathic organics (compounds with a hydrophilic and hydrophobic ends) can also be quite toxic to microbes as well as these partition into the cell membrane. The toxicity of an organic pollutant to bacterial cells is typically proportional to how easily the compound can be degraded by the bacterium through metabolic pathways. Degradation through metabolism would decrease the intracellular (and eventually external) concentration and thus lead to less effect on the cell. There are some general trends of organic compounds that inhibit a bacteria's ability to degrade them and thus are more toxic. To be more toxic to bacteria, the organic toxin would *generally* have:

- Increased branching
- Shorter aliphatic chain length
- Few polar groups
- Few hydrolysable groups
- Unsaturated aliphatic chains
- Double bonds to nitrogen atoms
- More rings in polyaromatics
- Increased halogen substitutions, particularly meta and ortho substitutions
- Quaternary bonded atoms.

Many xenobiotics also have hetero atom substitutions, particularly S, N, B, P. The research is thin on these, but it is clear that the presence of a heteroatom can signif
icantly change how well the compounds are acted on by bacteria and thus how toxic a compound is. This can lead to increased or decreased toxicity depending on the compound and the particular microbial strain.

These comments are highly generalized. As mentioned above the diversity of bacteria in the tree of life defines extreme diversity. Given that we have only studied a small percentage of all microbial life on the planet, there are sure to be exceptions to the general rules stated here. There is a consideration in the environmental microbiology field that somewhere on the planet there will be a bacteria that would be able to take a new toxin and degrade it.

3.4.1.2 Sites of organic toxin interactions within bacteria

Due to the hydrophobic and amphipathic nature of organics, the primary target is the bacterial cell membrane. As with the general discussion in Chapter 2, depending on the nature of the organic compound, it will either be transported into the cytoplasm or it will partition into the lipid bilayer and affect membrane-related functionalities. Compared to eukaryotic cells, the bacteria cell membrane is also the bioenergetics engine, where the electron transport chain (ETC) is contained (compared to the mitochondria and chloroplast organelles – which are in fact ancient prokaryotes engulfed by early cells). The result is that additional toxin events are possible at the bacterial cytoplasmic membrane compared to the eukaryotic cell membrane.

The difference in cell structure between Gram-positive and -negative bacteria can determine which cell type and cell region is preferentially targeted by a particular toxin. Pending the presence of any charge on the organic toxin, it may get bound up by the negatively charged LPS, such as with quaternary cation compounds. The peptidoglycan cell wall also contains charges and thus can also interact with incoming organics and pending the nature of the toxin-cell wall interaction can lead to a strengthening or weakening of the wall.

The bacteria cytoplasmic membrane (inner membrane for gram-negative), like the eukaryotic membrane, will have solute and ion concentration differences across it. But because there is also electron flow that "pumps" protons or ions out, there is a substantial proton electrochemical gradient (or proton motive force; PMF) and a resulting pH differential across this membrane. Thus organic compound partitioning can lead to reactive oxygen species (ROS) produced right at this critical membrane leading to lipid peroxidation resulting in alterations in membrane integrity. The lipid head group spacing changes can lead to ion permeability changes and subsequently uncoupling of the PMF or electrochemical gradient and loss of cell energetics. For this reason, many organic-based antimicrobials are very effective as they damage the energetic system of bacteria directly, whereas in eukaryotic cells the compounds have to travel a longer path to eventually partition to the mitochondrial matrix membrane to uncouple the PMF and or produce ROS from the ETC.

Once in the cytoplasm, of course, various vital cellular processes can be targeted as discussed in Chapter 2. A clear difference from eukaryotes is there is typically a lack of compartmentalization in a bacterial cell. Since there is no nucleus protecting the DNA in bacteria, the nucleic acids are immediately exposed to the toxin once in the cytoplasm. Many PAH will partition between nucleic acid bases leading to genotoxic effects. Similarly, all enzymes of metabolic systems are fully exposed.

3.4.1.3 Adaptation and resistance of bacteria cells to organic pollutants

We will now discuss a bacteria's abilities to tolerate organic pollutant stress. Given the remarkable diversity of the microbial world it is important to point out that unique biochemical mechanisms are involved. We will not review all, but highlight a few relevant examples to illustrate how microbes can respond to toxins. Overall, exposure to hydrophobic toxins such as organic solvents can result in changes in the cell membrane lipid composition [16]. Examples of such toxin solvents would be benzene, toluene, xylene, phenol, cyclohexane, ethers, long-chain alcohols (e.g., octanol), etc. The first exposure of a naïve bacteria to a pollutant can result in changes to the lipid composition in a temporal fashion.
Short-term:
- Cis to trans isomerization of double bonds in fatty acid chains.
- Hydration reactions to increase fatty acid saturation.

Longer-term:
- Exchange of lipids with increased polarity or change in head group charge.
- Change out lipids with longer or shorter acyl chain lengths.

During these changes, one may see an increase or decrease in membrane lipid content. An extreme example is the process of shedding lipid bilayer vesicles from the cell into the external environment around the cells in order to act as a sponge to mop up the toxin in order to dilute out what might finally reach the live cell. Shedding of membrane vesicles is a common stress response in bacteria [17]. Another approach is to increase the membrane lipid content that leads to wrinkling of the surface but provides more lipids to adsorb the shock of the toxin. Of course, the opposite would be to remove lipids so there is no "home" for the organic to partition into. Many of these findings were discovered using the Gram-negative *Pseudomonas putida* or the Gram-positive *Bacillus subtilis,* which are both fairly ubiquitous prokaryotes that occupy many different niches. *P. putida* is an organism of biotechnology importance in agriculture and for the bioremediation of organic pollutants.

Beyond lipid changes, other long-term adaptations may also occur. One sees a down-regulation and decreased accumulation of outer membrane (OM) porins in Gram-negative bacteria [18]. Porins are protein channels in the OM that allow the movement of solute molecules from the extracellular environment to the periplasm. Thus a reduction of these will lead to decreased toxin making it to the cell membrane. In part, this comes from their partition to the OM lipid bilayer and subsequently becoming trapped there. As the OM has no absolute critical function, this tends to be less damaging.

As opposed to the response of reducing porins, a similar outcome to reduce the amount of toxin getting into the cytoplasm, there can be an increase in expression and accumulation of multidrug resistance efflux pump (MDREP) transporters and other small molecule transporter systems [20]. These are multi-substrate, promiscuous transporters that can expel foreign organic pollutants from the lipid bilayer, cytoplasm, and periplasm. Thus, any organic toxin that is able to still get to the cell membrane or cytoplasm can be expelled to the external environment subsequently keeping the cytoplasmic membrane and internal cell concentration of the toxin quite low.

Other proteome changes include the upregulation of oxidative stress response, general stress response, energetic metabolism, fatty acid biosynthesis, cell envelope biosynthesis. Proteins downregulated are those involved in nucleotide biosynthesis and cell motility [19].

In summary, the bacterial cells' strategy to survive organic pollutant exposure is to adjust its membrane lipids to keep functionality, and then make changes to not let the toxin inside. And finally, if the organic toxin does get in, spit it back out again by a MDREP.

3.4.1.4 Biodegradability of organic pollutants (*Bioremediation*)

An important sidebar to the discussion on how organic pollutants are toxins to microbes is a discussion about the potential of microbes to degrade the pollutants. As discussed in Chapter 2, bioavailability is a measure of potential toxicity to an organism. In the case of the ability of a microbe to degrade a toxin, bioavailability becomes an important variable as typically the compound must be able to enter the cell to be metabolized (i.e., degraded). The term "bioremediation" refers to the use of biological organisms to biodegrade pollutants in order to remediate (return to natural state) marine, aquatic, or soil environments. Most bioremediation is performed by microorganisms (see also reviews [21–23]).

There are a number of factors that affect the biodegradability of an organic pollutant, many of which also relate to toxicity (Chapter 2). Although many of the factors have been mentioned previously, they are worthwhile to repeat here in this context. The factors can be a combination of the chemistry of the molecule itself and the physicochemistry of the environment. Considerations around the molecule include chemical structure, concentration, solubility, and physical properties (charge, hydrophobicity, functional groups). Physicochemical differences in the environment can significantly affect a microbes' physiology and fitness. These include the pH, temperature, redox potential, ionic strength and composition (types of ions), presence of appropriate organic and inorganic nutrients (e.g., nitrogen, sulfur, phosphorus) for growth, and in aquatic environments pO_2 and amounts and types of other gasses that may be available as terminal electron acceptors or carbon sources (e.g., H_2S, CO_2, CO, CH_4, H_2). Finally, for effective bioremediation, there must be the presence of the appropriate microbial species present in the population with the genetic potential to code for enzymes that will break down the pollutant.

Inevitably, the sudden addition of an organic pollutant toxin to an environment leads to a rapid decrease in the biodiversity of the organisms in that environment. This is the result of the killing off of any species that did not have the physiological fitness to tolerate the stress. Very quickly the bacteria population stabilizes to a community that is a mixture of tolerant and pollutant metabolizing species. In a diverse population, it is not necessary for a single species to completely degrade the toxin, as syntrophy is often employed [24]. One may imagine one bacteria starting the degradation (such as a dehalogenation reaction) then secreting for the next organism to take in and perform the ring cleavage reaction, and then a third organism carries out the final mineralization, with perhaps energy molecules like H_2 feeding back to the first in the series. This is Syntrophic metabolism.

Tolerant organisms are able to adapt to the toxin, and may or may not also contain genes that code for degradative enzymes. Those with specific genes present for metabolizing an organic pollutant may be able to mineralize the pollutant completely (conversion to CO_2 and H_2O). One of the early studied aromatic pollutant degradation metabolic pathways is that of the TOL plasmid from *Pseudomonas putida* for xylene and toluene catabolism pathways, the *xyl* operons [25]. Other tolerant species may not have specific toxin degradative genes, but have a metabolic reaction or pathway that the compound can be co-metabolized. In this case, an enzyme or metabolic pathway can accept a toxin that closely resembles a natural metabolite in chemical form. An example of this is the degradation of chlorinated solvents [26]. Although biotechnologists have explored specific species to target the degradation of specific pollutants, in natural environments organic pollutant degradation is typical via a combination of specific genes, syntropy, and co-metabolism.

Above, for the most part, one is considering natural compounds that are at high concentrations or spilled into an environment where the compound is not typically seen. To bacteria, various organic compounds are frequently seen in natural environments as plants release them as they decompose (in part by bacterial and fungal action). Molecules such as alkanes from fatty acids, alkenes, aromatics such as benzene, phenol, benzoic acid are quite common and thus bacteria have taken advantage and found ways to use these molecules as carbon sources. On the other hand, Xenobiotics, compounds that are produced by chemical synthesis for specific properties of interest in our industrial age, may not have any counterparts in the natural world.

Subsequently, xenobiotics are not only more problematic to higher organisms, but bacteria have a more difficult time degrading these molecules and as such are more toxic and subsequently they are recalcitrant in the environment.

3.4.1.5 Biochemistry of bacterial bioremediation

The key biochemical enzymes involved in organic degradation, particularly for aromatics, are oxygenases [27]. Monooxygenases catalyze the incorporation of a single atom of molecular oxygen into the substrate. There are two common types: aromatic hydroxylase and alkane hydroxylase. Dioxygenases catalyze the incorporation of both atoms of O_2 into the substrate. Some common dioxygenases are: benzene 1,2-oxygenase, catechol 1,2-dioxygenase, catechol 2,3-dioxygenase. The biodegradability of a substituted aromatic compound is affected by the ability to modify or remove one or more substitute groups in order to allow for the hydroxylation of two adjacent carbons. For aromatics, this gives an ortho di-hydroxylated ring (catechol). This is followed by the ring cleavage. This allows for ortho or meta fission of the ring. From this reaction, the now carboxylated molecule can be fed into the tri-carboxylic acid (TCA) cycle for energy and carbon utilization. Thus, the ability to degrade an aromatic pollutant is reliant on a bacterial species in the community to have enzymes that can produce the catechol intermediate.

   Halogenated compounds are quite toxic, as their presence makes the adjacent carbon-carbon bonds more difficult to hydrolyze. But since halogenated compounds do exist in nature, particularly chlorinated, bacteria have evolved to deal with these as well. Anaerobic reductive dechlorination with H_2 is quite thermodynamically favorable and thus can be used in a process referred to as dehalo respiration [28]. Dehalogenation reactions for bromine and particularly fluorine are energetically unfavorable and thus these reactions are much rarer in nature. This leads to the recalcitrant persistence of perfluorinated compounds. Generally, to date, for the organisms studied, dehalogenation is quite rare and only a few strains have been isolated capable of such reactions.

3.4.1.6 Summary of organic pollutant interactions with microbes

Most bacteria are remarkably resilient to the challenge of organic pollutants. As organics tend to be quite hydrophobic they partition into bacterial membranes. The bacteria respond by changing their lipids to maintain the functionality of membrane constituents. Many bacteria have multi-ligand binding efflux transporters that can remove persistent organics from the cell if they cannot be metabolized. The diversity of prokaryotic and archaea species gives huge metabolic potential to metabolize a wide variety of natural and anthropogenic organic compounds. Oxygenase enzymes are important in facilitating the breakdown to change a toxic organic compound into a carbon/energy source that is fed into the TCA cycle of central metabolism. This metabolism is exploited for bioremediation of polluted sites.

3.4.2 Toxic metal and metalloid ions

Microbes have evolved with metal ions in their environment and have biochemical processes in place to manage their stress at background concentrations. However, high concentration challenges can occur naturally such as a volcanic eruption or through anthropogenic activities such as exposure to mine leachates or tailings releases leave bacteria challenged to a pulse of metal ion stress. Another source of exposure comes from metals, as metal salts, alloys, or nanomaterials are increasingly being used as antimicrobials in infection control for humans, animal husbandry, livestock, and crop agriculture. Also in agriculture, we see their use as both dietary supplements and prophylactics of disease control. In the context of bacteria in the medical setting, metal ions have been used since antiquity for the treatment of many ailments. It was not until Sir Alexander Fleming introduced penicillin and the era of antibiotics in 1928 that led to a temporary decrease in their use as antimicrobials. However, in today's antimicrobial resistance (AMR) era, we see a huge resurgence in their use. Here in this section, we will overview the biochemical mechanisms in bacteria responsible for metal(loid) ion toxicity, tolerance, and resistance.

3.4.2.1 Metal(loid) toxicity

Generally metal(loid) toxicity to bacteria follows the hard-soft acid-base (HSAB) theory as discussed in Chapter 2. We see a trend to increased toxicity for metals acting as soft acids compared to harder acids. This leads to reactivity with biomolecules with bases as thiols or thioethers. The borderline metal acids have a pretense toward the imine-containing biochemical bases.

Considering the physicochemical properties of the metal(loids) (discussed in Chapter 2), the covalent index ($X^2_m\ r$) and the hydrolysis potential (pK_{OH}) were the best two variables modeled that correlated relative metal ion toxicities to bacteria [29], at least for a single marine organism growing planktonically. However, the more extensive study did not see any correlations with pK_{OH} for the soil organism *Pseudomonas fluorescens* growing either planktonically or as a biofilm [30]. For the soil bacterium, correlations of metal(loid) toxicity followed σ_p, ΔE^o, X_m, Log *AR/AW*, $X^2_m\ r$, and pK_{sp} for values plotting against the minimal inhibitory concentration (MIC) of planktonic bacteria. It was found that as in other studies, the regression analysis works better for class 2 elements that have partially or completely filled d-orbitals (the so-called heavy metals). When the bacteria were explored growing as a biofilm, few correlations were found with only slight correlations to the minimal biocidal concentration (MBC) to Log AR/AW or $X^2_m\,r$.

Recent studies have shed some important light onto the important effect that the media conditions that bacteria are grown in strongly influence the toxicity of metals to microbes. The growth of microbes in the lab and the environment relies on pH, osmolarity, ionic strength, and ion types, temperature, redox potential, carbon source and energy sources (electron donors and acceptors), and trace elemental nutrients (N, P, S, Ca, Mg, Mn, Fe, Cu, etc.). Differences in environmental conditions can strongly influence the overall metal atoms' charge and the thickness of the hydration layer around the ion, as well as coordination to various organic molecules; all of which will change the bioavailability to the bacteria. Additionally, such environmental differences strongly influence the genomic response mediating their physiological fitness state and thus their ability to tolerate the metal ion challenge.

Another consideration is whether a single type of bacteria is present in the environment or if a mixture of bacterial species are present. Striking differences are seen between a pollutant exposed microbial community and the metal tolerant strain *Cupriavidus metallidurans* [31]. In this study, inhibition of biofilm formation was differently correlated to σ_p, ΔE^0, X_m, and pK_{sp}. Even two non-related bacterial species growing together can strongly influence each other's tolerance to a metal challenge [32, 33]. Overall, prokaryotes growing as a biofilm are considered tolerant to antimicrobial challenge, and, in the case of metals, a biofilm can be as much as 10 to 100 times more tolerant than the planktonic form. However, this metal ion tolerance wears off and metals can still kill cells in a biofilm in a time-dependent fashion [34, 35].

3.4.2.2 Microbial biochemical targets of toxicity

First, a comment on the ability of silver, copper, and their alloys (e.g., brass) to control microbial growth. This is still very much a poorly understood phenomenon. Karl Nägeli in 1893 referred to what was called an "oligodynamic effect" [36], where it is considered that a few atoms of the metal are released from solid alloys and are able to "force" the cell to die. We will see in the discussions below that potentially the positive-charged metal ions distort the cell wall by bonding to negatively charged lipids. This allows the metal ion to enter into the cell. Once the metal ions are in the cell, they can subsequently bind to DNA, RNA, enzymes, and cellular proteins which eventually causes cell damage and/or death. Moving forward we will now discuss metal toxicity in the context of the metal atom or ion.

Bacteria can be damaged by a metal(loid) ion challenge very similar to what happens in eukaryotic cells. However, the bacterium cell structure is less compartmentalized and thus once inside the cell metals can access nucleic acids directly. Also, their bioenergetics are more vulnerable as their electron transfer chain (ETC) is contained in their cytoplasmic membrane and not compartmentalized in an organelle, although some prokaryotes do have cytoplasmic bodies for unique functions including biogenic processes such as photosynthesis.

The biochemical challenges and subsequent physiological damages that can occur in microbes as a result of metal ion exposure have already been outlined in Chapter 2 (see for more detailed description). Metal(loid) toxicity mechanisms to prokaryotes and archaea include the interaction of the metal(loid) with lipids; the exchange of the catalytic metal with a meta(loid) in the active site of an enzyme; genotoxic effects from reactive oxygen species (ROS) or metal ions binding to the DNA backbone; ROS damage to protein amino acids; metal ion coordination to functional/structural amino acids; exchange of a structural or regulatory metal in an protein/enzyme; solute carrier competition and inhibition; oxidation of R-SH containing biomolecules. The difficulty of delineating the mechanisms that are involved in metal(loid) toxicity is that metal(loid)s tend to have a variety of biochemical targets that can lead to multifactorial phenotypic symptoms.

The case of lipids is of particular interest. Prokaryotic membrane lipid content is considerably different from eukaryotic membranes and thus tend to be comparatively more negatively charged. This allows for metal cations to interact preferentially with lipids that can lead to the disruption of the fluidity and the dynamics of the membrane and thus change the functionality of integral membrane proteins. Also for gram-negative cells, the outer membrane outer leaflet is made primarily of lipo-polysaccharides (LPS), where the saccharo lipid A core is phosphorylated and carboxylated, and the inner core sugars can also be phosphorylated. This gives a very negatively charged surface for what would be the first interaction for a metal cation with the microbe. Thus, it is considered that metal cations get trapped and concentrated on the cell surface that can potentially magnify the adverse effects on the bacteria.

The oxidation of R-SH groups is another unique damage to prokaryotes compared to eukaryotes. The cytoplasm of prokaryotes tends to be far more reduced than that of eukaryotic cells, in part due to the ETC being directly exposed to the cytoplasm. The typical concentration of the thiol oxidative buffering compounds (glutathione, bacilithiol, redoxins) in prokaryotes is 10 mM and bacteria use NAD(P)H reductases to maintain their reduced state. Many metals react and oxidize these thiols (Ag^+, Cu^{1+}, Cu^{2+}, Se (IV), etc.) others will use R-SH as ligands and thus poison them (As(III), Te(IV)). The effect is raising the redox poise in the cytoplasm, which will subsequently affect metabolism and the thermodynamic efficacy of the electrochemical gradient for energy production.

A common observation upon exposure of bacteria to metal(loid)s is an observation of the production of ROS. As discussed in Chapter 2, there are a number of pathways to this outcome depending on the metal challenge. Fenton chemistry from Fe(III) can also be catalyzed biologically by Cu(II). But many other metals can lead to ROS through disruption of Fe homeostasis, particularly through the release of Fe from cofactors. Gallium can easily replace Fe^{3+} in cytochromes, thus releasing the iron for Fenton reactions. Several metals can also out-compete Fe from iron-sulfur centers [Fe-S], again leading to free iron release to catalyze ROS production.

Various metal(loids) including Ag(I), Cd(II), Co(II), Zn(II), Cr(VI), As(III), Se(IV) and Te(IV) can deplete the thiol pool through catalyzing oxidation (2 R-SH -> R-S-S-R) [33]. With less oxidation buffering, the naturally produced ROS from the ETC will increase in concentration and lead to irreversible damage. Thus, we see ROS can be produced indirectly through metals poisoning the microbial reduced thiol pools (glutathione and redoxins).

An example study evaluated 8 metal(loids) effects on *E. coli* for their ability to produce ROS vs thiol oxidation [37]. In this study, they found that Cu(II) and Cr(VI) are primarily ROS producing, whereas Co(II), Ag(I), Zn(II), and Te(IV) are primarily oxidative to the cellular RSH pool. As(III) and Se(IV) were found to be equally RSH reactive and ROS producing. The study also looked for changes in gene expression in biomarkers of cellular stress (*sodA* or *sodB*, *soxS*, *oxyR*, *rpoS*). Co(II) and Ag(I) did not induce any of these, Zn tended to repress these genes, Se(IV) upregulated them all, whereas the closely related Te(IV) only induced an *oxyR* response. Cr(IV) and Cu(II) both induced *sodA* and *soxS*. A curious observation was that if either of the general regulator genes *soxQ* or *marR* being present on the genome, there was increased lethality for several of the metal challenges but in different growth stages of planktonic vs biofilm. This study illustrates nicely that the different metals lead to remarkably different responses of the same bacterium to different metal(loid)s. Figure 3.5 summarizes sources of RSH and ROS reactions.

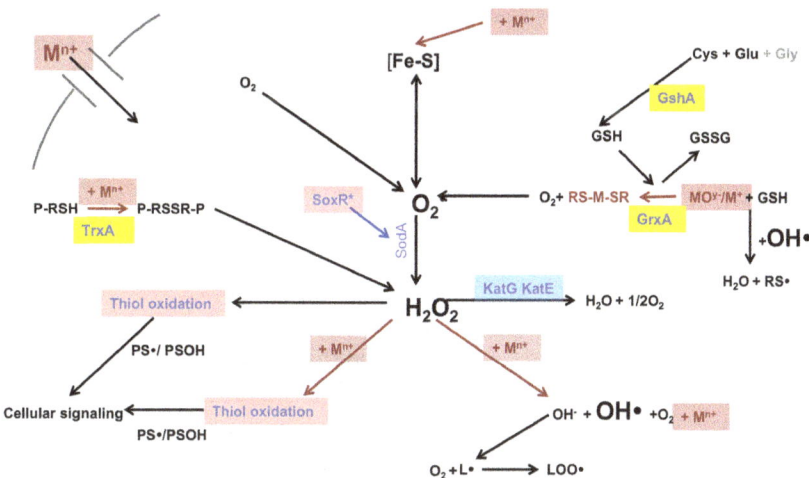

Figure 3.5: Biochemical processes altered by metal(loid) exposure. SoxR: general oxidative response regulator, particularly to superoxide dismutase (Sod) and catalase (Kat) and thioredoxin (TrxA), glutathione reductase (Grx), and glutathione synthesis (Gsh). Figure inspired from [37]. "P" in this figure refers to Protein, where "L" is Lipid, M^{n+} represents any metal cation, and MO^{y-} represents a metalloid oxyanion.

3.4.2.3 Systems biology understanding of metal(loid) challenge

Our understanding of key systems and genes found to be important upon metal challenge is remarkably incomplete. Most research looks for genes and systems involved in providing resistance as elaborated on below. However, the advent of various omics studies that survey all systems at once are beginning to provide us with lists of specific targets for various metals to bacteria, rather than the general view described above. Here a few examples will now be discussed briefly.

A metabolomics approach established that planktonic and biofilm growth states of bacteria have completely different physiological states and that these respond very differently to metal challenges [38]. The data suggested that biofilm physiology is more poised to deal with metal stress, with coordinated pathways, whereas the planktonic response was more chaotic. A proteomics-based study provided a model for the response of *Deinococcus radiodurans* to tellurite exposure [39]. This study validated that TeO_3^{2-} exposure led to ROS and that the ROS detoxification enzymes increased, but it also highlighted that protein folding chaperones also increased along with several key metabolic enzymes perhaps to deal with tellurite damage. Additionally, they observed that the cellular E^0 substantially decreased. This newer system study supports earlier work that showed tellurite uncoupled the ΔpH across the inner membrane of *E. coli* as well as depleted the ATP levels [40].

Chemical genomics (also called toxicogenomics) approach was used to find systems enriched upon Ag^+ [41], Cu^{2+} [42], and Ga^{3+} [43] challenge (Figure 3.6). The outcome of their study provides a systematic comparative view of three metal cations used as antimicrobials and demonstrates that these metal cations affect cellular systems very differently. Exploring the datasets, it was found that very few genes for tolerance or sensitivity were shared for all three metal ions. In fact, only two genes were found to be key for tolerance to all three. The folate binding protein, *ygfZ*, is considered to be involved in [Fe-S] cluster assembly, thus it was inferred to be critical to repair [Fe-S] damage from toxic metals. The other was *tolC*, the channel protein component of several MDREPs, thus implied critical for mediating the removal of metal ions and damaged biochemicals from the cell. Another impressive approach is metalloproteomics, where one identifies all proteins with a metal bound. This was performed in combination with metabolomics and gene expression on an Ag^+ challenge to *E. coli* [44]. This is an impressive study that noted that some biochemicals such as citrate and other tricarboxylic acid cycle metabolites actually enhance the bactericidal effect of Ag^+. The study identified that several proteins in the tricarboxylic acid cycle, glyoxylate cycle, glycolysis, translation, and regulation of intracellular pH were being specifically poisoned by Ag(I). These approaches are helping to build a comprehensive view of metal challenge to the model organism; however, there is still variation in findings, likely due to different media and strains used between groups.

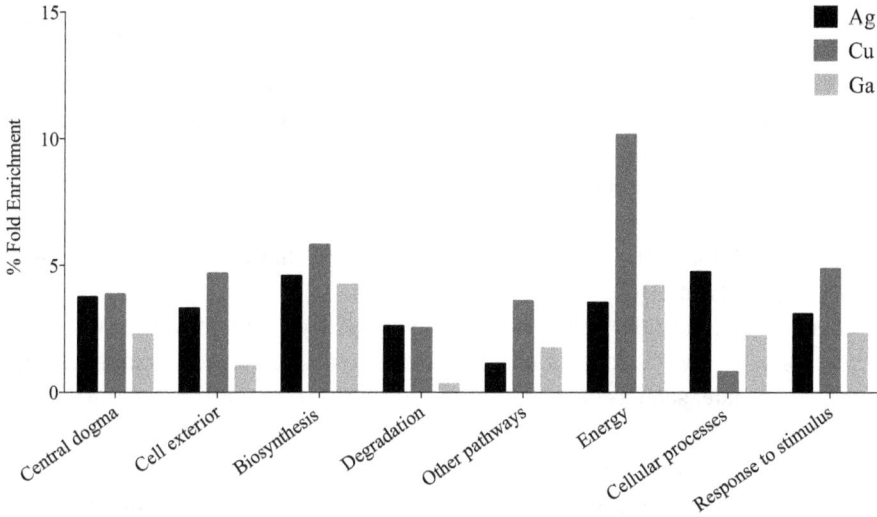

Figure 3.6: Percent enrichment for the Ag(I), Cu(II), and Ga(III) sensitive hits in *Escherichia coli* grown on Glucose minimal media with a metal challenge below the minimum inhibitory concentration. Enrichment was performed using Omics Dashboard from EcoCyc which calls attention to pathways and processes whose changes are statistically different; the significance value was $p < 0.05$. Adapted from [43]. (Credit. Dr. N. Gugala, Turner Research Group).

3.4.2.4 Resistance and tolerance toward metal(loid)s

Here we need to first define the difference between tolerance and resistance. Tolerance comes from the specific physiological state of the bacteria, whereas resistance is derived by specific mechanisms coded for by genes. We will speak of these separately, but they are in fact intertwined.

3.4.2.4.1 Resistance

The study of metal resistance mechanisms in bacteria traces back to the 1970s and through to the late 1990s with studies still being reported today. During these early years, specific metal or metalloid ion resistance determinants, consisting of genes or groups of genes (metal(loid) resistance genes; MRGs) were being isolated, sequenced, and mechanistically characterized. The MRGs are often organized into operons and were often identified on large conjugative plasmids and other mobile genetic elements. This early work performed in the pre-omics' era made great strides in exploring bacteria response to silver, nickel, cadmium, mercury, copper, arsenite/arsenate, and tellurite [45, 46]. This work led the field to appreciate that metal(loid) resistance in bacteria essentially follows a limited number of biochemical processes (Figure 3.7) that includes:

- Prevention of uptake. This may include proteins blocking transporters or channels, or performing biochemistry on lipids or carbohydrates to change binding and passive diffusion.
- Efflux. If the metal ion gets into the cell, an efflux transporter is coded for on the MRG to pump the metal back out again. These may be specific to one element (CadA for Cd(II) resistance) or multiple (Czr that mediates resistance to Co(II), Zn(II), Cd(II)).
- Chemical modification. The most common of this type is the oxidation-reduction reactions. Here MRGs code for oxidoreductases to act on the metal(loid) to change the redox state, such as changing the more toxic Cu(I) to Cu(II) or Cr(VI) to Cr(III). In some rare cases, there is a chemical modification to add an organic moiety. This may be methyl group(s) or alkylation or thiolation, where through the reactions with the cysteine of glutathione or similar molecules gives a metal-S covalent bonded complex. In the case of glutathione, this could be then transported out of the cell by the glutathione transporters or MDREPs. Another form of chemical modification is to reduce the metal(loid) ion to elemental form, this often leads to nano precipitate or crystallites inside or outside of the cell.
- Biomolecular repair and metabolic bypass. MRGs or native genes can code for oxidative repair mechanisms to repair: damaged biomolecules, particularly proteins and nucleic acids from ROS, or direct metal-catalyzed reactions. Additionally, an enzyme alternate can be coded for by an MRG that allows a replacement of a metal-sensitive enzyme in a metabolic pathway. This then provides a bypass reaction in the pathway around the damaged enzymatic step. The alternative enzyme may not have equivalent kinetics to the regular enzyme but allows the organism to sustain life under toxic metal stress.
- Sequestration and sorption. This is also referred to as metal sorption. The biochemistries of sequestration can be mediated through a variety of biomolecules. The charged EPS will act as an ion exchange matrix, coded MRGs can produce modified EPS and increase the amount of EPS found on cell surfaces. MRGs can code for specialized metal-binding organic molecules that can sequester metals and protect against their toxicity mechanisms. Specifically coded binding-proteins (metallothionein or similar) are overexpressed to act as a sponge to decrease the free metal ion content. Sequestration may also be achieved through the reduction of the metal ion to its elemental form leading to metal nano crystalline precipitates in or on the surface of cells.

Although many MRGs follow using a single mechanism for resistance, there are examples of MRGs that mix and match biochemical processes. Resistance to mercury includes MRGs that actually facilitate uptake into the cell, but at the same time never letting the metal ion be free (i.e., it is always protein bound). After uptake Hg^{2+} is

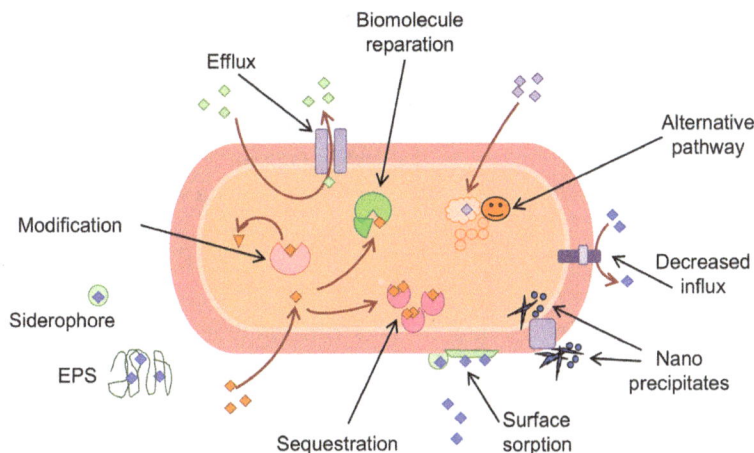

Figure 3.7: Summary representation of resistance mechanisms. See text for information. (Credit. Dr. N. Gugala; Turner Research Group).

reduced to Hg^0, which is actually volatile and easily percolates out of the cell. Arsenate enters the cell via phosphate transporters and then is reduced to arsenite by *arsC*, where the ion is then effluxed via the ATP-dependent transporter component (*arsAB*).

Finally, MRGs for tellurite continue to baffle microbiologists and biochemists. When a resistance mechanism is described for all other metals, a single biochemical mechanism is defined and that mechanism is found to be shared throughout all microbes and kingdoms of life. For tellurite, there are at least five different genetic determinants with completely different approaches [47]; unfortunately, most of the Te MRG's mechanisms still elude researchers.

3.4.2.4.2 Tolerance

A remarkable amount of work has been put forward to understand such processes of toxicity and resistance at the genetic, biochemical, and structural biology levels. Yet, even with the power of various omics approaches, there are still many metal-microbe interaction puzzles left to solve. Certainly, we understand a lot less of the general physiological changes that lead to tolerance toward metal (loid)s than we do about specific MRGs. Some clues are provided by those organisms that live in extreme ecosystem environments with comparatively high toxic metal loads. These include acid mine drainage, metal contaminated soil or aquatic systems, and deep-sea Volcanic vents. Some of these organisms continue to be studied, including *C. metallidurans*. This organism displays a wide spectrum of resistance mechanisms to many metals. Some of the resistance comes from MRGs on accessory genomes of plasmids, but others are built into the general physiology of the organism with enzymes that are not susceptible to ROS or metal-binding damage. There can also be unique features of the cell wall/membrane, where metal ions are unable to penetrate or become sequestered on the surface by metal-binding proteins or bound to secreted molecules.

Microbes in a biofilm have been found to be susceptible to metal ion toxins but only in a time-dependent fashion [35]. The slow rate of metal killing of biofilms vs planktonic cells may be due to a reaction-diffusion phenomenon attributed to the extracellular matrix. Both single and mixed-species biofilms are equipped with a huge degree of physiological variation that provides tolerance to metal(loid) challenges. Secreted metabolites from the unique physiology can catalyze bioinorganic reactions. The metabolic stratification will lead to changes in respiration, redox states and pH, and thus metal ion speciation. Many different types of microbial EPS molecules possess functional groups that can bind the metal ions like an ion-exchange matrix and thus facilitate sequestration of the metals within the biofilm matrix. Functional groups on cell surfaces also can sequester certain metal species [48], thus cells on the periphery of the biofilm can restrict the penetration of metals to the biofilm core. Metals bound to cell surface sites may also serve as nucleation centers for mineralization, enabling reactions with cell metabolites to form metal precipitates [49, 50]. Such sorption is also relevant to organic toxins.

Similar to sorption by proteins, bacteria secrete small molecular weight metal-binding molecules called metallophores. A specific type of metallophore is a siderophores specifically to capture Fe atoms to bring to the cell for import. Iron is essential for all life on earth and in many niches it is a limiting element. In the case of bacteria living in heavy metal loaded environments, the siderophores also bind other metal atoms, but the organisms also produce a variety of other metallophores to capture and limit their bioavailability. An interesting recent survey of metallophores is that for an interesting group of bacteria referred to as aerobic anoxygenic phototrophs [51].

The variety and amount of signaling molecules between cells are increased in biofilms vs planktonic. The molecular and metabolite signaling can also be between species and secreted molecules from one can protect other species from a metal ion challenge or make them even more susceptible [32]. Genetic rearrangements of phase variants can lead to the removal of a metal target and replacement with one that is not affected or produce a metal-binding protein or efflux system to reduce metal ion concentrations. Finally, so-called persister cells have time-dependent tolerance to metal ions in part due to physiological targets that are not active [52]. We should also not forget that the cells in the biofilm may also carry the genetic resistance MRG determinants and general MDREPs that can also efflux metal ions.

3.4.2.5 Summary of metal(loid) interactions to microbes

Metals can kill bacteria by damaging proteins and DNA, loss of membrane/cell wall integrity, and arresting cell metabolism where parts of this killing may be direct metal contact or indirectly mediated by ROS. It is important to stress that not all metal(loids) kill bacteria the same way and that different metal(loids) will stress different species' unique biochemistry in different ways. The toxicity can be dependent on the physiological state. There are a lot of parameters to consider around the speciation of metals and the physiological state. Bacteria have evolved a wide variety of methods to deal with metal(loid) toxicity either through specific metal resistance genes or specific physiological states. Toxicity of metal(loids) generally follows the Hard-soft acid-base theory by Pearson with increasing toxicity seen for the softer acid metals of lower oxidation state and larger polarizability. Resistance is mediated either by species' unique genetics or by metal(loid) resistance genes. Bacteria growing as a biofilm are physiologically different and are often affected differently by metals depending on the unique species and species present.

3.4.3 Antibiotics and antiseptics

To a bacterium cell, antibiotics, antiseptics, and disinfectants are toxins. These are toxins either created in nature or anthropogenically to control microbial growth to prevent and cure infections. Here we will overview the cellular targets of antibiotics as well as mechanisms of resistance. The importance of the antimicrobial resistance (AMR) problem will be brought to light, as medical issues from AMR will soon rival cancer-inducing pollutants as a major form of mortality.

First, we need to define some terms that are associated with antimicrobial agents. Disinfectants are antimicrobial compounds that are applied to non-living objects and surfaces with the purpose of killing any bacteria present. Antiseptics are applied to the surface of living organisms and tissues to prevent bacterial colonization. Antibiotics are compounds that are naturally produced by a microorganism (mostly the soil bacteria and fungi, particularly *Streptomycetaceae*) that are antagonistic to the growth of other microorganisms at low concentrations. Antibiotics can be administered outside and inside the body. Beyond the naturally produced "first-generation antibiotics," subsequent derivatives are produced by purposely chemically modifying these natural compounds. The activity of antimicrobial agents is either bacteriostatic (inhibits bacterial growth but does not kill so when the antimicrobial is removed the bacteria recover) or bactericidal (able to kill bacterial cells).

3.4.3.1 Antibiotic mode of action

Overall, each antibiotic type has different and highly specific modes of action. They also have different activity; the types of microbes they target, because of their unique biochemistries. As discussed in Chapter 2, toxins affect central processes of biology, and antibiotics are no different. Most antibiotics attack a specific step/

process of the central dogmatic flow of molecular biology. Examples of types and targets of antibiotics include:

Cell wall synthesis
- Beta lactams, Vancomycin; leads to osmotic lysis due to no mechanical strength
Cell Membrane
- Polymixins; uncoupling of membrane proton motive force.
Nucleic acid synthesis
- Folate synthesis
  - Sulfonamides; affects DNA replication
- DNA gyrase
  - Quinolones; affects DNA replication
- RNA Polymerases
  - Rifampin; prevents mRNA synthesis and thus protein production
Protein Synthesis
- RNA Polymerases (30S ribosomal subunit)
  - Tetracyclines, Aminoglycosides; inhibits amino acyl-tRNA binding
- RNA Polymerases (50S ribosomal subunit)
  - Macrolides, chloramphenicol (primarily bacteriostatic)

3.4.3.2 Antibiotic resistance

The action of the antibiotics may also be resisted by different mechanisms. There are many different antibiotic resistance mechanisms, each with its specific biochemical activities. Let's define some terminology first. Antimicrobial Resistance (AMR) typically refers to bacteria that are resistant to more than two antibiotics. The resistance is mediated by specific genetic determinants called antibiotic resistance genes (ARG). Multidrug resistance (MDR) is the term used for the resistance toward multiple types of antibiotics, antiseptics, antimicrobials, and or biocides. This term is also used to describe resistance to chemotherapy drugs, such as anti-cancer treatments. This resistance is typically mediated through multidrug efflux pump transporters (MDREPs) [20]. "Multidrug tolerant" (MDT) is a newer term that is used to describe the lack of killing from a variety of biocides and is a result of biofilm physiology.

One of the hallmark phenotypes of a biofilm is its high tolerance to antimicrobial agents. A primary general mechanism of bacterial cells to protect themselves from antimicrobials is to produce a robust extracellular matrix limiting penetration and diffusion of the antimicrobial into the biofilm. Thus, deeply buried cells will be exposed to sublethal concentrations – a serious concern regarding the use of antibiotics in the treatment of an infection. Beyond the physical structure of the biofilm, resistance is provided through different physiological states of bacteria. Phenotype variants (often seen as differential colony phenotypes on laboratory agar media) arise from differentially expressed operons and are considered to be regulated by *gacSA*. These variants can have markedly different gene complements and as such may not have the target present in the cell for a given antibiotic, or a resistance system now expressed. As previously mentioned, persister cells are dormant cell states that are neither dead nor alive [53], as such, they have no biochemical activity and thus have no antibiotic targets. The nature of biofilms leads to an increased frequency of both phase variants and persister cells.

Antibiotic resistance mechanisms from ARGs follow a defined set of biochemical mechanisms. Preventing antibiotic access to the biochemical target is key, thus reducing antibiotic uptake into the cell through either extracellular degradative enzymes or changes to porins and import transporters is found. If the antibiotic does get in, effluxing it back out or degrading it is key to keeping the concentration inside the cell below toxic levels (ability to repair damage faster than damage is done). Another resistance mechanism involves alterations to the antibiotic target so that the antibiotic does not bind, which can occur via a gene coding for a protein that binds the target or allosterically changing the conformation of the target protein. Finally, if an antibiotic destroys a key enzyme, a resistance gene that codes for a different enzyme with the same activity can provide an alternative pathway(s) to maintain the biochemistry of the cell.

3.4.3.3 Multidrug resistance efflux pumps (MDREP)

Multidrug resistance is an evolving issue around AMR [54]. This resistance is mediated through multidrug-resistant efflux pump (MDREP) transporters, which are prevalent in microbes in the environment [55]. These are of particular concern as they do not follow the biochemical principle of one enzyme for one substrate (biochemical specificity of an enzyme). But a single protein complex is able to accept and expel a wide variety of substrates across the membrane. These MDREP were initially discovered in the eukaryotic medical field in the 1970s providing resistance to cancer drugs. They were recognized further in the 1990s to contribute to bacterial multidrug resistance. A large number of bacterial drug exporters have now been characterized and fall into six superfamilies: small multidrug resistance (SMR); multidrug and toxin extruder (MATE); major facilitator superfamily (MFS); resistance nodulation and cell division (RND); ATP-binding cassette (ABC); and the most recent addition which is far less characterized, the proteobacterial antimicrobial compound efflux (PACE). Of course, many members of these families are involved in moving solutes as part of the bacterial physiology for metabolite export, cell homeostasis, and intercellular signal trafficking. But a fortuitous outcome of their promiscuous substrate selectivity leads them to accept substrates from organic and inorganic pollutants, pesticides, antiseptics (quaternary cation compounds), antibiotics, and various other biocides. MDREPs allow bacteria to live in highly toxic environments by maintaining the internal toxin concentration within repair levels. In the context of antibiotic resistance, they work synergistically with other tolerance/resistance mechanisms to further enhance the resistance levels leading to clinically relevant resistant profiles. MDREPs are often coded on mobile genetic elements such as plasmids or transposons and are easily horizontally transferred within and between species in various environments. These characteristics provide huge challenges to antimicrobial development and use.

3.4.3.4 Evolution of resistance

It is worth reviewing how new biochemical activities for resistance phenotypes evolve. If a bacterial population without resistance is exposed to biocide "X" at a lethal concentration, the bacteria die. However, random mutations occur through the normal error rate of DNA Polymerases (~1 per 10^7–10^9 bases). If a mutation gives greater fitness to the bacteria under the biocide stress, bacteria can survive and the strain with the resistance will multiply passing on the resistant trait. Thus the next time they are exposed to the X compound, the population survives. Such processes are amplified under sub-lethal concentrations where the bacteria are not killed, yet are struggling and thus there is pressure for selection for any mutation that provides even the slightest increase in fitness toward biocide X. This gives many incremental changes toward high resistance rather than the need for a dramatic change in a gene to acquire a high level of resistance right away.

Alternatively, a bacteria carrying ARG can join a population of sensitive strains. These strains can acquire the ARG through a number of means (transformation, transduction, conjugation) facilitating horizontal gene transfer (HGT). Those that are able to acquire the donated ARG, are able to survive biocide X and thus multiply and propagate. All these processes are far more efficient with bacteria cells in a biofilm due to the local high cell density which mediates HGT and provides a lower biocide penetration leading to sub-lethal concentrations.

3.4.3.5 Collateral effects of antimicrobial use

Side effects of antimicrobial use, or also called collateral resistance, is a phenomenon that has only recently become recognized that can contribute significantly to AMR. Oddly, some collateral effects can also lead to increased sensitivity to an antimicrobial. Collateral resistance arises from exposure of a bacteria species to one type of compound (antimicrobial, pollutant, or other toxin) and subsequently develops resistance to the compound. But the resistance mechanism is also capable of endowing resistance or sensitivity to other pollutants or biocides the cell has not previously been exposed to [56]. An example of this is the observation that using metal-based antimicrobials can induce MDREPs. Since both metals and antibiotics as well as other toxins can be effluxed by many MDREPs, one obtains cross-resistance of metals, antibiotics, and organic pollutants toxins. This has been seen by the use of copper or zinc for control of bacterial diseases in an agricultural setting and the metal exposed bacteria subsequently transferred to urban and/or medical settings leading to antibiotic resistance contributing to the AMR problem. If we consider the long-time misuse of antibiotics in the animal husbandry and fish farming industries, the potential of collateral transfer is becoming very concerning. A similar problem being ignored is the overuse of antiseptic biocides in commercial cleaning and anti-biofouling uses. Various antiseptics (quaternary ammonium/cation compounds) induce MDREPs as well providing a pool of conditioned bacteria for either HGT of MDREP genes or collateral resistances against antibiotics used to treat infectious diseases.

AMR issues are also arising from antibiotic manufacturing where the industry will dispose of poor or expired batches into aquatic systems leading to the destruction of local periphytons and/or the ability to generate a reservoir of evolved ARGs [57]. A similar issue around antibiotic pollution is the inadequate disposal practices and overdosing. Poor disposal is where an individual may not use their full prescription regime and dump the pills into the sewer system. This provides low antibiotic concentrations to give stress pressure to the environmental bacteria and subsequently evolve resistance. Overdosing also leads to unprocessed antibiotics secreted by the individual and again finding themselves in the wastewater. Thus, antibiotics like many other pharmaceuticals are turning up in wastewater treatment plants at increasing concentrations leading to poisoning of the bacteria in the treatment facility and downstream toxicity to aquatic/marine/soil periphyton.

3.4.3.6 Summary of antibiotics to microbes

Antibiotics tend to target very specific biochemical processes in a cell, typically inactivating a specific enzyme critical to growth or survival. Antibiotic resistance may be mediated by antibiotic resistance genes specific to a single antibiotic or class, or through more promiscuous multi-antibiotic resistance efflux pumps (for review see [58]). MDREPs also provide tolerance to antiseptics, various organic pollutants, and metal(loids). The biofilm growth state of bacteria are particularly tolerant to antibiotics and is one of their definitive phenotypes. AMR is predicted to lead to over 10 million deaths a year by 2050, and the WHO considers it to be a greater threat to the lifespan of humans than climate change.

3.5 Summary

Overall, our knowledge level of how various toxins effect prokaryotes is still rather superficial; however, we are seeing increased molecular and genetic mechanistic information that has become available over the past decades, and this trend will increase as we learn more about the importance of the microbial world to life on the planet. Issues in studying the toxicity of various compounds against prokaryotes and comparing results arise due to the vast differences in strains and species used in the studies as well as media type and growth conditions. Different combinations of genetics and fitness from environmental living conditions generate unique physiologies in prokaryotes, where each state will respond differently to toxic compounds and reflect in differing stress responses in each cell.

Suggested readings

Alexander M. Biodegradation and Bioremediation 2nd edition. Academic Press, 1999.
Das S, Dash HR. (editors) Handbook of Metal-Microbe Interactions and Bioremediation. CRC Press, 2017,(*collection of specific articles covering many different topics around metal-microbe interactions related to the topic of this textbook*)
Das S, (editor) Microbial Biodegradation and Bioremediation. Elsevier, 2014.
Ghannoum M, Parsek M, Whiteley M, Mukherjee PK. Microbial Biofilms. 2nd edition. ASM Press, 2015.
Gualerizi CO, Brandi L, Fabbretti A, Pon CL, (editors) Antibiotics; Targets, Mechanisms and Resistance. Wiley-VHC, Germany, 2014.
Harrison JJ, Turner RJ, Marques LLR, Ceri H. Biofilms. Am Sci. 2005, 93, 508–515.
Jin Y, Wu S, Zeng Z, Fu Z. Effects of environmental pollutants on gut microbiota. Environ Pollut. 2017, 222, 1–9.

Bibliography

[1]   Hug L, Baker B, Anantharaman K, *et al.* A new view of the tree of life. Nat Microbiol 2016, 1, 16048.
[2]   Editorial, Microbiology by numbers. Nat Rev Microbiol 2011, 9, 628.
[3]   Jolivet-Gougeon A, Bonnaure-Mallet M. Biofilms as a mechanism of bacterial resistance. Drug Discov Today Technol. 2014, 11, 49–56.
[4]   Yan J, Bassler BL. Surviving as a Community: Antibiotic Tolerance and Persistence in Bacterial Biofilms. Cell Host Microbe. 2019, 26, 15–21.
[5]   Hall CW, Mah TF. Molecular mechanisms of biofilm-based antibiotic resistance and tolerance in pathogenic bacteria. FEMS Microbiol Rev. 2017, 41, 276–301.
[6]   Penesyan A, Paulsen IT, Gillings MR, Kjelleberg S, Manefield MJ. Secondary Effects of Antibiotics on Microbial Biofilms. Front Microbiol. 2020, 11, 2109.
[7]   Kumar M, Jaiswal S, Sodhi KK, Shree P, Singh DK, Agrawal PK, Shukla P. Antibiotics bioremediation: Perspectives on its ecotoxicity and resistance. Environ Int. 2019, 124, 448–461.

[8]   Mukherjee S, Bassler BL. Bacterial quorum sensing in complex and dynamically changing environments. Nat Rev Microbiol. 2019, 17, 371–382.

[9]   Jenal U, Reinders A, Lori C. Cyclic di-GMP: Second messenger extraordinaire. Nat Rev Microbiol. 2017, 15, 271–284.

[10] Waters CM, Bassler BL. Quorum sensing: Cell-to-cell communication in bacteria. Annu Rev Cell Dev Biol. 2005, 21, 319–346.

[11] Irving SE, Choudhury NR, Corrigan RM. The stringent response and physiological roles of (p)ppGpp in bacteria. Nat Rev Microbiol. 2021, 19, 256–271.

[12] Lapouge K, Schubert M, Allain FH, Haas D. Gac/Rsm signal transduction pathway of gamma-proteobacteria: From RNA recognition to regulation of social behavior. Mol Microbiol. 2008 Jan;67(2): 241–253.

[13] Lewis K. Persister cells. Annu Rev Microbiol. 2010, 64, 357–372.

[14] Yachi S, Loreau M. Biodiversity and ecosystem productivity in a fluctuating environment: The insurance hypothesis. Proc Natl Acad Sci (USA). 1999, 96, 1463–1468.

[15] Koshlaf E, Ball AS. Soil bioremediation approaches for petroleum hydrocarbon polluted environments. AIMS Microbiol. 2017, 3, 25–49.

[16] Murínová S, Dercová K. Response mechanisms of bacterial degraders to environmental contaminants on the level of cell walls and cytoplasmic membrane. Int J Microbiol 2014, 873081.

[17] Mozaheb N, Mingeot-Leclercq MP. Membrane Vesicle Production as a Bacterial Defense Against Stress. Front Microbiol. 2020, 11, 600221.

[18] Roma-Rodrigues C, Santos PM, Benndorf D, Rapp E, Sá-Correia I. Response of Pseudomonas putida KT2440 to phenol at the level of membrane proteome. J Proteomics. 2010, 73, 1461–1478.

[19] Santos PM, Benndorf D, Sá-Correia I. Insights into Pseudomonas putida KT2440 response to phenol-induced stress by quantitative proteomics. Proteomics. 2004, 4, 2640–2652.

[20] Li X-Z, Elkins CA, Zgurskaya HI, (editors) Efflux-mediated Antimicrobial Resistance in Bacteria: Mechanisms, Regulation and Clinical Implications. Springer Publishing, Switzerland, 2016.

[21] Abbasian F, Lockington R, Mallavarapu M, Naidu R,. Comprehensive A. Review of Aliphatic Hydrocarbon Biodegradation by Bacteria. Appl Biochem Biotechnol. 2015, 176, 670–699.

[22] Gkorezis P, Daghio M, Franzetti A, Van Hamme JD, Sillen W, Vangronsveld J. The Interaction between Plants and Bacteria in the Remediation of Petroleum Hydrocarbons: An Environmental Perspective. Front Microbiol. 2016, 7, 1836.

[23] Varjani SJ. Microbial degradation of petroleum hydrocarbons. Bioresour Technol. 2017, 223, 277–286.

[24] Morris BE, Henneberger R, Huber H, Moissl-Eichinger C. Microbial syntrophy: Interaction for the common good. FEMS Microbiol Rev, 2013, 373, 384–406.

[25] Ramos JL, Marqués S, Timmis KN. Transcriptional control of the Pseudomonas TOL plasmid catabolic operons is achieved through an interplay of host factors and plasmid-encoded regulators. Annu Rev Microbiol. 1997, 51, 341–373.

[26] Semprini L. Strategies for the aerobic co-metabolism of chlorinated solvents. Curr Opin Biotechnol. 1997, 8, 296–308.

[27] Broderick JB. Catechol dioxygenases. Essays Biochem. 1999, 34, 73–89.

[28] Fincker M, Spormann AM. Biochemistry of Catabolic Reductive Dehalogenation. Annu Rev Biochem. 2017, 86, 357–386.

[29] Newman MC, McCloskey JT. Predicting relative toxicity and interactions of divalent metal ions: Microtox bioluminescence assay. Environ Toxicol Chem 1996, 15, 275–281.

[30] Workentine ML, Harrison JJ, Stenroos PU, Ceri H, Turner RJ. *Pseudomonas fluorescens* view of the periodic table. Environ Microbiol. 2008, 10, 238–250.

[31] Frankel ML, Demeter MA, Lemire J, Turner RJ. Evaluating metal tolerance capacity of microbial communities isolated from the Alberta oil sands processed water. PLOS One. 2016, 11, e0148682.

[32] Monych NK, Turner RJ. Multiple compounds secreted by *Pseudomonas aeruginosa* increase the tolerance of *Staphylococcus aureus* to the antimicrobial metals copper and silver. mSystems. 2020, 5, e00746–20.

[33] Lemire J, Harrison JJ, Turner RJ. Antimicrobial activity of metals: Mechanisms, molecular targets and applications. Nat Rev Microbiol. 2013, 11, 371–384.

[34] Harrison JJ, Ceri H, Stremick CA, Turner RJ. Biofilm Susceptibility to Metal Toxicity. Environ Microbiol. 2004, 6, 1220–1227.

[35] Harrison JJ, Ceri H, Turner RJ. Multimetal resistance and tolerance in microbial biofilms. Nat Rev Microbiol. 2007, 5, 928–938.

[36] Nägeli K. Wilhelm Über oligodynamische Erscheinungen in lebenden Zellen. Neue Denkschriften der Allgemeinen Schweizerischen Gesellschaft für die Gesamte Naturwissenschaft, XXXIII. 1893.

[37] Harrison JJ, Tremaroli V, Stan MA, Chan CS, Vacchi-Suzzi C, Heyne BJ, Parsek MR, Ceri H, Turner RJ. Chromosomal antioxidant genes have metal ion-specific roles as determinants of bacterial metal tolerance. Environ Microbiol 2009, 11, 2491–2509.

[38] Booth SC, Workentine ML, Wen J, Shaykhutdinov R, Vogel H, Ceri H, Turner RJ, Weljie AM. Differences in metabolism between the biofilm and planktonic response to metal stress. J Proteome Res 2011, 10, 3190–3199.

[39] Anaganti N, Basu B, Gupta A, Joseph D, Apte SK. Depletion of reduction potential and key energy generation metabolic enzymes underlies tellurite toxicity in *Deinococcus radiodurans*. Proteomics. 2015, 15, 89–97.

[40] Lohmeier-Vogel EM, Ung S, Turner RJ. In vivo 31P nuclear magnetic resonance investigation of tellurite toxicity in Escherichia coli. Appl Environ Microbiol. 2004, 70, 7342–7347.

[41] Gugala N, Lemire J, Chatfield-Reed K, Yan Y, Chua G, Turner RJ. Using a chemical genetic screen to enhance our understanding of the antibacterial properties of silver. Genes 2018, 9, 344.

[42] Gugala N, Chatfield-Reed K, Turner RJ, Chua G. Using a chemical genetic screen to enhance our understanding of the antimicrobial properties of gallium against *Escherichia coli*. Genes 2019, 10, 34.

[43] Gugala N, Salazar- Alemán DA, Chua G, Turner RJ. Using a chemical genetic screen to enhance our understanding of the antimicrobial properties of copper. Metallomics. 2022, 14, mfab071.

[44] Wang H, Yan A, Liu Z, Yang X, Xu Z, Wang Y, et al. Deciphering molecular mechanism of silver by integrated omic approaches enables enhancing its antimicrobial efficacy in *E. coli*. PLoS Biol. 2019, 17, e3000292.

[45] Silver S, Phung LT. Bacterial heavy metal resistance: New surprises. Annu Rev Microbiol. 1996, 50, 753–789.

[46] Hobman JL, Crossman LC. Bacterial antimicrobial metal ion resistance. J Medl Microbiol. 2014, 64, 471–497.

[47] Taylor DE. Bacterial tellurite resistance. Trends Microbiol. 1999, 7, 111–115.

[48] Fein JB, Daughney CJ, Yee N, Davis TA. A chemical equilibrium model for metal adsorption onto bacterial surfaces. Geochim Cosmochim Acta. 1997, 61, 3319–3328.

[49] Flemming H-C, Sorption sites in biofilms. Water Sci Technol. 1995, 32, 27–33.

[50] Golby S, Ceri H, Marques LLR, Turner RJ. Mixed species biofilms cultured from oil sands tailings pond can biomineralize metals. Microbiol Ecol. 2014, 68, 70–80.

[51] Kuzyk SB, Hughes E, Yurkov V. Discovery of Siderophore and Metallophore Production in the Aerobic Anoxygenic Phototrophs. Microorganisms. 2021, 9, 959.

[52] Harrison JJ, Turner RJ, Ceri H. Persister cells, the Biofilm Matrix, and Tolerance to Metal Cations in Biofilm and Planktonic *Pseudomonas aeruginosa*. Environ Microbiol 2005, 7, 981–994.

[53] Lewis K. Persister cells. Annu Rev Microbiol. 2010, 64, 357–372.

[54] Piddock LJ. Clinically Relevant Chromosomally Encoded Multidrug Resistance Efflux Pumps in Bacteria, Clin Microbiol Rev. 2006, 19, 382–402.

[55] Brown D, Demeter M, Turner RJ. Prevalence of multidrug resistances efflux pumps (MDREP) in Environmental Communities. in Microbial Diversity in Genomic Era Academic Press, Elsevier, USA, (Editors: Das S, Dash HR) Chapter 31. 545–557, 2019.

[56] Roemhild R, Andersson DI. Mechanisms and therapeutic potential of collateral sensitivity to antibiotics. PLoS Pathog 2021, 17, e1009172.

[57] Nijsingh N, Munthe C, Larsson DGJ. Managing pollution from antibiotics manufacturing: Charting actors, incentives and disincentives. Environ Health. 2019, 18, 95.

[58] Blair JM, Webber MA, Baylay AJ, Ogbolu DO, Piddock LJ. Molecular mechanisms of antibiotic resistance. Nat Rev Microbiol. 2015, 13, 42–51.

Maryam Doroudian and Jürgen Gailer

Chapter 4
Toxic metal(loid) species at the blood–organ interface

4.1 Introduction

Although environmental pollution has been known since Hippocrates, it is somewhat surprising that two closely related issues have only recently been recognized. The first is the relatively common misperception that environmental pollution is only an environmental problem, while the second is the enormous impact of environmental pollution on human health, the magnitude of which has only recently been quantified [1]. Both issues are directly related to the fact that over 140,000 new chemicals have been produced on an industrial scale since 1950. Today, still only half of the 5,000 high-volume chemicals have undergone any testing for safety or toxicity [1]. While Rachel Carson's book *Silent Spring*, which was published in 1962, identified the exposure of organisms to pesticides as an important health issue, certain human populations, including children, are today exposed to a cocktail of these and a variety of other environmental chemicals.

In terms of human exposure to environmental chemicals, one needs to be aware of two fundamental differences between organic pollutants (e.g., polycyclic aromatic hydrocarbons) and inorganic pollutants (e.g., As). The first difference is that inorganic pollutants are elements that cannot be broken down and are therefore inherently persistent in the environment, while most organic pollutants can be degraded by ultraviolet (UV) light and/or microorganisms into water and carbon dioxide (see Chapter 3). The second difference between these pollutant categories is that ~90 elements are present in the earth's crust where they are "locked up" in form of minerals (e.g., realgar or αAs_4S_4) and geological formations. Thus, if elevated concentrations of any inorganic pollutant are detected, for example, in a ground water sample, it is critical to first consider if the inorganic pollutant has a "natural origin" (i.e., mobilization from a mineral/geological formation in the subsurface into the drinking water where it can affect millions of people [2]) before blaming "industrial sources."

In this chapter, we will focus entirely on the toxic metal(loid)s As, Cd, Pb, and Hg or rather their corresponding toxic metal(loid) species (TMS), which can bioaccumulate in certain food crops [3] and thus eventually in human tissues [4]. From a public health

Acknowledgments: The author would like to acknowledge funding from the Natural Sciences and Engineering Research Council of Canada.

Maryam Doroudian, Jürgen Gailer, Department of Chemistry, 2500 University Drive NW, University of Calgary, Calgary, AB, Canada, e-mail: jgailer@ucalgary.ca

https://doi.org/10.1515/9783110626285-004

point of view, the chronic exposure of humans to TMS represents a considerable long-term problem for future populations. Before we confront the status quo, however, it is instructive to provide a brief historical perspective to understand how environmental pollution with TMS has become a truly global problem. Organic pollutants must be considered as well, but are not the main focus here.

4.1.1 Historical perspective on human exposure to toxic metal(loid)s

Humans have been unwittingly exposed to TMS since we first learned to use metals, in fact, since we first used fire [5]. While ancient civilizations already knew lead (Pb) and mercury (Hg), arsenic (As) was not discovered until 1250 AD by Albertus Magnus and cadmium (Cd) in 1817 by Friedrich Strohmeyer. As early as 400 BC, Hippocrates recognized that human health – and, by extension, the health of all organisms on Earth – is inextricably linked to the geochemistry of their surrounding environment. From a purely inorganic chemistry point of view, this observation can be readily explained by the fact that every organism needs to continuously exchange chemical matter contained in food and drinking water with its environment in order to maintain its health and wellbeing [6]. In this context, it is crucial to remember that the onset of the industrial revolution (i.e., the manufacturing of a wide range of consumer goods on a massive scale) coincided with a dramatic increase in the emission of Pb^{2+}, Hg^{2+}, As^{III}, and Cd^{2+} into the environment for two reasons. To begin with, fossil fuels, including toxic metal(loid)-laden coal, were needed to power various manufacturing processes. Second, the manufacturing of consumer goods required large quantities of chemical elements as building blocks, which had to be mined on an industrial scale. The elements Cd and Pb, for example, were utilized to produce the paint pigments CdS (yellow) and $PbCO_3$ (white) [4], while elemental Hg was used to manufacture barometers, mirrors and thermometers. The metalloid As served as a building block to synthesize a variety of pesticides, wood preservatives and salvarsan, the first chemotherapeutic drug (also referred to as a "magic bullet") to treat patients who suffered from syphilis, a bacterial infection. Early on, however, it was recognized that industrial workers in plants that manufactured As, Cd, Pb, and Hg and/or their compounds developed cardiovascular disease and neurodevelopmental disorders [4], but because only a comparatively small number of people were affected, the danger that is associated with human exposure to these remained concealed.

It took four major poisoning incidents for humankind to learn that chronic human exposure to comparatively low daily doses of Pb^{2+}, Cd^{2+}, Hg^{2+}, and As^{III} is associated with toxic effects [4]. The first one happened between 1850 and 1930 when millions of people were exposed to Pb^{2+} from water pipes, which caused "plumbism" and eclampsia (i.e., seizures and/or unexplained coma during pregnancy or postpartum in women). Another incident involved Cd^{2+} and took place in Japan, where people used water from a river downstream of a mine to irrigate rice. This event resulted in severe kidney disease predominantly in women and was referred to as Itai-Itai disease in Japanese, which means "ouch-ouch." Yet another incident occurred in the late 1950s in Japan, where fishermen in Minamata ingested CH_3Hg^+-contaminated fish, which triggered a neurological disorder named "Minamata disease." The last incident – the biggest mass poisoning incident in history – started to unfold in the late 1980s in West Bengal/India and Bangladesh and involved unsafe levels of As^{III} [i.e., $As(OH)_3$] in drinking water. The resulting disease is called "arsenicosis," which currently afflicts an estimated 77 million people in this part of the world [7].

4.1.2 Present exposure of human populations to toxic metal(loid)s

Human activities, such as fossil fuel consumption and the mining industry, emit quantities of As, Cd, Pb, and Hg into the environment that rival or exceed those released from natural sources (volcanism, chemical weathering, etc.) and thus profoundly affect entire ecosystems in certain parts of the world [8]. While these elements naturally occur in the earth's crust, their multiple medical, industrial, domestic, agricultural, and technological applications have led to their dispersal in the global urban environment. The adverse health effects associated with humans' exposure to comparatively low doses of these prompted the World Health Organization (WHO) to include them into their "Ten elements of major public health concern."

Relatedly, some scientists claim that we live in "the Anthropocene," defined as a geological age in which human activities have a measurable impact on the global environment [9]. Due to the continuous global emission of toxic metal(loid)s into the biosphere, certain human populations are therefore inadvertently exposed to higher daily doses of TMS [10] and other industrial chemicals [11] than ever before. In principle, humans can be chronically exposed to TMS via three exposure pathways: inhalation, ingestion, and dermal contact. We also need to differentiate between environmental exposure (e.g., the ingestion of contaminated food with one or more TMS) and/or occupational exposure (e.g., the inhalation of toxins at the workplace for up to 8 h/day). Factors that significantly affect the outcome of the chronic human exposure to TMS include the molecular form of the toxic metal (e.g., Hg^{2+} is predominantly toxic to the kidneys, while CH_3Hg^+ adversely affects the brain), the dose (acute vs chronic exposure), the age (vulnerability: fetus > child > adult), the sex [12], the health status as well as the length of exposure (weeks, months, years, and/or decades).

4.1.3 Public health concerns of chronic human exposure to toxic metal(loid)s

There are five conceptual reasons why chronic human exposure to TMS matters:

– Past pollution incidents have demonstrated that human exposure to small daily doses of TMS (30–600 µg range/day) – which are known to bioaccumulate in tissues over time – is associated with severe health effects (e.g., Minamata disease, Itai-Itai disease, arsenicosis, plumbism).

– Blood and urine analysis has verified that humans are exposed to TMS, but we don't know what these concentrations mean in terms of their health relevance [13].

– The prevalence rate of certain grievous human diseases is skyrocketing, but what causes their onset is often unknown. Type 2 diabetes, for example, is projected to rise from 220 million people that are affected today to 300 million people by 2025 [14]. Also, autism spectrum disorders are now recognized to affect up to 1% of the population and are expected to be a major future public health concern due to their early onset, lifelong persistence, and high levels of associated impairment [15].

– 9 million people died in 2015 because of exposure to environmental pollutants, corresponding to 15% of all deaths worldwide [1]. This estimate likely represents an underestimation, as the cancer risk is often calculated only using individual contaminants, but not their combination [16].

– Pollution is costly. Global welfare losses attributed to pollution are estimated at US $4.6 trillion per year, and pollution mitigation can yield large net gains for human health and the economy [1].

Taken together, it appears timely to address the profound knowledge gaps that exist in terms of assessing how chronic TMS exposure in human populations allows one to determine the true global disease burden and to develop low-cost palliative treatments.

4.1.4 Knowledge gaps regarding chronic human exposure to toxic metal(loid) species

It is instructive to identify knowledge gaps that pertain to Cd^{2+}, yet similar knowledge gaps exist for numerous other TMS. Human exposure to Cd has long been linked to lung cancer, osteoporosis, renal dysfunction [17], and to exert adverse effects on the testes [18]. More recent studies have suggested that this metal species also adversely affects the cardiovascular system [19], the endothelium [20], and is implicated in the etiology of stroke [21], atherosclerosis [22–24], age-related macular degeneration [25, 26], cataract formation in the eyes [27], neurodevelopment in children [28], anemia [29], amyotrophic lateral sclerosis (ALS) [30], immunotoxicity [31], spontaneous abortion [32], renal cancer [33, 34], low birth weight [35], and diabetes [36].

Some biochemical aspects which have not received the attention they deserve in terms of delineating how individual TMS adversely affect target organs are:

i)   determine the chemical nature of all TMS-derived metabolites that are formed in the bloodstream (e.g., complexes with plasma proteins [37] and up to 400 small molecular weight metabolites [38]), which will then impinge on target organs [39, 40].

ii) identify all uptake mechanisms that translocate individual TMS metabolites into target organs [41, 42].

iii) identification of all dietary components (e.g., vitamin C [43]) which modulate the metabolism of the TMS in the bloodstream (i.e., to enhance the excretion of a TMS via the kidneys) and may therefore result in a decreased uptake of essential trace elements, such as Se into organs [6, 44, 45].

Therefore, establishing causal links between TMS exposure and specific adverse health effects crucially requires a comprehensive understanding of the underlying biomolecular mechanisms that happen in the bloodstream and within organs (Figure 4.1).

4.1.5 Biochemical mechanisms involved in the chronic toxicity of TMS in humans

Mammalian organisms continually exchange chemical elements with their environment by ingesting food/drinking water and excreting waste products. This dynamic exchange is mediated by the gastrointestinal (GI) tract, which effectively absorbs nutrients, such as minerals, vitamins, amino acids, fatty acids, and sugars from the diet. The GI tract, however, also absorbs environmental pollutants from the diet to various degrees, as is evidenced by the presence of a variety of organic and inorganic pollutants that have been accurately quantified in the bloodstream of the average population [13]. Therefore, the GI tract effectively represents a double-edged sword as it absorbs both nutrients and TMS from the diet into the bloodstream, from where the latter can then be delivered to organs (Figure 4.2).

To unravel toxicologically relevant biochemical mechanisms involved in the chronic exposure to pollutants, we need to recognize that nutrients *and* TMS dynamically "flow" through organisms. Accordingly, it is important to characterize their interaction with biomolecules/cells of an organism if we want to better understand their

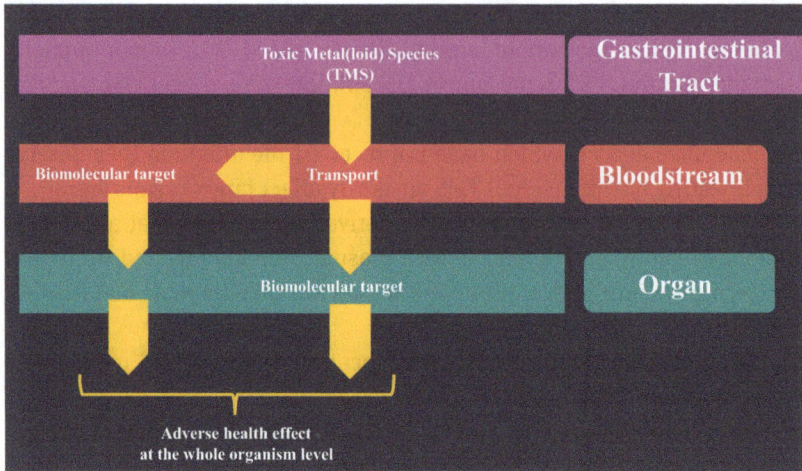

Figure 4.1: Simplified scheme of principal interactions of a toxic metal(loid) species (TMS) with biomolecules during its metabolism in mammals. While these interactions can unfold in the bloodstream and/or in organs, it is important to distinguish between two types of interactions that can unfold therein. The first kind (downward orange arrows) refers to the transport of the TMS to a particular organ where organ damage happens. The second kind (orange arrow pointing to the left) is associated with the TMS targeting a biomolecule in the bloodstream which will eventually result in organ damage over time.

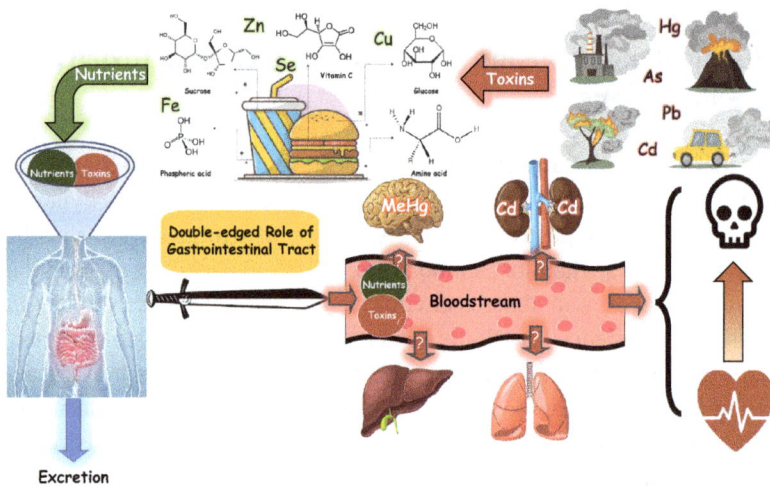

Figure 4.2: Conceptual depiction of the "flow" of nutrients and toxins through a human organism, which includes their absorption from the GI tract into the bloodstream, their subsequent distribution to organs followed by their eventual excretion.

chronic toxicity. Based on the "flow" of nutrients and TMS through organisms, we need to distinguish between interactions that occur in the bloodstream (primary interaction) and those that unfold within organs (secondary interaction). Thus, uncovering the mechanism of chronic toxicity thus requires us to probe the interactions between nutrients and TMS at the blood–organ interface and within target organs. Notably, the biochemical transformations of TMS that unfold in the bloodstream must be integrated with the biochemical events that unfold within organs to establish the mechanism of chronic toxicity in the whole organism. Before we even consider how to probe these highly complex TMS interactions, it is important to think about the experimental design to obtain meaningful results [8].

4.1.5.1 The dose–response approach

To evaluate the effect of any TMS on a mammalian organism, a relationship must be established between the dose (i.e., µg of TMS per kg body weight) and the response (e.g., the number of individuals suffering from of an adverse outcome) in any given population. Ideally, individuals should be exposed to a single TMS. An unfortunate poisoning incident in Iraq in the 1970s involved the exposure of thousands of people to a defined dose of CH_3Hg^+ per day via food (this compound was used as a seed dressing to prevent spoilage during transport) [46]. While a linear dose–response relationship could be established, there are three reasons why this general approach to assess the risk associated with human exposure to a toxin (e.g., a TMS) is not feasible:
1. Based on the obtained dose–response relationship, one cannot extrapolate what would happen in a real-life situation if lower doses of CH_3Hg^+ were ingested.
2. Humans are exposed to a mixture of organic and inorganic pollutants (multi-pollutant exposure), and different individuals may absorb different fractions of pollutants depending on their eating habits.
3. One must also consider an individual's genetic make-up, gender, and age, as these factors may significantly influence the outcome [47].

Taken together, it is evident that the dose–response relationship does not represent a feasible approach to assess the risk that is associated with the exposure of humans to different TMS. Thus, a better approach must be developed in order to gain new insights.

4.1.5.2 The 'molecular toxicology' approach

There is an alternative approach which requires probing all biochemical interactions of a TMS and its metabolites within an organism at the molecular level to identify all toxicologically relevant biomolecular mechanism(s) of action that may result in an adverse health effect [48]. Determining the daily dose of all TMS in an exposed human individual is practically impossible due to the prohibitively expensive costs associated with the fact that several exposure pathways would have to be considered [4]. Nevertheless, so-called biomonitoring studies can provide relevant information about the exposure of human populations to toxic metal(loid)s [4]. These studies report on the concentration of environmental chemicals in easily accessible fluids, such as human blood and/or urine, that are collected from people of different age, sex, and ethnicity. While this approach is of practical use because it provides background concentrations of chemicals in the average population, we do not know what these concentrations (which can be exceedingly low) mean. Pb is a notable exception as a blood threshold concentration exists, and if it is exceeded, treatment (e.g., chelation therapy) is usually initiated right away. Another flaw of biomonitoring studies is that they do not provide any information about the concentration of TMS within organs (e.g., Cd is known to accumulate in the kidneys, which is linked to kidney failure). The most significant deficiency of many biomonitoring studies is that the total concentration of toxic metal(loid)s in human blood and/or urine does not provide any information about which molecular form(s) (i.e., the TMS and/or its metabolites) are present. The analysis of hair, however, represents a notable exception as it could be demonstrated that human dietary exposure to CH_3Hg^+ could be accurately assessed [49].

To make further progress in molecular toxicology, four fundamental components must be considered: three that are related to the experimental design to probe the interaction of a TMS with relevant biomolecules and one that is related to the interpretation of the result(s) (Figure 4.3):

1. Selecting an "analytical tool" which is readily available and appropriate to probe the fate of any given TMS in a complex biological tissue/fluid.
2. Identifying a "biological fluid/tissue" where novel and toxicologically relevant interactions of the TMS with biomolecules and/or nutrients are likely to occur.
3. Deciding on a "conceptual approach" to probe a TMS-biomolecule interaction either bottom-up (reductionistic approach) or top-down (addressing biological complexity head-on) to ultimately explain how a small dose can result in a big adverse/toxic effect at the organ level (Figure 4.4) [22, 50, 51].
4. Integrating the obtained results (from step 3) with what is known about the metabolism of the TMS to establish a comprehensive "toxicological mechanism of action" in the whole organism.

We will now point out how each of these problems can be addressed and share some recent results from our research efforts.

4.1.6 Selecting an analytical tool

To reveal the fate of a TMS in any biological fluid (e.g., blood plasma, cerebrospinal fluid) and/or solid tissue (e.g., the brain) one needs to "observe" it and its metabolites by using appropriate analytical tools. What these biological fluids/tissues have in common is their biological complexity, which means that they contain thousands of biomolecules (proteins, amino acids, sugars, peptides, etc.). Therefore, one must select appropriate tools to observe the TMS in these complex "molecular habitats." This task is equivalent to find a needle (a TMS) in the haystack (thousands of biomolecules), which translates into a considerable separation problem.

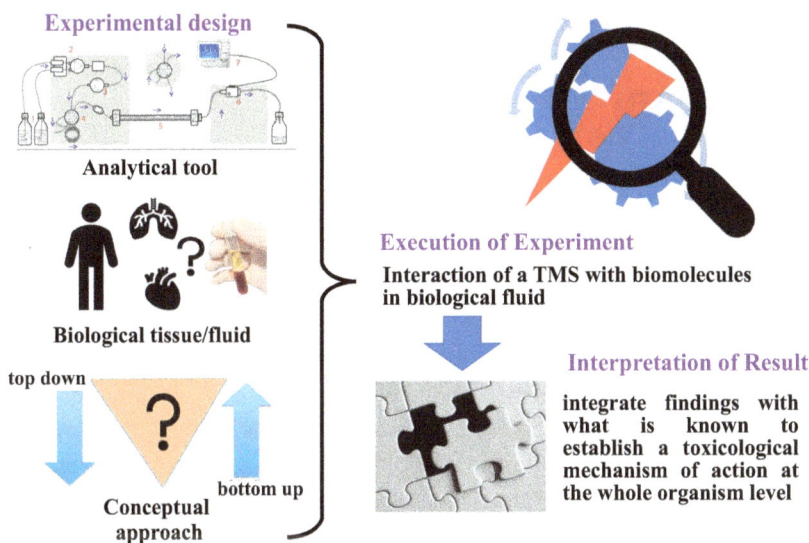

Figure 4.3: Major aspects that one needs to consider in the context of advancing our understanding of the molecular toxicology of TMS. The three aspects on the left are related to the design of experiments to better understand the interaction of a TMS with relevant biomolecules. The one aspect on the right has to do with the interpretation of the result(s) to establish the 'holistic' biomolecular mechanism of chronic toxicity in a mammalian organism.

Analytical chemists have tackled this fundamental challenge – the separation of an exceedingly complex mixture of biomolecules – by developing two principal experimental approaches (Figure 4.5, a + b), each of which is associated with inherent equipment costs and analysis times. There is another approach, however, that considerably simplifies the analytical separation problem: a "metallomics approach" (Figure 4.5c). The latter involves coupling a separation technique with a detection system to specifically observe the toxic metal(loid) of interest, including its metabolites.

In this context, it is pertinent to point out that the integrity of the metal(loid)-protein bond (i.e., the species we want to observe) must be maintained during sample

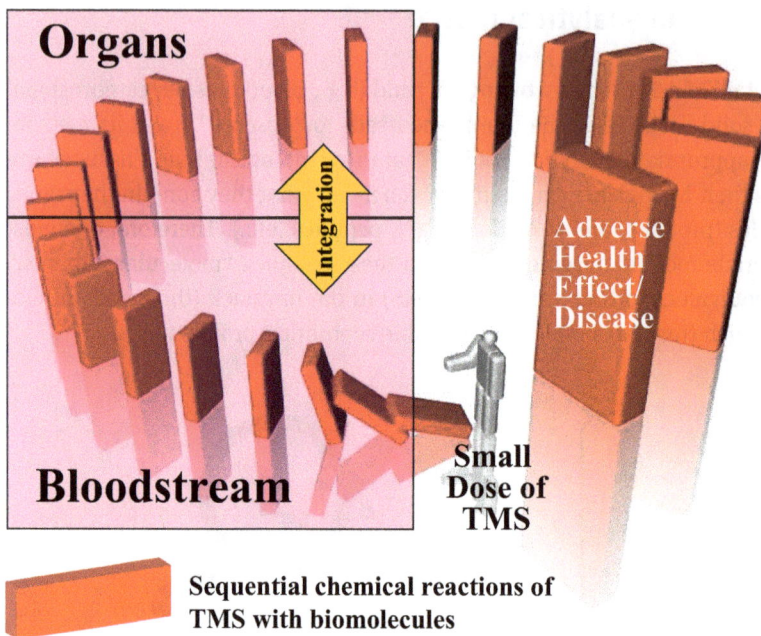

Figure 4.4: Conceptual depiction of biochemical mechanisms which can explain how a small dose of TMS may eventually result in adverse health effects over time.

collection/storage and the entire analysis (Figure 4.6, top) to avoid the generation of artifacts. We note that using metallomics tools simplifies the analytical separation problem in general since rather than separating all biomolecules contained in a biological fluid (this is referred to as the "proteome," which includes thousands of proteins), only the toxic metal(loid) and its metabolites need to be separated from one another! Metallomics tools are therefore uniquely suited to directly analyze complex biological fluids (e.g., blood plasma, cerebrospinal fluid) for toxic metal(loid)s (e.g., Cd and Hg) and/or their metabolites. However, one needs to be aware that one inherent limitation of applying metallomics tools is that no structural information about the species that contain the toxic metal(loid) in a biological fluid is obtained, whatsoever. Thus, tools must be employed in conjunction with metallomics tools to acquire this important information (see Chapter 5). While it is beyond the scope of this book chapter to provide a detailed overview of the various separation and detection methods that comprise a metallomics tool to analyze any given biological fluid, a conceptual metallomics system is depicted in Figure 4.6 (bottom). For more information, the interested reader is referred to comprehensive reviews [52, 53].

4.1.7 Choice of biological tissue/body fluid

Any TMS first enters the bloodstream (primary interaction), after which it and/or its metabolites may infiltrate organs (Figure 4.2) as well as other body fluids (secondary interaction) (Figure 4.7).

Figure 4.8 depicts the variety of primary interactions between any given TMS and biomolecules within the bloodstream, which may involve plasma proteins as well as red blood cells. Primary interactions have not been thoroughly studied predominantly because of a lack of appropriate tools and because they are logistically complex to execute. However, these very interactions require more attention as these interactions will determine the organ availability of a TMS (i.e., which organ is targeted and to what extent [6]). Secondary interactions of the TMS and/or its metabolites within organs must be better understood as they determine the organ-based toxic effects (Figure 4.8). Crucially, however, the integration of primary and secondary interactions is necessary to establish the biochemical mechanism of chronic toxicity in the whole organism. Thus, the application of metallomics tools enables obtaining fundamentally new insights into the biomolecular mechanisms of a TMS in any given organism.

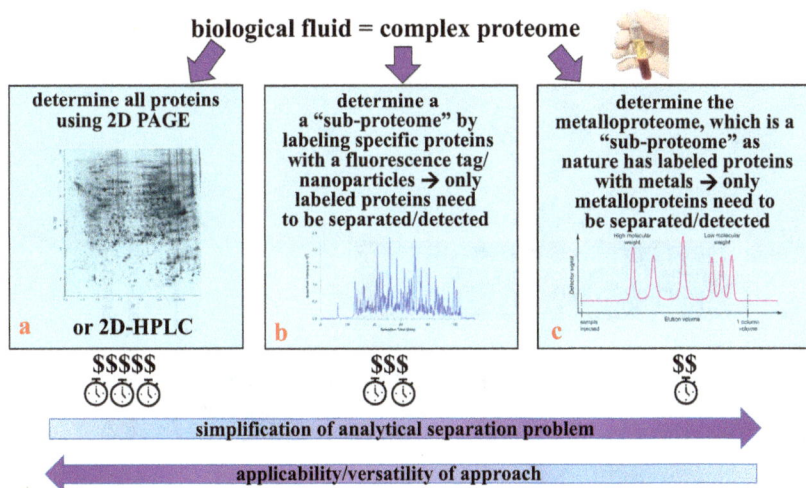

Figure 4.5: Available instrumental analytical approaches that can be employed to observe the fate of a TMS in a complex biological fluid. Each approach allows to tackle the inherent biological complexity in different ways that is associated with costs and time.

a)

Integrity of the TMS-protein bond <u>must</u> <u>not</u> be compromised during sample collection and analysis!

b)

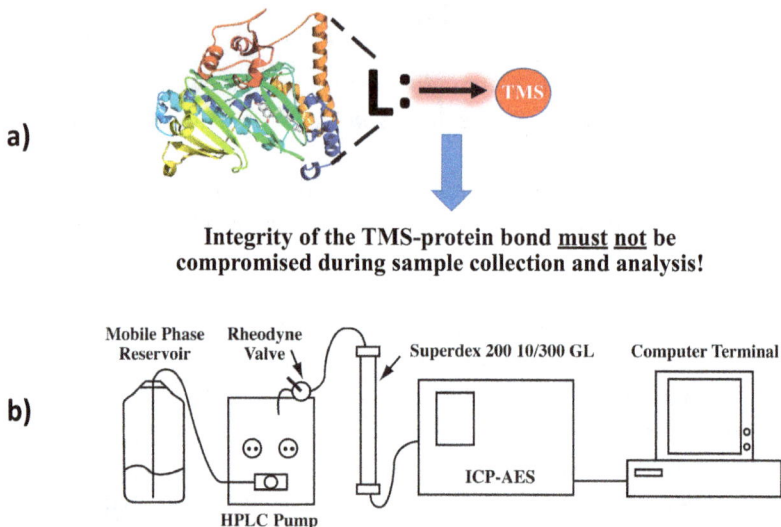

Figure 4.6: Illustration of the need to maintain the integrity of any given TMS-protein bond during a metallomic analysis to avoid the generation of artefacts (a). Conceptual depiction of a metallomics tool which is comprised of a separation system (e.g., HPLC) and a detection system (e.g., ICP-AES) (b).

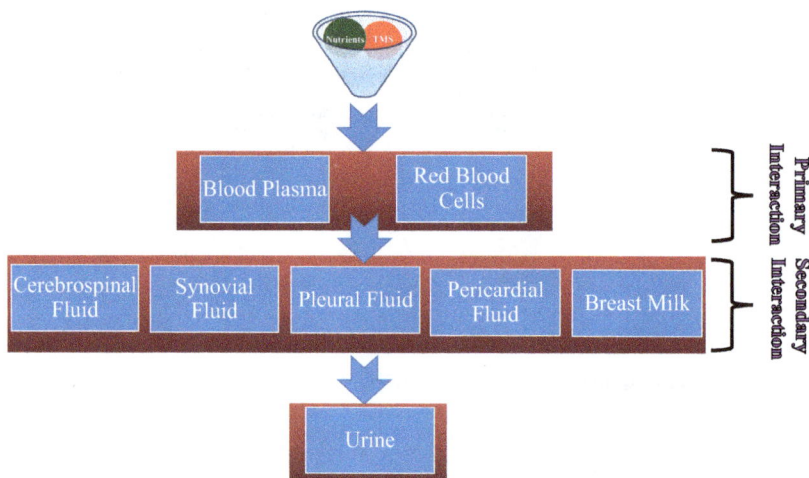

Figure 4.7: Illustration of the most important biological fluids that are amenable to analysis with metallomics techniques. Organs are not shown for clarity. Note that based on the flow of chemical matter through an organism, interactions of a TMS with biomolecules will first unfold in primary fluids (i.e., blood) followed by interactions within organs and then in secondary fluids (e.g., cerebrospinal fluid).

4.1.8 Bottom-up vs top-down

After choosing an appropriate metallomics tool and a biological tissue/fluid, one needs to decide how to conceptually study the interaction of a TMS with a complex biological fluid (e.g., its binding to a specific protein/peptide). One can either employ a bottom-up approach to probe the binding of a TMS with a single biomolecule (e.g., a plasma protein) using appropriate analytical techniques (e.g., ^{1}H-NMR, ^{119}Cd-NMR) at near physiological conditions (e.g., in phosphate buffered saline at pH 7.4) [54]. One may also employ a top-down approach, which involves the direct analysis of a complex biological fluid (e.g., plasma) with an appropriate metallomics tool (e.g., analysis of plasma to which the TMS has been added for TMS-protein complexes). Ideally, the bottom-up and the top-down approach results should mutually validate each other.

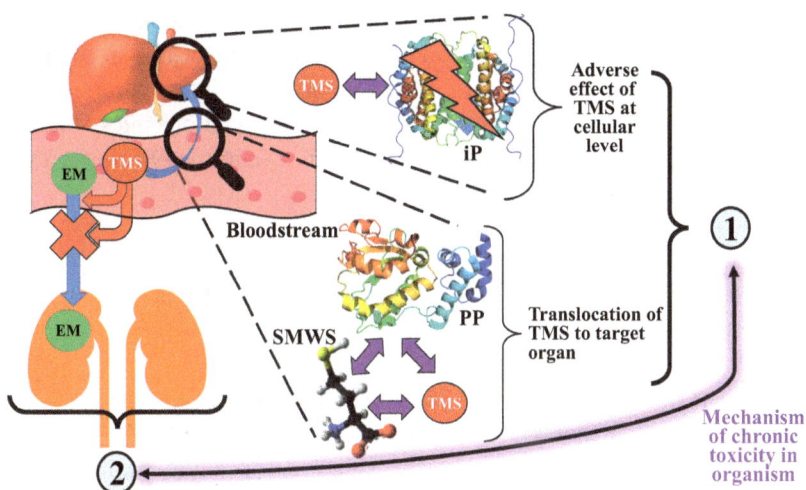

Figure 4.8: Conceptual depiction of toxicologically relevant interactions of a toxic metal species (TMS) in a mammalian organism. Biochemical interactions of the TMS in the bloodstream are important as they determine which organs are targeted while its interactions with biochemical targets within an organ determines the extent of the organ damage over time (mechanism 1). TMS may also adversely affect the transport of essential metals (EM) in the bloodstream, which may perturb its uptake into organs and thus result in dyshomeostasis over time (mechanism 2). Importantly, mechanism 2 provides the required leverage (see Figure 4.4) to identify biochemical mechanisms which can explain how human exposure to a comparatively small TMS dose may result in an adverse health effect and/or a disease. Abbreviations: IP intracellular protein, PP plasma protein, SMWS small molecular weight species.

4.1.9 Integration

Integrating the obtained result(s) with what is currently known about the metabolism
and the biochemistry of a TMS is critical to advance our understanding of its biochemical mechanism of chronic toxicity (Figure 4.8). Thereafter, one must critically
evaluate if the uncovered mechanism of action can explain how the exposure of an
organism to a small dose of a TMS can explain the development of an adverse
health effect at the whole organism level (Figure 4.4).

4.2 Application of metallomics tools to probe TMS-induced biochemical mechanisms of action

Here we will provide a specific example to illustrate how the application of a metallomics tool can provide relevant toxicological insight into the fate of TMS in complex
biological fluids. Although this example illustrates one specific metallomics approach, it can be coupled with other omics approaches, such as proteomics and metabolomics to provide more comprehensive information toward the effect of a TMS
exposure in any biological system.

4.2.1 Different Hg species target hemoglobin in red blood cell (RBC) cytosol

RBCs are one of the first cell types a TMS encounters in an exposure event. Despite
this fact, the toxicology of metal(loid)s in RBCs is not well understood [55]. Since ma-
ture RBCs are non-nucleated, protein synthesis cannot be carried out, and protein
damage will inevitably result in loss of cell function. The rupture of RBCs (i.e., hemo-
lysis) is potentially fatal as free hemoglobin in the bloodstream interferes with the
function of the kidneys and can result in thrombosis. A metallomics approach was
recently employed to probe the comparative fate of Hg^{2+}, CH_3Hg^+, and thimerosal in
rabbit RBC cytosol. RBC lysate was spiked with Hg^{2+}, CH_3Hg^+, and thimerosal and
analyzed using size-exclusion chromatography coupled on-line to an inductively
coupled plasma atomic emission spectrometer (SEC-ICP-AES) over a 6 h period. Notably, a mobile phase containing 2.5 mM glutathione (GSH) was used to simulate the
conditions of erythrocyte cytosol to protect the integrity of potential metal-protein
adducts that had formed in the RBC cytosol after spiking with different Hg species
[56]. Simultaneous monitoring of the elution of Hg and hemoglobin (Hb) via its Fe-
emission line revealed that >92% of total Hg^{2+} eluted as a GS_xHg complex, and that a
small fraction (<1.5%) eluted bound to a cytosolic protein (2 h and 6 h), which was
also observed for Cd^{2+}. Interestingly, 4.8–6.9% of Hg^{2+}, 4.3–5.8% of CH_3Hg^+, and

4.8–11.8% of $CH_3CH_2Hg^+$ (derived from thimerosal, which is a mercury based bactericidal preservative found in some vaccines) co-eluted with Hb, revealing new insight into its molecular toxicology in RBCs (Figure 4.9). Subsequent analysis of collected column fractions which contained the Hg^{2+}-Hb-adduct with an advanced spectroscopic technique (EXAFS, see Chapter 5), showed that Hg was coordinated to two sulfur-ligands with a Hg-S distances of 2.316 Å. These results implied the binding of Hg^{2+} to one Cys residue from Hb, while the nature of the second Cys residue is unknown. The reason for the unexpectedly strong binding of CH_3Hg^+ and the thimerosal-derived $CH_3CH_2Hg^+$ species to Hb – in the presence of 2.5 mM GSH – remains unknown. However, these experiments demonstrate that the interaction of all investigated Hg-species with Hb within RBC cytosol is considerably stronger than that of Cd^{2+}. These findings are of direct toxicological relevance as it has been recently shown that the binding of $CH_3CH_2Hg^+$ to Hb directly affects its O_2-binding capacity [57].

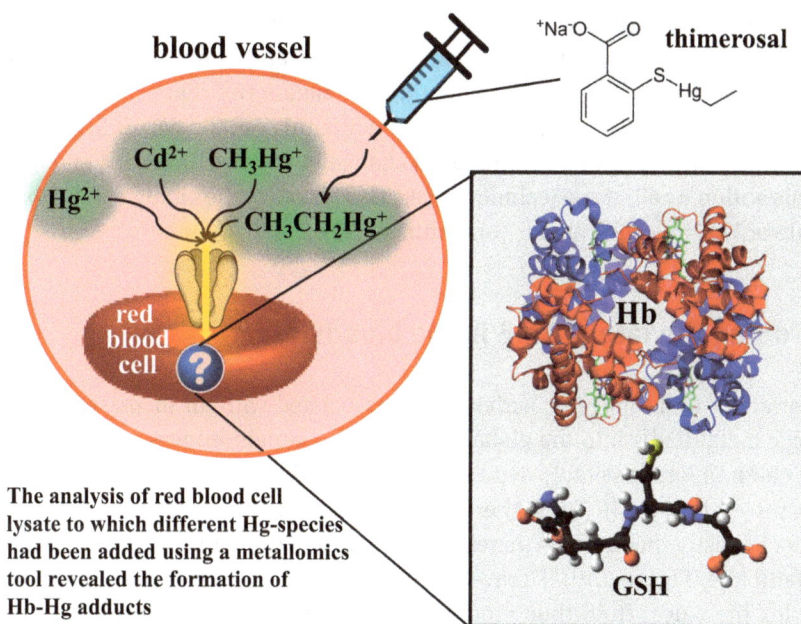

Figure 4.9: The application of metallomics tools combined with XAS provided structural information about different Hg-species within red blood cells. All investigated Hg-species strongly bound to hemoglobin (Hb) in the presence of 2.5 mM glutathione (GSH), while the binding of Cd^{2+} to Hb was comparatively weak. The binding of $CH_3CH_3Hg^+$ to Hb has been shown to decrease its O_2-binding capacity [57].

4.2.2 Probing the biochemical fate of a TMS within cells/organs

The biochemistry of individual TMS that unfolds within target organ cells is incompletely understood. Cd, for example, is known to target the kidneys [58–60], but the biomolecular mechanisms by which Cd damages this and other organs are multifactorial [61] involving the disruption of physiological signaling cascades [62], oxidative stress [63], the activation of cell death pathways [64], the reprogramming of developmental signaling pathways,[65] metallohormone effects [65], carcinogenesis [65–67], endocrine disruption [68], as well as epigenetic effects [69, 70]. Mitochondria and particularly complex III of the electron transfer chain therein has been identified as a critical intracellular target of Cd toxicity [71]. While the aforementioned results were predominantly obtained using classical biochemical techniques, the application of metallomics tools has the potential to provide new information about the adverse effect(s) of toxic metal(loid)s within target organ cells at the metalloprotein level. Synchrotron-based metallomics techniques (see Chapter 5) are particularly useful in this context as they allow one to localize multiple metal(loid)s within organ tissue slices (e.g., from the brain) to generate two-dimensional metal distribution maps (e.g., of Fe, Cu and Zn) [72]. While this unique capability alone allows one to directly observe a TMS-induced change in the distribution of endogenous metals within a cell, these techniques can also visualize the change of the oxidation state of TMS in a sub-cellular compartment [73].

4.2.3 Probing the fate of TMS in the bloodstream

Since a variety of anthropogenic activities are associated with an increased emission of toxic metal(loid)s into the global environment, certain human populations, including children, are inevitably exposed to multiple TMS via the diet and/or new consumer products [74, 75]. Thus, the influx of TMS into the human bloodstream and the biochemistry that unfolds therein is directly implicated in the chronic toxicity of metal(loid)s (Figure 4.10). Unravelling the biochemical events that unfold in this complex biological fluid thus represents a new frontier in the 'toxicology of metals'.

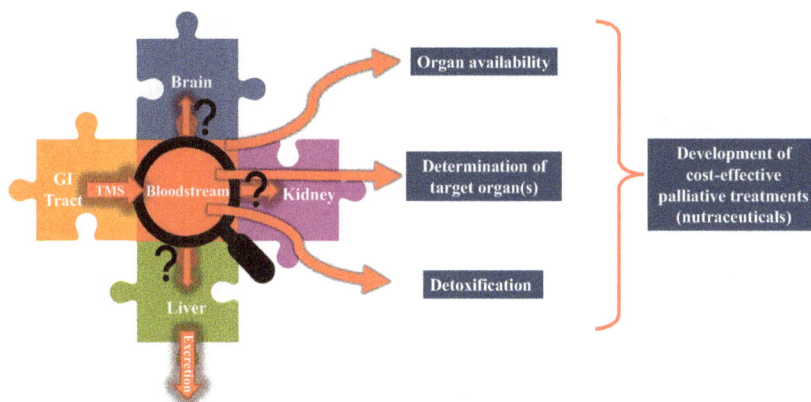

Figure 4.10: Illustration of the critical role that the biochemistry of TMS that unfolds in the bloodstream has in terms of better understanding the mechanisms of their organ-based chronic toxicity in humans.

Summary

One of the biggest knowledge gaps in terms of understanding organ-based human diseases that do not have a genetic cause [76] is the lack of causal links between human exposure to pollutants and disease etiology in the post-genomic era [76–78]. For example, knowledge about the metabolites that are formed as a result of the interactions between TMS, plasma proteins, SMW metabolites, and essential trace elements in the bloodstream (i.e., plasma and RBCs) is limited. To that end, the competitive interaction of a TMS with plasma proteins, SMW metabolites, and ions (homocysteine, HCO_3^-, Cl^-) plays an important role in directing them to their toxicological target cells/organs (i.e., RBCs, endothelial cells, cardiomyocytes, kidneys, liver). In addition, the relative flux of a TMS to individual organs is unknown, which ultimately determines the toxicological effect in target organs over time ("the dose still makes the poison"). Furthermore, single pollutant exposure rarely occurs in our overly industrialized world and, as such, dealing with the multipollutant exposure problem remains difficult to tackle experimentally [79]. Perhaps most importantly, all interactions of any given TMS that "flows" through an organism need to be assembled into a concise sequence of biochemical events, which should ultimately allow one to causally link exposure to an adverse health effect and/or a particular organ-based disease. Thus, the concerted application of modern analytical techniques [40, 52, 56, 80] is expected to obtain fundamentally new insights into the biomolecular processes which govern the exposure-disease link. Only if human exposure to multiple TMS can be causally linked to the origin of a human diseases will it be justified to impose stricter regulations on the anthropogenic emission of toxic metal(loid)s into the environment. Establishing the mechanisms of the chronic toxicity of TMS is an important goal not only to better protect all mammals and their future generations from avoidable diseases [81–85] but also to significantly reduce the global health care costs [86].

Suggested readings

Bridle, TG, Doroudian M, White W, Gailer J. Physiologically relevant hCys concentrations mobilize MeHg from rabbit serum albumin to form MeHg-hCys complexes. Metallomics 2022, 14, mfac010.
Sarpong-Kumankomah S, Miller K, Gailer J. Biological Chemistry of Toxic Metals and Metalloids, Such as Arsenic, Cadmium and Mercury, Encyclopedia of Analytical Chemistry. John Wiley & Sons, Ltd. 2006–2020
Maret W. Metallomics: A Primer of Integrated Biometal Sciences. Paperback 156 pages, Imperial College Press, 2016.
Nordberg G, Fowler BA, Friberg LT. Handbook on the Toxicology of Metals, Third Edition. 2008.
Izatt RM. Metal Sustainability, Global Challenges, Consequences and Prospects. Wiley, 2016.
Sigel A, Sigel H, Sigel RKO. Cadmium: From Toxicity to Essentiality, Metal Ions in Life Sciences. 11, Springer, 2013.
Sturla SJ, Boobis AR, Fitzgerald RE, Hoeng J, Kavlock RJ, Schirmer K, Whelan M, Wilks MF, Peitsch MC. Systems toxicology: From basic research to risk assessment. Chem Res Toxicol. 2014, 27, 314–329

References

[1] Landrigan PJ, Fuller R, Acosta NJR, Adeyi O, Arnold R, Basu N, Balde AB, Bertollini R, Bose-O'Reilly S, Boufford JI, Chiles T, Mahidol C, Coll-Seck AM, Cropper ML, Fobil J, Fuster V, Greenstone M, Haines A, Hanrahan D, Hunter D, Khare M, Krupnick A, Lanphear B, Lohani B, Martin K, Mathiasen KV, McTeer MA, Murray CJL, Ndahimananjara JD, Perera F, Potocnik J, Preker RB, Ramesh J, Rockstroem J, Salinas C, Samson LD, Sandilya K, Sly PD, Smith KR, Steiner A, Steward RB, Suk WA, van Schayck OCP, Yadama GN, Yumkella K, Zhong M. Lancet. 2018, 391, 462–512.
[2] Flanagan SV, Johnston RB, Zheng Y. B. World Health Organ. 2012, 90, 839–846.
[3] Coulthard TJ, Macklin MG. Geology. 2003, 31, 451–454.
[4] Campbell PGC, Gailer J, Effects of non-essential metal releases on the environment and human health (Chapter 10). Metal Sustainability: Global Challenges, Consequences and Prospects, Izatt, RM (Editor), John Wiley & Sons Ltd., Chichester, United Kingdom, 2016, 221–252.
[5] Nriagu JO. Science. 1996, 272, 223–224.
[6] Sarpong-Kumankomah S, Gibson MA, Gailer J. Coord Chem Rev. 2018, 374, 376–386.
[7] Rahman MM, Chowdhury UK, Mukherjee SC, Mondal BK, Paul K, Lodh D, Biswas BK, Chanda CR, Basu GK, Saha KC, Roy S, Das R, Palit SK, Quamruzzaman Q, Chakraborti D. Clin Toxicol. 2001, 39, 683–700.
[8] Bridle TG, Kumarathasan P, Gailer J. Molecules. 2021, 26, 3408.

[9]   Steffen W, Crutzen PJ, McNeill JR. Ambio. 2007, 36, 614–621.
[10]   Karri V, Schumacher M, Kumar V. Environ Toxicol Pharmacol. 2016, 48, 203–213.
[11]   Grandjean P, Landrigan PJ. Lancet. 2006, 368, 2167–2178.
[12]   Vahter M, Akesson B, Liden C, Ceccatelli S, Berglund M. Environ Res. 2007, 104, 85–95.
[13]   Fourth National Report oh Human Exposure to Environmental Chemicals. Centers of Disease
       Control and Prevention, GA, USA, 2009, http://www.cdc.exposurereport/pdf/FourthReport.pdf.
[14]   Agardh E, Allebeck P, Hallqvist J, Moradi T, Sidorchuk A. Int J Epidemiol. 2011, 40, 804–818.
[15]   Simonoff E, Pickles A, Charman T, Chandler S, Loucas T, Baird G. J Am Acad Child Psy. 2008,
       47, 921–929.
[16]   Mitchell E, Frisbie S, Sarkar B. Metallomics. 2011, 3, 874–908.
[17]   Nordberg GF. Toxicol Appl Pharmacol. 2009, 238, 192–200.
[18]   Parizek J, Zahor Z. Nature. 1956, 177, 1036–1037.
[19]   Kopp SJ, Glonek T, Perry HM, Erlanger M, Perry EF. Science. 1982, s217, 837–839.
[20]   Lukkhananan P, Thawonrachat N, Srihirun S, Swaddiwudhipong W, Chaturapanich G,
       Vivithanaporn P, Unchern S, Visoottiviseth P, Sibmooh N. J Toxicol Sci. 2015, 40, 605–613.
[21]   Peters JL, Perlstein TS, Perry MJ, McNeely E, Weuve J. Environ Res. 2010, 110, 199–206.
[22]   Revis NW, Zinsmeister AR, Bull R. Proc Natl Acad Sci USA. 1981, 78, 6494–6498.
[23]   Knoflach M, Messner B, Shen YH, Frotschnig S, Liu G, Pfaller K, Wang X, Matosevic B, Willeit
       J, Kiechl S, Laufer G, Bernhard D. Circulation J. 2011, 75, 2491–2495.
[24]   Bergstroem G, Fagerberg B, Sallsten G, Lundh T, Barregard L. Plos One.2015, e0121240.
[25]   Park SJ, Lee JH, Woo SJ, Kang SW, Park KH. Ophthalmology. 2015, 122, 129–137.
[26]   Wills NK, Kalariya N, Ramanujam VMS, Lewis JR, Abdollahi SH, Husain A, van Kuijk FJGM. Exp
       Eye Res. 2009, 89, 79–87.
[27]   Harding JJ. Br J Ophtalmology. 1995, 79, 199–201.
[28]   Ciesielski T, Weuve J, Bellinger DC, Schwartz J, Lanphear B, Wright RO,.Environ Health
       Perspect. 2012, 120, 758–763.
[29]   Raval G, Straughen JE, McMillin GA, Bonhorst JA. Clin Chem. 2011, 57, 1485–1489.
[30]   Bar-Sela S, Reingold S, Richter ED. Int J Occup Environ Health. 2001, 7, 109–112.
[31]   Bigazzi P. Lupus. 1994,3, 449–453.
[32]   Saad AA, Gaber K, Youssef AI, Amer NM, Ashour MN, Farag MK, Diab NAM. Hum Ecol Risk
       Assess. 2014, 17, 906–914.
[33]   Il'yasova D, Schwartz GG. Toxicol Appl Pharmacol. 2005, 207, 179–186.
[34]   Godt J, Scheidig F, Grosse-Siestrup C, Esche V, Brandenburg P, Reich A, Groneberg DA.
       J Occup Med Toxicol. 2006, 1, 22.
[35]   Rahman A, Kumarathasan K, Gomes J. Sci Total Environ. 2016, 569–570, 1022–1031.
[36]   Edwards JR, Prozialeck WC. Toxicol Appl Pharmacol. 2009, 238, 289–293.
[37]   Johansson A, Enroth S, Palmblad M, Deelder AM, Bergquist J, Gyllensten U. Proc Natl Acad Sci
       USA. 2013, 110, 4673–4678.
[38]   Shin S-Y, Fauman EB, Petersen A-K, Krumsiek J, Santos R, Huang J, Arnold M, Erte I, Forgetta
       V, Yang T-P, Walter K, Menni C, Chen L, Vasquez L, Valdes AM, Hyde CL, Wang V, Ziemek D,
       Roberts P, Xi L, Grundberg E, The Multiple Tissue Human Expression Resource (MuTHER)
       Consortium. Waldenberger M, Richards JB, Mohney RP, Milburn MV, John SL, Trimmer J, Theis
       FJ, Overington JP, Suhre K, Brosnan MJ, Gieger C, Kastenmueller G, Spector TD, Soranzo
       N. Nat Genet. 2014, 46, 543–550.
[39]   Morris TT, Keir JLA, Boshart SJ, Lobanov VP, Ruhland AMA, Bahl N, Gailer J. J Chromatogr
       B. 2014, 958, 16–21.
[40]   Sagmeister P, Gibson MA, McDade KH, Gailer J. J. Chromatogr. B. 2016, 1027, 181–186.
[41]   Limaye DA, Shaikh ZA. Toxicol Appl Pharmacol. 1999, 154, 59–66.
[42]   Wong C-H, Mruk DD, Siu MKY, Cheng CY,.Endocrinology. 2005, 146, 1893–1908.

100 — Maryam Doroudian and Jürgen Gailer

[43] Kim H, Lee HJ, Hwang J-Y, Ha E-H, Park H, Ha M, Kim JH, Hong Y-C, Chang N. J Nutr. 140, 2010, 1133–1138.
[44] Sasakura C, Suzuki KT. J Inorg Biochem. 71, 1998, 159–162.
[45] Bao R-K, Zheng S-F, Wang X-Y. Environ Sci Pollut R. 24, 2017, 20342–20353.
[46] Bakir F, Damluji SF, Amin-Zaki L, Murtafdha M, Khalidi A, Al-Rawi NY, Tikriti S, Dhahir HI, Clarkson TW, Smith JC, Doherty RA. Science. 181, 1973, 230–241.
[47] Langlois MR, Delanghe JR. Clin Chem. 42, 1996, 1589–1600.
[48] Josephy PD, Mannervik B. (Ed.). Molecular Toxicology, 2nd Edition, 2006, Oxford University Press, 608 pages.
[49] Kempson IM, Lombi E. Chem Soc Rev. 40, 2011, 3915–3940.
[50] Kepp KP. Chem Rev. 112, 2012, 5193–5139.
[51] Bomer N, Beverborg NG, Hoes MF, Streng KW, Vermeer M, Dokter MM, Ijmker J, Anker SD, Cleland JGF, Hillege HL, Lang CC, Ng LL, Samani NJ, Tromp J, van Veldhuisen DJ, Touw DJ, Voors AA, van der Meer P. Eur J Heart Failure. 22, 2019, 1415–1423.
[52] Gomez-Ariza JL, Jahromi EZ, Gonzalez-Fernandez M, Garcia-Barrera T, Gailer J. Metallomics. 3, 2011, 566–577.
[53] Sarpong-Kumankomah S, Miller K, Gailer J. Encyclopedia of Analytical Chemistry. John Wiley & Sons, Ltd.. 2006–2020, 2020.
[54] Gailer J, Lindner W. J Chromatogr B. 716, 1998, 83–93.
[55] Hill A, Gailer J. J Inorg, Biochem. 216, 2021, 111279.
[56] Gibson MA, Sarpong-Kumankomah S, Nehzati S, George GN, Gailer J. Metallomics. 2017, 9, 1060–1072.
[57] de Magalhanes Silva M, de Araujo Dantas MD, de Sila Filho RC, Dos Santos Sales MV, de Almeida Xavier J, Leite ACR, Goulart MOF, Grillo LAM, de Barros WA, de Fatima A, Figureiredo IM, Santos JCC. Intern J Biol Macromol. 2020, 154, 661–671.
[58] Buchet JP, Lauwerys R, Roels H, Bernard A, Bruaux P, Claeys F, Ducoffre G, De Plaen P, Staessen J, Amery A, Lijnen P, Thijs L, Rondia D, Sartor F, S.r. A., Nick L. Lancet. 1990, 336, 699–702.
[59] Kägi JHR, Vallee BL. J Biol Chem. 1960, 235, 3460–3465.
[60] Thevenod F, Wolff NA, Metallomics. 8, 2016, 17–42.
[61] Yang H, Shu Y. Int J Mol Sci.2015, 16, 1484–1494.
[62] Thevenod F. Toxicol Appl Pharmacol. 2009, 238, 221–239.
[63] Casalino E, Sblano C, Landriscina C. Arch Biochem Biophys. 1997, 346,171–179.
[64] Pi H, Xu S-W, Reiter RJ, Guo Pan, Zhang L, Li Y, Li M, Cao Z, Tian L, Xie J. Autophagy. 2015, 11, 1037–1051.
[65] Thevenod F, Lee W-K, Toxicology of cadmium and its damage to mammalian organs (Chapter 14). Sigel A, Sigel H, Sigel RKO. (Ed.). Cadmium: From Toxicity to Essentiality. 11, Springer Science + Business Media Dordrecht, New York, 2013, 415–490.
[66] Rani A, Kumar A, Lal A, Pant M. Int J Environ Health Res. 2014, 24, 378–390.
[67] Hartwig A. Biometals. 23, 2010, 951–960.
[68] Johnson MD, Kenney N, Stoica A, Hilakivi-Clarke L, Singh B, Chepko G, Clarke R, Sholler PF, Lirio AA, Foss C, Reiter R, Trock B, Paik S, Martin MB. Nat Med. 2003, 9, 1081–1084.
[69] Sanders AP, Smeester L, Rojas D, DeBussycher T, Wu MC, Wright FA, Zhou Y-H, Laine JE, Rager JE, Swamy GK, Ashley-Koch A, Miranda M-L, Fry RC. Epigenetics. 2014, 9, 212–221.
[70] Skipper A, Sims JN, Huang J, Guo N, Shi J, Lin Y, Lin Z. Int J Environ Res Public Health. 2016, 13, 1–10.
[71] Wang Y, Fang J, Leonard SS, Rao KMK. Free Rad Biol Med. 2004, 36, 1434–1443.
[72] Popescu BFG, Robinson CA, Chapman LD, Nichol H. Cerebellum. 2009, 8, 340–351.
[73] Harris HH, Pickering IJ, George GN. Science. 2003, 301, 1203.

[74] Guney M, Zagury GJ. Environ Sci & Technol. 2013, 47, 5921–5930.

[75] Liu S, Hammond SK, Rojas-Cheatham A. Environ Health Perspect. 2013, 121, 705–710.

[76] Gailer J, Inorg J. Biochem. 2012, 108, 128–132.

[77] Buenzli J-CG. Front Chem. 2013, 1, 1–3.

[78] Lahner B, Gong J, Mahmoudian M, Smith EL, Abid KB, Rogers EE, Guerinot ML, Harper JF, Ward JM, McIntyre L, Schroeder JI, Salt DE. Nat Biotechnol. 2003,21, 1215–1221.

[79] Rider CV, McHale CM, Webster TF, Lowe L, Goodson WH, La Merrill MA, Rice G, Zeise L, Zhang L, Smith MT. Environ Health Perspect. 2021, 129, 035003–035001.

[80] Hoffman AS, Sokaras D, Zhang SB, Debevre LM, Fang C-Y, Gallo A, Kroll T, Dixon DA, Bare SR, Gates BC. Chem Eur J. 2017, 23, 14760–14768.

[81] Gunn SA, Gould TC, Anderson WAD. J Reprod Fertil. 1968, 15, 65–70.

[82] Interdonato M, Pizzino G, Bitto A, Galfo F, Irrera N, Mecchio A, Pallio G, Ramistella V, De Luca F, Santamaria A, Minutoli L, Marini H, Squadrito F, Altavila D. Clin Endocrinol. 2015, 83, 357–362.

[83] Sebastian A, Prasad MNV. Agron Sustain Dev. 2014, 34, 155–173.

[84] Greenberg MI, Vearrier D. Clin Toxicol. 2015, 53, 195–203.

[85] Landrigan PJ, Sly JL, Ruchirawat M, Silva ER, Huo X, Diaz-Barriga F, Zar HJ, King M, Ha EH, Asante KA, Ahanchian H, Sly PD. Ann Global Health. 2016, 82, 10–19.

[86] Trasande L, Liu Y. Health Aff. 2011, 30, 863–870.

Graham N. George, Ben Huntsman, Olena Ponomarenko,
Emérita Mendoza Rengifo, Monica Y. Weng,
Julien, J. H. Cotelesage, Natalia V. Dolgova, and Ingrid J. Pickering

Chapter 5
Structural and chemical aspects of the molecular toxicology of heavy metals and metalloids

5.1 Introduction

The chemical compounds of heavy metals and metalloids are among the most toxic species to which humans are commonly exposed. Indeed, many such compounds known in antiquity were variously used as poisons and medicines. The sixteenth-century physician and alchemist Paracelsus (1493–1541) is widely acknowledged as the father of toxicology. Paracelsus, whose full name was Theophrastus Philippus Aureolus Bombastus von Hohenheim, was a controversial and outspoken individual and a thorn in the side of the medical establishment; our word "bombastic" derives from his name. Paracelsus was frequently contemptuous of his peers and was given to exhibitions such as publicly burning revered writings of antiquity, upon which sixteenth-century conventional medicine relied heavily, such as those of Galen and Avicenna. Instead, Paracelsus emphasized experiment and evidence [1].

Acknowledgments: The authors gratefully acknowledge the use of photographs from the Stanford Synchrotron Radiation Lightsource (SSRL) and the Canadian Light Source (CLS). The experimental data that we have shown were collected at SSRL, CLS and the Advanced Photon Source. The authors also acknowledge funding from the Canada Research Chairs program (IJP, GNG), the Natural Sciences and Engineering Research Council of Canada (IJP, GNG), the Sylvia Fedoruk Canadian Centre for Nuclear Innovation Inc. and Innovation Saskatchewan (GNG, IJP), the Ataxia Charlevoix-Saguenay Foundation (GNG, IJP), the Richardson Research Fund (GNG, IJP) and the University of Saskatchewan (GNG, IJP).

Graham N. George, Ingrid J. Pickering, Molecular and Environmental Sciences Group, Department of Geological Sciences, University of Saskatchewan, 114 Science Place, Saskatoon, Saskatchewan S7N 5E2, Canada; Toxicology Centre, University of Saskatchewan, Saskatoon, Saskatchewan, S7N 5B3, Canada; Department of Chemistry, University of Saskatchewan, Saskatoon, Saskatchewan, S7N 5C9, Canada, e-mail: g.george@usask.ca
Ben Huntsman, Olena Ponomarenko, Emérita Mendoza Rengifo, Monica Y. Weng,
Julien, J. H. Cotelesage, Molecular and Environmental Sciences Group, Department of Geological Sciences, University of Saskatchewan, 114 Science Place, Saskatoon, Saskatchewan S7N 5E2, Canada
Natalia V. Dolgova, Calibr - California Institute for Biomedical Research, Scripps Research, La Jolla, California, 92037, USA

https://doi.org/10.1515/9783110626285-005

One of his most famous quotes can be translated as "solely the dose determines that a thing is not a poison", meaning that nearly anything can be toxic if taken in sufficiently high dose, and that some toxic substances may be medically useful at low doses. Thus, arsenic trioxide, As_2O_3, long-known as a deadly poison, has been used to treat leukemia since 1878 [2], and is now an established treatment for acute promyelocytic leukemia [3, 4]. We note in passing that arsenic compounds have been used in China as part of traditional medicines for considerably longer than this, but in some cases their use is without valid scientific basis, and consequently deaths have resulted [5]. We now understand that Paracelsus, while a remarkable thinker and substantially ahead of his time, was only partly correct; it is not only the dose but also the *speciation* of a metal or metalloid that makes the poison. Chemical speciation defines how an element is presented at the molecular level. It is this speciation that controls how a metal or metalloid is transported in the environment and taken up into living organisms, and once taken up whether it is toxic, benign or beneficial. Thus, while arsenic is notorious for the toxic nature of its compounds, these have widely varying toxicities, with some being deadly poisons and others completely benign.

This chapter reviews the scientific toolbox available to a modern molecular toxicologist to determine speciation and understand toxicity at the molecular level. Our intent is not to provide a comprehensive review of the entire field, but instead to present a focused examination of selected areas to illustrate the application of various techniques. We discuss techniques for understanding structure and chemistry, ranging from computational chemistry and quantum mechanical approaches to advanced in situ probes for speciation such X-ray spectroscopy and imaging, illustrated by specific examples of how these methods provide toxicological insights.

5.2 Formal and actual charge and oxidation state

Before we proceed to discuss the main topics of this chapter we will review some basic concepts, which are charge, oxidation state, and the effects of chemical bonding. Oxidation state (or oxidation number) can be defined as the charge that an atom would possess if the compound containing it were composed of pure ionic species. We will use mercury as an example. Thus, $Hg(OH)_2$, a species that exists in solution but not as a solid, has mercury in the formal +2 oxidation state, which is often written as Hg^{2+}, $Hg(II)$ or Hg^{II}. A formal view of $Hg(OH)_2$ would be a single Hg^{2+} bound by two HO^- groups; however, the electronic charges are altered by chemical bonding which distributes charges over the whole molecule. In the case of $Hg(OH)_2$ the Hg–OH bonds possess covalency, so that the actual charge on mercury in this compound is +0.695. The charges on the oxygens of the two HO^- groups are similarly affected by the bonding so that these are both −0.766, with +0.419 on each hydrogen. We can consider the compound series H_3CE–Hg–ECH_3, where E is any of the chalcogens (periodic table group 16), O, S, Se and Te, and examine the trends in charge going down the group. Here the charge on mercury is +0.694, +0.266, +0.185, +0.013 for E =O, S, Se and Te, respectively. Put very simplistically, going down the group the number of electrons available for covalent bonding increases, which means that the charge can more readily spread out, and the charge on the metal decreases nearly to zero for tellurium. For the chalcogen, the charge again decreases with increasing atomic number, with −0.595, −0.288, −0.185 and +0.022 for E =O, S, Se and Te, respectively. Extending the series to a hypothetical polonium[1] compound with E = Po ($H_3CPo-Hg-PoCH_3$) is instructive; polonium is formally classified as a metal because its electrical conductivity decreases with increasing temperature (a characteristic of all metals), whereas tellurium and selenium can be semiconductors, and sulfur and oxygen are non-metals. For the polonium compound, the charge on the mercury, formally Hg^{2+}, is actually negative at −0.032 and positive on the Po at +0.091, so that in this (hypothetical) compound mercury would behave as a non-metal. In all of these compounds mercury has a formal charge of Hg^{2+} with a formal oxidation state of Hg^{II}, which is reflected in the total electron count of the molecules, but in reality the charge on mercury will in all cases be very much less than 2 +. Formal charge is distinct from oxidation state in that it can be used for groups, rather than just atoms, for example arsenobetaine (Section 5.5) possesses an arsonium group with a formal +1 charge $[-CH_2(CH_3)_2As]^+$, and a carboxylate group with a formal −1 charge $[-CO_2]^-$, while the actual charges are ±0.467 with near-neutral methyl groups (H_3C-) and arsenic itself possessing a near-unit charge of +0.721. To give an additional example, we can compare the charges on the species $[(CH_3)_3E]^+$ where E again is a chalcogenide and which are known for oxygen, sulfur, selenium and tellurium. These compounds have an overall charge of +1 with a chalcogen formal oxidation state of −II. Here, as we go down group 16, the charge on the central chalcogen is −0.470, −0.249, −0.102, +0.103 and +0.186, for oxygen, sulfur, selenium, tellurium and polonium, respectively. Clearly, the *formal* charge and the *formal* oxidation state have little to do with the *actual* charge on individual atoms; the effects of chemical bonding are very much more important. Nonetheless, the concepts of formal oxidation state and charge are useful because they allow us to formulate the overall electron count in a site or a molecule, which is key to the chemical bonding and properties. The perceptive reader may realize that more than one of the compounds that we have just discussed are unknown. So, how can we possibly know the charges on the various atoms? The answer is that they can be reliably calculated using density functional theory, which we will discuss in Section 5.4.

---

1 Polonium would be an obscure element to the general public but for its use in a 2006 poisoning [6], in which the victim died from radiation poisoning because of the alpha decay of ^{210}Po (polonium has no non-radioactive isotopes) that had been administered in a cup of tea.

Throughout this chapter we will use the convention of superscript Roman numerals when referring to the formal oxidation state of an element denoted by its symbol (e.g., As^{III}), in parentheses when the full name is spelled out (e.g., mercury (II)), and superscript Arabic numerals when referring to formal charge (e.g., Hg^{2+}).

5.3 X-ray-based in situ speciation techniques

In this section we will provide an overview of X-ray spectroscopic techniques that can be used to study the chemical speciation of metals and metalloids giving insights that in many cases are unique. With a view to accessibility from a range of toxicological backgrounds, we have attempted to restrict our discussions largely to concepts and have omitted discussion of theory. Commercial bench-top instruments for these methods are now available but typically a synchrotron radiation facility must be used, together with other specialized equipment. Because of this they may not be as routine as many other techniques, but the reader should not consider them inaccessible. Several decades ago, access was limited to specialists who would often have designed and built the equipment used, but now facilities have become refined so that even the novice spectroscopist might use the methods.

5.3.1 X-ray absorption spectroscopy

X-ray absorption spectroscopy (XAS) is a tool that can provide information upon both physical and electronic structure. Originating in the early 1970s, with time the method has become increasingly applied to a very wide variety of fields. One of the strengths of the technique is that it can be used with little pre-treatment of samples, whether they be purified biomolecules, tissue samples or even intact whole (small) organisms. An early weakness was the lack of availability of synchrotron radiation X-ray sources which are required for all except the most concentrated of samples, but with increasingly wide availability of such sources, along with their constant improvement, accessibility is no longer a major limitation.

X-ray absorption spectra arise from excitation of core-level electrons through absorption of an X-ray photon. A schematic of the underlying physics is shown in Figure 5.1. Spectra are thus dependent upon the ionization energies of the core electrons; if the energy of incident X-rays is scanned, a sharp rise is observed at this threshold energy, which is called the absorption edge. The nomenclature typically used is known as X-ray notation, with K, L, M and N denoting excitation of electrons with principal quantum numbers n of 1, 2, 3 and 4, respectively. Thus, K-edges correspond to a 1s excitation, L_I to a 2s, L_{II} and L_{III} respectively to $2p_{1/2}$ and $2p_{3/2}$, M_I to 3s, M_{II} to $3p_{1/2}$, M_{III} to $3p_{3/2}$, M_{IV} to $3d_{3/2}$ and M_V to $3d_{5/2}$ excitation. Almost all XAS studies

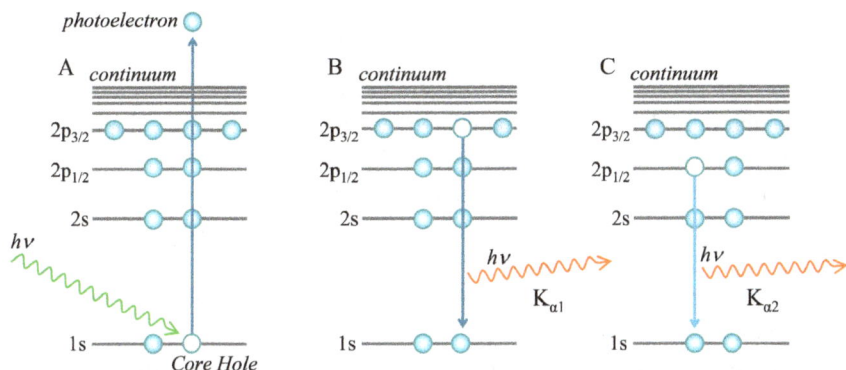

Figure 5.1: Physics of X-ray spectroscopy. A shows the primary excitation of a core electron, in this case a 1s electron corresponding to a K-edge, by a photon ($h\nu$) yielding a photoelectron and a core-hole. B and C show the decay of an outer electron to fill the core hole with concomitant emission of X-ray fluorescence photons; $K_{\alpha1}$ (B) and $K_{\alpha2}$ (C) from the $2p_{3/2}$ and $2p_{1/2}$ levels, respectively.

applicable to molecular toxicology use K-edges for lighter elements and L-edges for heavier elements, with the choice being typically driven by experimental considerations. XAS spectra are separated into two different regions: the extended X-ray absorption fine structure or EXAFS, which occurs at energies higher than the absorption edge; and the near-edge region (Figure 5.2A, inset), often called the X-ray absorption near-edge structure or XANES, which consists of features lower in energy than the major rise of the absorption edge plus any at higher energies which are not part of the analyzable EXAFS. Here we include a word on nomenclature – unfortunately the acronym XANES has been used by some researchers to mean different parts of the spectrum; in some cases, XANES has been used to include the entire near-edge region and in others, the post-edge region excluding pre-edge features. For this reason, we prefer not to use XANES, and instead refer to this region as the near-edge spectrum.

XAS is element-specific and is applicable to a very wide range of elements; it can be used to investigate solids, liquids (including solutions), gaseous materials, and probes all occurrences of an element within a sample with moderate sensitivity, so that no major fraction of a mixture (at least in principle) can elude detection by the experimenter. The sensitivity of XAS will depend upon the experimental setup used, including the specific synchrotron beamline, the type of detectors that are available, and whether facilities for maintaining samples at low temperatures such as a helium cryostat are used. With a state-of-the-art facility it should be possible to measure near-edge spectra at the micro-molar level, with several studies of samples at these levels having been reported.

The EXAFS part of the spectrum arises from backscattering, by nearby atoms, of the photoelectron emitted following an absorption event. It can, with appropriate analysis, give a local radial structure around the absorber. Figure 5.2 illustrates this process.

The near-edge part of the spectrum arises from transitions to bound states, unoccupied molecular orbitals that can be populated by the excited electron at highly characteristic energies. Intense features arise from dipole-allowed $\Delta l = \pm 1$ transitions, such as 1s→4p transitions for a K-edge spectrum of a p-block element, or 2p$_{3/2}$→5d for the L$_{III}$ edge of a heavy metal. The short wavelength of X-rays means that formally dipole-forbidden, quadrupole-allowed $\Delta l = \pm 2$ transitions are also weakly observed when such levels are available, such as 1s→3d transitions for the K-edge spectrum of a transition metal ion in a centrosymmetric environment.

The experimental setup for XAS of bulk samples has evolved from the early days to a fairly standard configuration, an example of which is shown in Figure 5.3. The X-ray beam enters the experimental setup on the right of the figure. Its intensity is first measured by a simple detector (a gas ionization chamber, or ion chamber for short) labeled I_0. The beam then passes through the sample within the cryostat with the downstream intensity of the beam then measured by another ion chamber I_1. The beam then passes through a foil made of the element being studied, and the intensity is again measured at I_2. The calibration foil is used to precisely calibrate the energy of the X-ray beam throughout the measurement.

Typically, samples are maintained at a low temperature, preferably within a suitable liquid helium cryostat, which are available commercially. Dual benefits in the use of low temperatures are both enhancing the EXAFS by freezing out vibrational modes and protecting the sample against radiation damage. XAS spectra, and in particular those of dilute samples, are frequently measured as fluorescence excitation spectra using the X-ray fluorescence that is generated on filling of the core-hole created through the primary photoexcitation event by decay of a higher electron. Intense fluorescence lines follow the same $\Delta l = \pm 1$ selection rules as absorption. Thus, the most intense fluorescence from a K-edge would be the K$_{\alpha 1}$ and K$_{\alpha 2}$ which arise from 2p$_{3/2}$→1s and 2p$_{1/2}$→1s transitions, respectively (Figure 5.1B and C), and typical XAS of a transition metal ion might monitor both of these transitions (unresolved) using a solid-state detector. The use of X-ray fluorescence means that the aforementioned cryostat must be equipped with suitably thin windows and large windows to transmit the X-rays. There are various commercial alternatives for X-ray fluorescence detectors, and pros and cons for each choice, that we will not discuss here. These devices are in effect large reverse bias diodes and are energy dispersive with an energy resolution that is at best about 150 eV (full width half-maximum). One limitation with solid state detectors is related to count-rates and readout electronics, which gives an upper-bound to the counts per second that can be accepted, above which the detector tends to saturate. The first specialized XAS detectors used arrays of individual diodes in combination with filters and Soller slits to preferentially eliminate some of the unwanted background signal from the sample, such as inelastic and elastic scattered X-rays. Modern detectors can operate at higher count

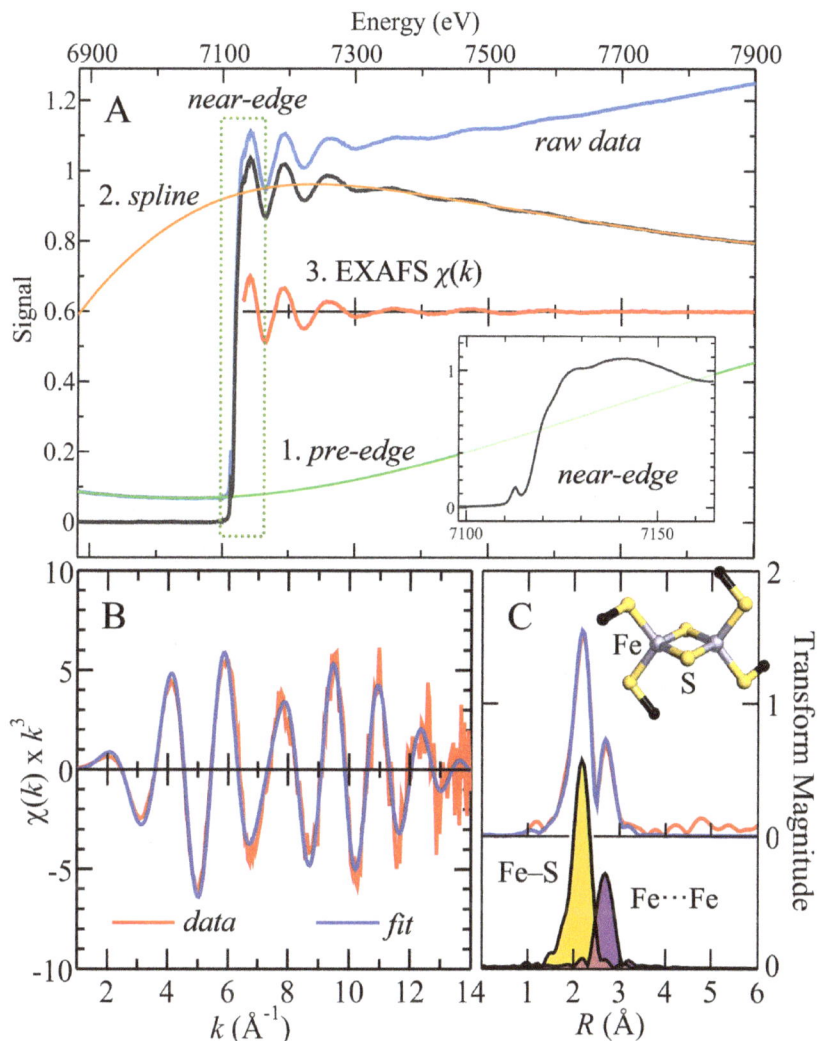

Figure 5.2: X-ray absorption spectroscopy data analysis. The figure shows the Fe K-edge XAS of a metalloprotein containing a two-iron [Fe$_2$S$_2$]$^{2+}$ cluster. Panel A shows the data reduction process, starting from the raw data (light blue curve), from which is subtracted a pre-edge function (green curve) yielding an iron-only XAS spectrum (gray curve). The near-edge (XANES) portion of the spectrum is outlined by the broken green line and is expanded in the inset in A. The EXAFS χ are the oscillatory part of the absorption, on the high-energy side of the absorption edge. To extract the EXAFS (red curve) a spline (orange curve) is fitted to the data and then subtracted from it. The EXAFS oscillations are then typically plotted as a function of photo-electron wave vector k, k^3 weighted (red curve, panel B) and are fitted with a theoretical model (blue curve, panel B) to obtain structural information. The EXAFS data are often viewed as the Fourier transform (panel C) which gives peaks at distances corresponding to the shells of neighboring atoms. Panel C also shows the core structure of the iron-sulfur site. The individual EXAFS components corresponding to the Fe–S and Fe⋯Fe interactions are shown in the lower part of C.

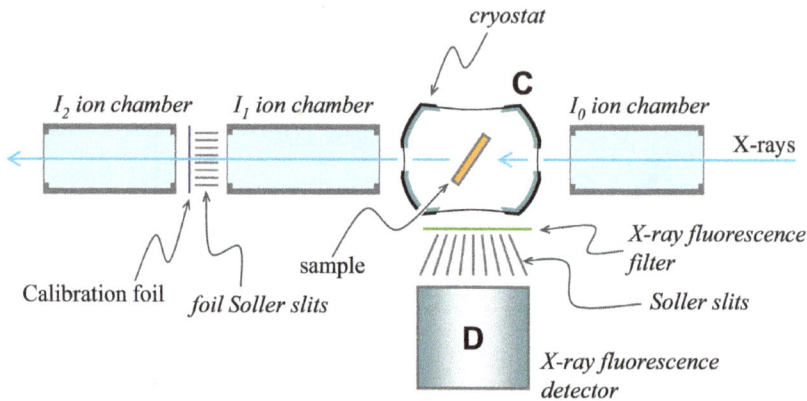

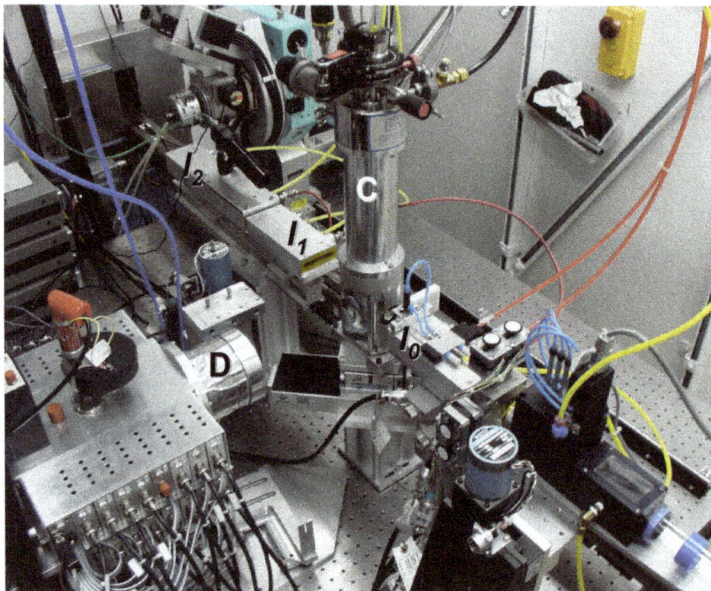

Figure 5.3: Experimental setup for XAS. The upper panel shows a schematic (viewed from above) of a typical setup, while the lower panel shows the experimental setup of beamline 9–3 at the Stanford Synchrotron Radiation Lightsource (photo courtesy Dr. Matthew Latimer), with ion chambers (I_0, I_1 and I_2), cryostat (C) and solid-state fluorescence detector (D) indicated. The detector D has been moved back from its normal position close to the cryostat to allow visualization of the cryostat windows, and the Soller slit/X-ray fluorescence filter assembly has been removed for the same reason. The setup includes a number of motorized stages to facilitate alignment and is enclosed inside a lead-lined radiation-proof "hutch".

rates than these early devices, but modern beamlines provide much greater photon fluxes, hence the strategy of filters and Soller slits is still required. Soller slits were invented by Walter Soller of the University of Cincinnati in 1924; a common misnomer that should be avoided is to call them "solar" slits. They are typically made of a metal that absorbs X-rays strongly, such as silver, and acts to eliminate X-rays coming from directions other than intended. Thus, in the case of the *foil Soller slits* (Figure 5.3, upper panel) the parallel slits eliminate much of the X-ray fluorescence from the energy calibration foil, preventing it from registering in the upstream I_1 ion chamber. In the case of the Soller slits separating the X-ray filter from the fluorescence detector they serve to discriminate only rays coming from the sample, eliminating X-ray fluorescence from the filter.

Modern theories of XAS are well developed, with methods of analysis of both the EXAFS and near-edge regions now well-established. The details of XAS analysis methods are outside the scope of this review, and the interested reader is referred to other more specialized texts [7, 8]. The EXAFS can be used to determine the approximate size (atomic number) and number of neighbor atoms to the absorber atom and can very accurately estimate mean inter-atomic distances for a particular neighbor type. The near-edge spectrum can give detailed information on electronic structure, which in turn informs on oxidation state, coordination environment and geometry. The near-edge spectrum can be used as an effective *fingerprint* of a particular chemical type, for which methods to analyze mixtures are well developed. The basis for this is that individual chemical species present subtly different spectra, which can be used to inform on species present in situ. For near-edge spectra of samples of unknown composition, the most commonly used methods of analysis employ linear combinations of spectra of standard compounds together with least-squares fitting, and principal component analysis. There is a common misconception that near-edge spectra give the oxidation state, while the spectra can inform on the oxidation state the specific chemical form is much more important. For example, Se^0 compounds include both the elemental forms and selenoxides, which have highly distinctive spectra. Near-edge spectra cannot provide absolute molecular identities, but instead probe the *local* speciation of a metal or metalloid. For example, the natural products L-selenomethionine and Se-methyl-L-selenocysteine have essentially the same local selenium environment with $CH_3-Se-CH_2-$, and as a consequence give near-identical near-edge spectra, and cannot be distinguished by XAS, so that other techniques must be used to provide this information. Another restriction in application of this type of analysis is that spectra of standard species having similar coordination to any unknowns must be available (collected using the same conditions), which may not be possible in the case of truly new or novel species. In such cases EXAFS spectra might be used to provide clues about the coordination environment, although this requires more lengthy data acquisition, and is challenging for mixtures. Despite these limitations, the use of XAS provides a powerful tool when an in situ probe of chemical form is required, with applications continuing to expand.

5.3.2 Overcoming energy resolution limitations – HERFD-XAS

All experimental methods have their inherent limitations, with XAS being no exception. A major limitation for near-edge spectra is associated with the poor spectroscopic energy resolution arising from core-hole lifetime broadening. This is a direct consequence of Heisenberg's uncertainty principal, which can be stated as $\Delta E \geq \hbar/2\Delta\tau$ where $\Delta\tau$ is the lifetime of the core-hole that is created by photoexcitation, $\hbar$ is Plank's constant divided by 2π, and ΔE is the resulting energy-broadening observed in the spectrum. Because XAS involves excitation of tightly bound core-electrons, the core-hole that is created by the excitation is very short-lived, typically with lifetimes that are a fraction of a femtosecond in duration. Thus, $\Delta\tau$ is very small, and as direct consequence ΔE is large, smearing out spectroscopic detail by including a Lorentzian broadening in the overall linewidth, which in the following is referred to as Γ. Recently, an experimental method that circumvents this substantial limitation has been coming to the fore. This is called high-energy resolution X-ray fluorescence detected X-ray absorption spectroscopy, or HERFD-XAS. How the method works is illustrated here by a K-edge experiment, though the method also applies at other edges. The natural linewidth of an X-ray fluorescence-detected K-edge measurement is the sum of the natural linewidths associated with the inner 1s shell containing the core-hole created by the primary photoexcitation, and the outer shell which decays to fill that core hole, which for the $K_{\alpha 1}$ fluorescence arises from a $2p_{3/2} \rightarrow 1s$ transition. The corresponding natural linewidth therefore is $\Gamma = \Gamma_{1s} + \Gamma_{2p_{3/2}}$. The HERFD-XAS method uses a special crystal analyzer-based spectrometer to measure the X-ray fluorescence line with much higher resolution than its natural linewidth. This high-resolution measurement defines the fluorescence photon energy, limiting the uncertainty to that of the outer-shell, and the resolution is now defined primarily by $\Gamma_{2p_{3/2}}$. Because the outer electron is much less tightly bound, it has a considerably longer lifetime than that of the primary core hole; consequently $\Gamma_{2p_{3/2}}$ is small and the observed HERFD-XAS shows a dramatic sharpening of the spectrum. Contributing to the overall lineshape are components that are due to the spectrometer and beamline optical resolutions, which provide Gaussian contributions to the overall linewidth, convoluted with the Lorentzian lifetime broadening to give a Voigt peak-shape. Both spectrometer and beamline optics are very important in performing an adequate HERFD-XAS measurement.

Figure 5.4 shows an example of spectroscopic improvement that is given by a HERFD-XAS measurement, relative to conventional XAS. The increase in detail in the HERFD-XAS spectrum is remarkable [9]. The method allows substantially enhanced speciation capabilities, with sensitivity that has been estimated (using Monte-Carlo simulations) as being better than a factor of two than conventional XAS.

Two pivotal experimental aspects are spectrometer resolution and beamline resolution. The preferred spectrometer for HERFD-XAS is a Johann geometry arrangement with a large Roland circle (*ca.* 1 meter), which uses Bragg diffraction

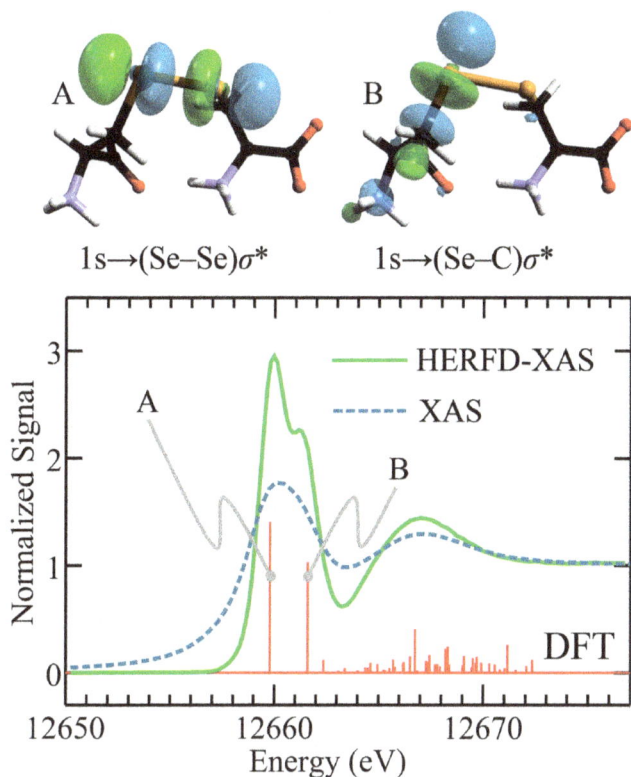

Figure 5.4: Comparison of conventional Se K XAS and Se $K_{\alpha 1}$ HERFD-XAS for an aqueous solution of L-selenocystine showing the remarkable improvement in spectroscopic resolution for the latter. A time-dependent DFT simulation (red line) of the spectrum is also shown, which indicates that the two intense transitions A and B populate excited states predominantly corresponding (Se–Se)σ^* and (Se–C)σ^*, respectively.

from a spherically bent analyzer crystal. As the Bragg angle θ_B of these crystal spectrometers approaches 90°, the resolution improves dramatically as a function of $\cot \theta_B$, becoming less sensitive to the size of the incident X-ray beam spot on the sample and approaching the Darwin-width of the analyzer crystal. Although $\theta_B=90°$ is not practical as sample and detector would need to be in the same place, $\theta_B=85°$ is certainly feasible, for example Se $K_{\alpha 1}$ HERFD-XAS uses a Si(844) analyzer which has $\theta_B=85.3°$ [9]. If the $\theta_B \to 90°$ condition is not satisfied, then enhanced resolution is not achieved, so that, for example, use of a Ge(844) analyzer with $\theta_B=73.1°$ at the Se $K_{\alpha 1}$ energy, gives spectra with energy resolution that is only very slightly better than conventional XAS [10]. Indeed, Pierre et al. [11] have reported that using a Si (555) analyzer for As $K_{\alpha 1}$ HERFD-XAS with $\theta_B= 69.7°$ actually gave *worse* spectroscopic resolution than was obtained with conventional As K-edge XAS. As Pierre et al. point out [11], the reason that the resolution was poorer than conventional XAS (rather than the same) was that the HERFD-XAS experiment employed a Si(111) double crystal monochromator which has an inherently poorer energy resolution than the Si(220) double crystal monochromator that was used for the conventional XAS.

In attempts to maximize sensitivity, conventional XAS experimental setups usually collect a large solid angle of X-ray fluorescence from the sample, which is uniform over all 4π steradians.[2] For this reason the use of arrays of solid-state detectors is now standard practice for XAS, with many beamlines equipped with monolithic pixel arrays comprised of 100 individual solid state pixels. HERFD-XAS also uses arrays, but in this case made up of crystal analyzers, each of which must be aligned individually with care; and at the time of writing the largest arrays in common use comprise 7 analyzer crystals and accept only $\sim^1/_{42}$ of the solid angle of a 100-pixel monolithic Ge detector. One initially unexpected result from early applications of HERFD-XAS to dilute samples was that the threshold for detection was lower than for conventional XAS, despite the lower solid angle accepted. With Se $K_{\alpha1}$ HERFD-XAS an adequate signal was obtained using samples of aqueous solutions containing only 100 nM of selenium, which would be impossible for conventional XAS. The reason for this is that the much higher resolution used in the HERFD-XAS experiment accepts substantially less background signal from inelastic X-ray (Compton) scattering, which for conventional XAS of very dilute samples would swamp the signal, making the HERFD-XAS practicable when the XAS is not.

There is a temptation to regard HERFD-XAS as simply a high-resolution version of XAS, but this can be an over-simplification and may give rise to errors. HERFD-XAS probes the diagonal of what is known as the resonant inelastic X-ray scattering or RIXS plane, in which the measured emission is plotted in three dimensions, usually as a function of incident energy and energy transfer (incident energy minus emission energy), often as a contour plot. The normal experimental procedure for recording a HERFD-XAS spectrum would be to move the incident energy far above the absorption edge and then record the X-ray fluorescence emission spectrum. This would then be used to determine the X-ray fluorescence peak centroid, and this centroid energy would then be used to monitor the fluorescence while the incident energy is scanned to record a HERFD-XAS spectrum. Problems may occur when there are chemical shifts in the fluorescence lines, such as the approximately 0.4 eV shift in the Se $K_{\alpha1}$ between Se^{-II} and Se^{IV}; this small energy difference is certainly large enough to pose substantial problems for the unwary experimenter. If the fluorescence energy is off-peak then the HERFD-XAS is shifted and distorted. If

---

2 A steradian (abbreviated sr) is the international system of units (SI) measure of three-dimensional solid angle, analogous to the two-dimensional angle radian. Just as one radian measured from the center of a circle of radius r describes an arc of length r on the circumference of the circle, for a sphere of radius r and area $4\pi r^2$ a solid angle of one steradian subtended at the center of the sphere and projecting a cone onto its surface will describe an area of r^2 on the surface of the sphere. The largest possible solid angle would thus be 4π sr.

chemical shifts are significant in a sample containing a complex mixture of species, then the experimenter must account for them either by collecting standards at the same energy as the mixture or by collecting a partial RIXS plane. Things can get even more complicated, since in some cases intensities will occur off-diagonal in the RIXS plane, for example for some 1s→3d transitions of first transition metal ions. These would be included in an XAS spectrum (at poor spectroscopic resolution) but not in a typical HERFD-XAS measurement. Because of these complications, and in general, HERFD-XAS should not be considered as just a high-resolution version of XAS; the interested reader is referred to the published literature for additional details [9].

5.3.3 X-ray fluorescence imaging

Many samples, such as biological tissues, exhibit an inherent and complex spatial structure. X-ray fluorescence imaging (XFI) allows such samples to be visualized in terms of the microscopic distribution of the constituent elements, their correlations with other chemical elements or anatomical features. XFI uses a micro-focused X-ray beam, which again is typically from a synchrotron radiation facility, to illuminate the sample which is raster scanned while the X-ray fluorescence is monitored using a solid-state detector. X-ray fluorescence from all heavier elements of interest within a sample can be monitored, providing that the energy of the incident X-ray beam is above that of the absorption edge of each of the elements concerned. An image or map of the sample in terms of its constituent elements can thus be built up. An example is shown in Figure 5.5. The spatial resolution of these methods ranges across many orders of magnitude, from more than 50 μm for large samples, to 20 nm or less. Like XAS, commercial devices are available, but the most advanced XFI experiments are conducted at synchrotron facilities. Specialized hardware is needed to generate the smallest X-ray beams, and not all synchrotrons can cater to all length-scales.

The chemical speciation of specific locations of interest on an XFI sample can be investigated by moving the sample so that the incident beam illuminates locations of interest and recording an XAS spectrum. This method is called micro-XAS, or μ-XAS. In some cases, one can go further and collect XFI data sets at different incident X-ray energies, chosen to preferentially excite particular chemical species, to give a chemically specific XFI data. Other variants of the method allow access to three dimensions. An X-ray fluorescence tomography setup is where a sample is both rotated and scanned relative to the incident X-ray beam while collecting the X-ray fluorescence, and even chemically selective X-ray fluorescence tomography have been reported. An attractive alternative method is provided by confocal XFI, where two sets of micro-focusing X-ray optics are used, one for the incident X-ray beam and one on the detector, acting in reverse, and focused on the focal point of the incident optic. This second optic serves to reject X-rays from all locations other than one specific volume element within a sample and the sample can be translated in three dimensions to collect XFI from different planes within it.

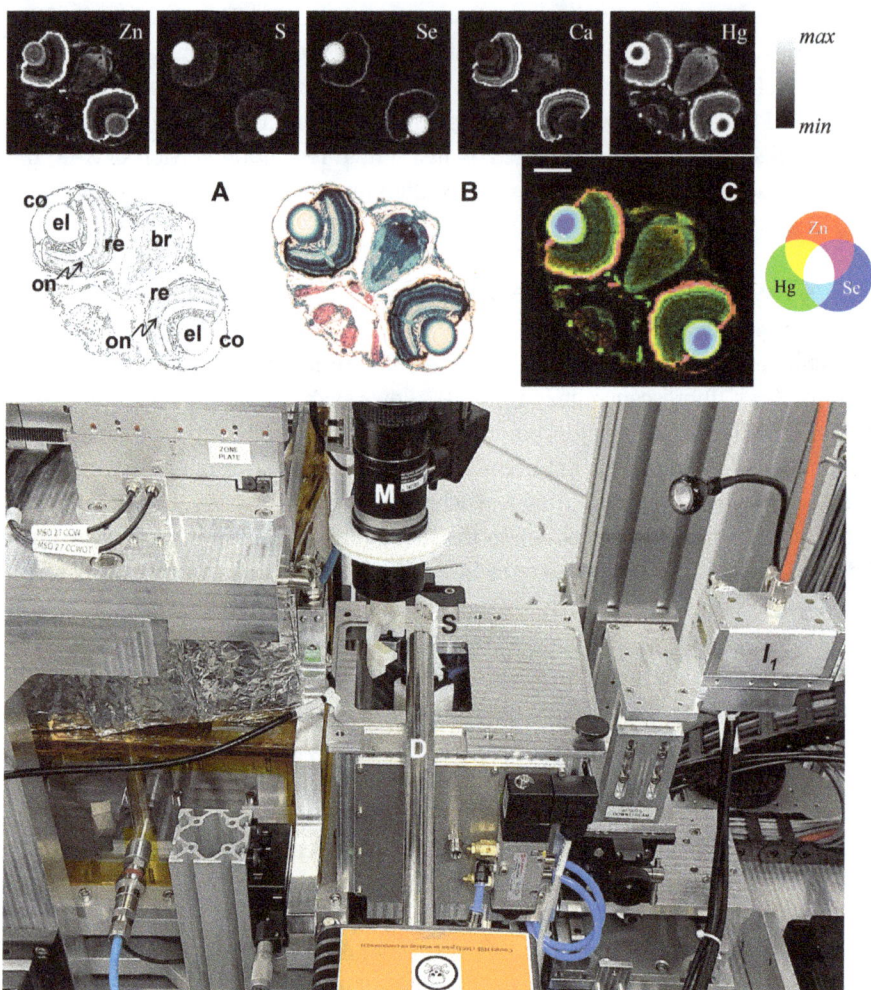

Figure 5.5: X-ray fluorescence imaging. The upper panel shows an example XFI data set; a head section of a larval stage zebrafish. A shows a line rendition of the histologically stained micrograph (B) showing the cornia (co), eye-lens (el), retina (re), optic nerve (on) and the fore-brain (br) (diencephalon). Images of individual elements are shown above, with Zn, Hg and Se superimposed in C. The scale bar in C shows 100 μm. The lower panel shows an experimental setup (BioXAS-imaging at the Canadian Light Source, courtesy of Dr. Malgorzata Korbas), with alignment microscope (M), sample (S) and solid-state X-ray fluorescence detector (D) indicated.

5.4 Computational chemistry

Recent increases in the power of readily accessible computer systems have been paralleled by development of increasingly available fast, highly accurate quantum chemical codes. This has revolutionized some approaches to structural chemistry. Density functional theory (DFT) is arguably the gold standard among these computational methods. Walter Kohn and John Pople shared the 1998 Nobel Prize in Chemistry for the development of DFT; since that time there has been a veritable explosion in applications of the method, which is evidenced from the number of publications using DFT listed in Chemical Abstracts, now totaling in excess of 307,000 with more than 28,000 in 2020 alone. Since there are many excellent introductions to DFT available [12, 13], what follows is intended as a superficial perspective.

A typical DFT calculation might begin with selection of a particular quantum chemical code, of which there are many, but all have much in common, each with its own specializations and particular strengths. All DFT codes use what is called the Kohn-Sham approach, in which the energy of a system is expressed as a function of electron density. The calculation requires the input of a starting model – a three-dimensional chemical structure with appropriate charge and interatomic distances – followed by choice of appropriate components of the calculation, such as basis sets and the exchange-correlation functional, after which the computation is initiated.

The DFT software will solve the Kohn-Sham equation iteratively until self-consistency is achieved, according to user-defined convergence criteria. The DFT calculation might be a geometry optimization, in which atomic positions are adjusted to minimize the energy using non-linear optimization methods, which can be followed by computation of properties, such as the partial charges discussed in Section 5.2 or the thermodynamic parameters discussed in Section 5.6.1. Figure 5.6 shows an example of a DFT geometry optimization, that of bis(cysteamine)-mercury(II) in comparison with the crystal structure (Figure 5.6a) [14]. Small molecule crystal structures are very accurate, with precisions in bond-lengths that are typically less than 0.01 Å, and hence present a good test of the abilities of DFT. When DFT is used to compute the structure in a vacuum (Figure 5.6b) or in a solvent, a different structure is obtained than that observed crystallographically. Some researchers have used comparisons of this sort to suggest that such differences correspond to a "failure" of the DFT method [15], a claim later refuted by others [16]. However, this is rather like comparing apples and oranges, since if the DFT method is applied to the crystal (Figure 5.6c and *d*) then the geometry optimization gives a very similar structure to that observed crystallographically.[3]

---

3 Of course, DFT does have its limitations; a well-known example is given by molecular sodium chloride (NaCl) which exists in the gas phase (along with the dimer Na_2Cl_2) as a covalently bound entity. If DFT is used to predict electron density of NaCl this works well with the actual bond-length of 2.36 Å, but if the atomic separation is increased to infinity then DFT still predicts that (small) partial charges exist, which is clearly a failure.

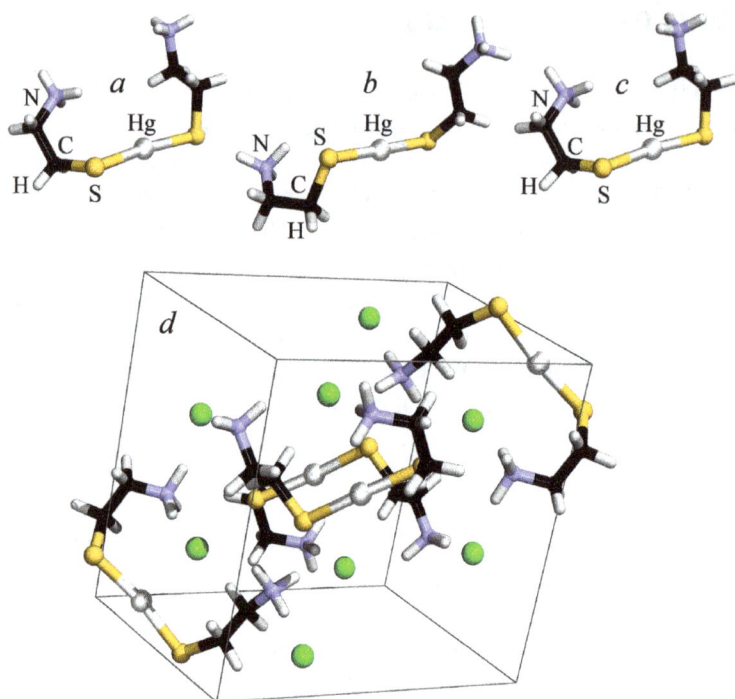

Figure 5.6: A comparison of density functional theory (DFT) geometry optimizations with small molecule crystallography: *a* shows [bis(cysteamine)-mercury(II)]$^{2+}$, as determined in the crystal structure. *b* shows the result of DFT geometry optimization (DMol3/GGA/PBE) of the same species in a vacuum, which is distinctly different from the structure (*a*) determined crystallographically. *c* shows a DFT geometry optimization of the crystalline structure (CASTEP/GGA/PBE) using the crystallographic cell with refinement of all crystal parameters. *c* can be seen to be very similar to the experimental crystal structure *a*. *d* shows the full DFT geometry optimized structure, with green spheres showing the location of the charge balancing Cl$^-$ in the crystal.

In the case of large molecules such as proteins and nucleic acids, crystal structures are underdetermined; so that to obtain a geometry-optimized solution, components of known structure (e.g., the amino acids in a protein) are used with bond-lengths held rigid, but with parts of the molecule with unknown local structure allowed to vary, in which case the errors can be greater than 0.1 Å. A similar level of caution to that for calculating small molecules with or without a crystal structure is needed when attempting to model the active sites of metalloproteins [17]. Thus, early work attempting to model active sites using truncated structures in a vacuum gave incorrect active site geometries, whereas later work using constraints developed from crystal structures gave more accurate results [17, 18], with insight into mechanism [19]. Even structures such as that of DNA, which depends upon relatively weak and difficult to model π-π stacking of bases within the double helix, can now be successfully modelled using DFT, with an appropriate choice of functional [20]. Structures calculated using DFT are thus expected to be reasonably accurate, with bond-lengths differing by less than 0.05 Å from the true values, with the only real limitations with modern codes relating to the speed of computation.

The computational burden of DFT scales non-linearly with the size of the model, increasing approximately as the cube of the number of basis functions. Because of this, large molecules can easily exceed the abilities of most desktop computers; one may need both a more powerful computer and a generous supply of patience, or to adopt more approximate methods. One such are so-called MM/QM approaches, in which molecular mechanics (MM) is used to model the bulk of a structure, typically a macromolecule such as a protein or a nucleic acid, and quantum mechanics (QM) and specifically DFT is used to model the more challenging parts of the molecule. A weakness of this approach is that the user must pre-define atoms to be treated quantum mechanically, and the quantum model will simply be a refinement of the starting model, with no substantial structural changes possible. Thus, in a recent study of mercury binding to DNA, the binding modes could be explored with DFT, such that if the Hg^{2+} was bound to oxygen atoms (for example) within the double helix, then DFT geometry optimizations allowed the Hg^{2+} to migrate to a lower energy location within the structure, which was found to be coordinated to N3 of cross-adjacent thymine bases (Figure 5.7) [20]. With the QM/MM method, on the other hand, the mercury would always stay bound to a QM atom, so no that large-scale changes are permitted by the method. Some DFT codes also give experimenters the capability to accurately simulate X-ray absorption near-edge spectra (e.g., Figure 5.4), closing a loop with experiment and giving insights into both physical geometry and electronic structure.

In a modern research setting, the power of DFT methods allows the user to conduct what are in effect in silico experiments at the molecular level, to test whether a proposed molecular arrangement is reasonable or not, and thus providing a computational reality check. In our view, the most important applications of DFT may be those that combine experimental measurements and computational studies. DFT also gives experimenters a convenient means to compute basic thermodynamic values, and thus develop an understanding of whether certain reactions are likely or unlikely. Having become an essential part of the toolbox for quantitative chemical research, we anticipate that future applications of DFT to the chemical problems of inorganic molecular toxicology will increase.

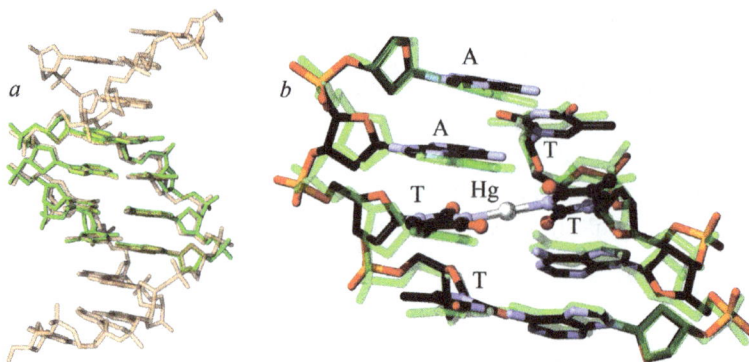

Figure 5.7: DFT geometry optimization of a four-base pair duplex DNA (dAATT) compared with a crystal structure containing the same central sequence (*a*). Green shows the DFT geometry optimized structure (DMol3/mGGA/M11L), and light brown the crystal structure. *b* shows a DFT simulation (geometry optimization) of Hg^{2+} interaction with DNA showing binding to cross-adjacent thymine bases (here, green shows the DFT structure without Hg^{2+}, as in *a*). In both *a* and *b* hydrogen atoms have been omitted to simplify the figure.

5.5 Chemical speciation and toxicology

As discussed above, understanding the chemical speciation of a toxic metal or metalloid is essential for understanding its roles in organisms, including any toxicity. Uncontaminated seawater naturally contains approximately 30 pM arsenic, but marine algae and a variety of marine animals accumulate arsenic by several orders of magnitude. In particular edible marine fish typically contain approximately 0.02% by weight of arsenic, but this poses no toxic hazard to humans, as it is present as the entirely non-toxic trivalent arsenicals arsenobetaine [21] and arsenocholine [22] (Figure 5.8). These As^{III} species contain the arsonium group (Figure 5.8) and are not substantially metabolized, although arsenocholine does appear to be metabolized to arsenobetaine [23]; such methylated arsonium species appear to present little or no toxic hazard to humans. They are completely absorbed from the gastrointestinal tract and are retained without eliciting any ill-effects until eventually excreted in the urine. Typical trends are that increasing methylation of arsenic lowers toxicity for a given arsenic formal oxidation state, and that As^{III} is more toxic than As^{V} in a similar coordination. Hence, in human cell lines [24] the toxicities of a series of simple arsenic compounds have been reported to decrease in the order methylarsonous acid [$CH_3As(OH)_2$] > dimethylarsinous acid [$(CH_3)_2As(OH)$] > arsenous acid [$As(OH)_3$] > arsenic acid [H_3AsO_4] > trimethylarsine [$(CH_3)_3As$] > dimethylarsinic acid [$(CH_3)_2AsO(OH)$] > methylarsonic acid [$CH_3AsO(OH)_2$] > trimethylarsine oxide [$(CH_3)_3AsO$] > arsenocholine [$(CH_3)_3As^+C_2H_4OH$] > arsenobetaine [$(CH_3)_3As^+CH_2CO_2^-$].Apart from the zwitterionic arsenobetaine, in the above list we have used formulae for the fully protonated acids, some of which are deprotonated at physiological pH values.

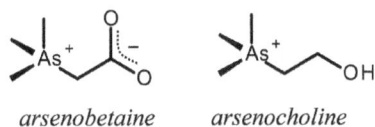

arsenobetaine   *arsenocholine*

Figure 5.8: Schematic structures of arsenobetaine and arsenocholine, with protonation states appropriate for physiological pH values.

Thus, arsenic species present a variety of toxic responses, critically depending upon the chemical speciation and the dose. When it comes to understanding toxicological molecular mechanisms, knowledge of the chemical form in solution, or even in the tissue itself, is very important, illustrating the need for in situ speciation tools.

5.6 Chelators and the chelate effect in toxicology

"Chelation therapy" is a term that has entered the public domain, with fringe medicine embracing such procedures as "heavy metal detox" with exaggerated claims of beneficial effects that have clouded the situation [25]. Despite this, chelation therapy can be highly effective when appropriately used, and embraces a range of different chelators. Chelators are chemical entities that bind a metal or metalloid by *more than one* coordinating group to form a chelate. The term chelator derives from the Greek *chelos* (χηλή), meaning pincer-like claws, as of a crab or lobster, which typically hold and object using two claws. Chemists speak of the *denticity* of coordinating groups; *monodentate* ligands, or alternatively *unidentate* ligands are coordinating groups with just one bonded atom (derived from the Latin, meaning one-tooth). Only *polydentate* species can be chelators, and, as we will see, not all of these actually are chelators in practice. Thus, the term *bidentate chelator* would mean a ligand having *two* coordinating atoms both of which bind a metal or metalloid. The denticity of a chelate complex is denoted by the use of the Greek letter κ, so that an As^{III} atom bound to both thiolates of the vicinal dithiol chelator British Anti-Lewisite [26] (abbreviated BAL) might be described as a $κ^2$-BAL complex. Denticity is distinct from hapticity, which refers to coordination of a metal or metalloid by an uninterrupted contiguous series of atoms. Hapticity, derived from the Greek word for touch, is denoted by the Greek letter η; an Fe^{3+} peroxide complex in which the metal is bound to both oxygens would have $η^2$ coordination.

Chelation therapy is the use of chelators to treat exposure to metal or metal ions by mobilizing in the tissues, and facilitating excretion. Before proceeding to discuss chelation therapy we will briefly discuss what is known as the *chelate effect*. This is observed when a true chelator interacts with a metal or metalloid, binding with greater tenacity than an equivalent monodentate ligand.

5.6.1 Thermodynamics of the chelate effect

Thermodynamically, the process of chelation is entropy driven. Relatively simple DFT calculations can be used to demonstrate this. If we consider the idealized reaction shown in Figure 5.9 which has a vicinal dithiol chelator reacting with monothiol coordinated trivalent arsenic, then DFT can give values for the Gibbs free energy change $\Delta G = \Delta H - T\Delta S$ using a frequency calculation to estimate both H and S for the various components and the summed values for products from those of the reactants. Since the computed As–S bond-lengths for the methanethiolate and ethanedithiolate chelate are very similar, in terms of bond-strengths alone there will be energetically little to choose between these alternatives. Using a temperature of 298.15 K, we can use DFT calculations to estimate enthalpy and entropy terms for both reactants and products and compute ΔG=−42.9 kJ·mol^{-1}, with ΔH=−0.7 kJ·mol^{-1}, and $T\Delta S$=42.2 kJ·mol^{-1}, showing that only minimal changes in enthalpy are expected and by far the largest thermodynamic driver is an increase in entropy which is due to an increase in degrees of freedom of the liberated monothiolate. These values yield an equilibrium constant for Figure 5.9 of 3.21×10^7 (very far to the right).

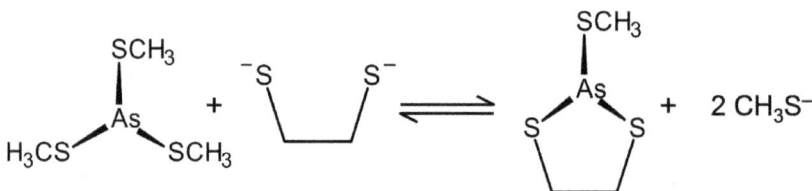

Figure 5.9: Schematic structures for hypothetical chelation chemistry discussed in the text.

5.6.2 Hard-ligand-based chelators

We would be remiss if we did not mention hard-ligand-based chelators, such as EDTA (ethylenediaminetetraacetic acid) or DTPA (diethylenetriaminepentaacetic acid) (Figure 5.10), both of which are approved for clinical use. EDTA can be found on the shelves of most laboratories, and is often used as a standard component of biochemical buffers in order to chelate any free metal ions. It is used clinically in chelation therapy, and comes as two different formulations which have unfortunately been given quite similar names, the disodium salt, which is called "edetate disodium" or the mixed calcium disodium form, which is called "edetate disodium calcium". Na$^+$ is not expected to be strongly bound by EDTA, but Ca^{2+} will bind strongly via four of the carboxylate oxygens and both nitrogens of a single EDTA, with two coordinated waters completing an overall 8-coordinate site (Figure 5.10) [27].

The formulation labeled "edetate disodium calcium" is approved for chelation therapy associated with heavy metal intoxication, such as Pb^{2+}, whereas that labeled

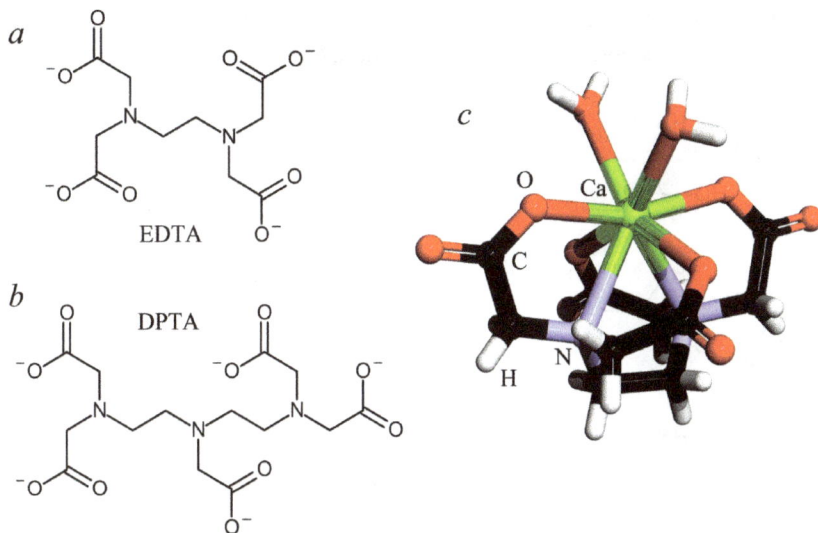

Figure 5.10: Schematic structures of a EDTA (ethylenediaminetetraacetic acid) and b DTPA (diethylenetriaminepentaacetic acid); charge states shown are those at physiological pH. c shows the crystal structure for $[CaEDTA(H_2O)_2]^{2-}$ [27].

"edetate disodium" is not. The "edetate disodium" formulation has been used to chelate Ca^{2+} for the emergency treatment of hypercalcemia (a calcium excess), to control of ventricular arrhythmias associated with digitalis toxicity, or in secondary prevention in post–myocardial infarct patients [28]. Giving "edetate disodium" under other circumstances carries a risk of inducing hypocalcemia (too little calcium) through chelation of blood Ca^{2+} after which coma and death rapidly follow. Unfortunately, in all probability it is the similarity in these drug names that has led to medical mistakes and some unnecessary deaths [29]. DTPA has substantially more potential coordinating groups and is approved for chelation therapy of actinides [30].

5.6.3 Vicinal dithiol-based chelators

We now turn to consider vicinal dithiol-based chelator drugs, some well-known examples of which are shown in Figure 5.11. The first of these to be developed was British anti-Lewisite (BAL) or 2,3-dimercaptopropanol which was developed for use against the arsenical agent of chemical warfare Lewisite (chlorovinyl arsine dichloride) [26, 31]. Trivalent arsenicals are lethal under conditions of acute exposure because of their strong binding to the dithiol of reduced lipoic acid in the pyruvate dehydrogenase complex and related enzymes, which forms a chelate with a six membered ring (Figure 5.12). This effectively blocks respiration, and thus can kill rapidly. Using similar DFT calculations to those discussed above we find that arsenical association with the propane-1,3-dthiol moiety of lipoic acid is again an entropy driven reaction, as is the displacement of the propane-1,3- dithiolate by alkane-1,2-dithiolate chelators.

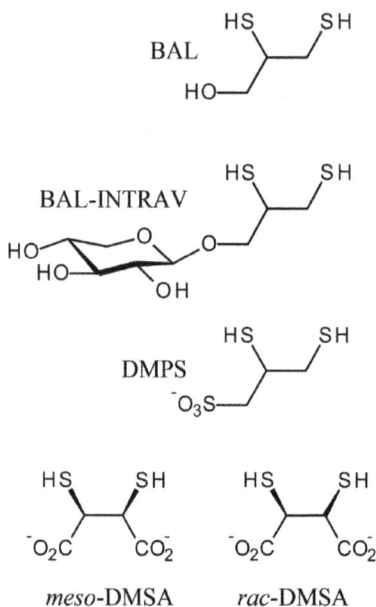

Figure 5.11: Schematic structures for vicinal dithiol chelators, BAL, BAL-INTRAV, DMPS, and *rac* and *meso*-DMSA. We note that DMPS is also chiral, and that commercial preparations contain the racemate.

BAL is the oldest of the vicinal dithiol chelator drugs; under the drug name Dimercaprol it remains a mainstay of chelation therapy and is included on the World Health Organization's list of essential medicines [32]. However, BAL suffers from a number of disadvantages; it has poor water solubility and a noxious smell. BAL is typically administered by intramuscular injection, and it has a range of adverse effects including pain at the injection site, vomiting, fever, raised blood pressure and tachycardia. Early work by Peter Mitchell (of chemiosmotic hypothesis fame) and co-workers considered the glucoside – which they called BAL-INTRAV (Figure 5.11) as a non-toxic alternative given by intravenous injection [33], but this has never seen clinical use. The related compound dimercaptopropane sulfonic acid (DMPS) was subsequently developed as a more water-soluble alternative to BAL; it has no perceptible smell, and is known by the drug names Dimaval or Unithiol. DMPS is usually administered by injection, either intramuscular or intravenous, although it can also be administered orally. Another water-soluble and odor-free vicinal dithiol is dimercaptosuccinic acid (DMSA) which has the drug names Dimaval and Succimer, and this is generally taken orally. DMSA occurs in the *rac* and *meso* diasteriomers (Figure 5.11), of which the *meso* form is that available for drug use. These chelation therapy drugs have been used for treatment of As^{3+}, Hg^{2+} and Pb^{2+} intoxication. In a ground-breaking study, BAL was also used as the first successful non-palliative treatment for Wilson's disease [34], which we will discuss in Section 5.6.4. Here, we will discuss the nature of the interaction of vicinal dithiol chelation therapy drugs with arsenic, mercury, cadmium and zinc (specifically, As^{3+}, Hg^{2+}, Cd^{2+} and Zn^{2+}).

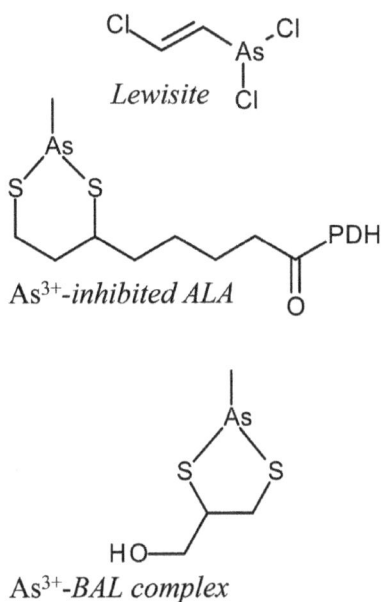

Figure 5.12: Schematic structures showing Lewisite, the As^{3+} chelate to reduced lipoic acid, where the pyruvate dehydrogenase complex is abbreviated as PDH, and the As^{3+} BAL chelate.

5.6.3.1 Vicinal dithiol chelators for treating Arsenic(III) intoxication

There are some remarkable accounts of successfully saving patients who have ingested otherwise massively lethal doses of arsenic compounds using vicinal dithiol chelation therapy, e.g., [35, 36]. In one case more than 9 grams of arsenic trioxide was orally ingested, and while at one point recovery seemed uncertain, the patient survived with only peripheral neuropathies. These cases illustrate the efficacy of correctly applied chelation therapy. We note one report in which intake of more than 1 gram of arsenic trioxide was successfully self-treated through the simple expedient of drinking large volumes of water (5 liters over 5 h), although subsequent chelation therapy may have helped the patient to recovery [37]. In all cases residual symptoms included a residual neuropathy, which improved with time. In December 2002 a woman was found guilty of murder by poisoning a communal pot of curry soup at a 1998 summer festival in the Sonobe district of Wakayama, Japan. A large quantity of arsenic trioxide had been added to the soup, and at least 67 people were poisoned. None of the patients received chelation therapy, and two adults and two children subsequently died. Interestingly, a recent study of data on the poisoned individuals showed that the children were more resistant to poisoning than were the adults, with a more rapid conversion to methylated forms, and much more

rapid excretion of the arsenic [38]. It remains unclear whether the victims who suc-
cumbed to the poisoning might have been saved by chelation therapy.

5.6.3.2 Interaction of vicinal dithiol chelation drugs with Hg²⁺

We now turn to consider the interaction of vicinal dithiol drugs with mercury. While
they cannot compare with other toxins, such as the organophosphorus nerve agents,
some of the compounds of mercury can be considerably more toxic than those of any
other non-radioactive heavy element. Figure 5.13 compares the toxicities of some environmentally problematic elements.

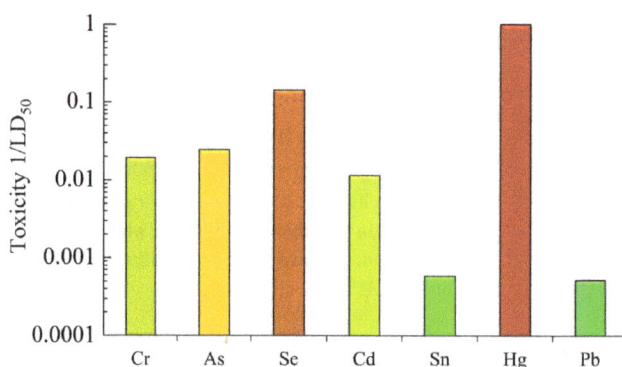

Figure 5.13: Toxicity comparison for some heavy elements that pose human health problems.
Toxicities are plotted on a $\log_{10}$ scale and are expressed as reciprocals of the LD_{50} (rat, oral,
mg/kg) taken from commercial safety data sheets (Sigma-Aldrich, Acros Organics) for inorganic forms
of each metal or metalloid, specifically, Na_2CrO_4, $NaAsO_2$, Na_2SeO_3, $CdCl_2$, $SnCl_2$, $HgCl_2$ and $PbCl_2$.

Mercury is a natural part of our environment, and in the metallic form is more mobile
than any other heavy element because of its high vapor pressure and correspondingly
effective atmospheric transport. Mercury pollution, however, is currently at an all-
time high, and the environment is showing signs of substantial increases that may be
problematic. In the environment, mercury occurs as the metal, inorganic forms, or organometallic forms, primarily as methylmercury compounds with some occurrence of
dimethyl mercury. Elemental mercury is mostly found either dissolved in waters, or
in the vapor phase in the atmosphere. Here, we will focus primarily on inorganic
forms. The chemistry of mercury is in some senses quite flexible, since Hg^{2+} can
adopt either two-coordinate digonal, three-coordinate trigonal or four coordinate tetrahedral type local coordination environments. Thus, at the time of writing, the
Cambridge Structural Database [39] contains some 2,796 four-coordinate, 595
three-coordinate and 1,333 two-coordinate Hg^{2+} species. Mercury is subject to
strong relativistic effects which together with a smaller contribution from the well-

known lanthanide contraction are responsible for characteristically shorter bond lengths. Mercuric ions are known for their affinity for chalcogenides, indeed the old name for thiols is mercaptan, which is derived from the Latin *mercurius captans* or mercury capturing, because of the strength of bonding between thiolates and mercuric ions. Restricting searches of the Cambridge Structural Database to only species with organic thiolate donors, the numbers of structurally characterized two, three and four coordinate species are 61, 9 and 66, respectively. The use of BAL, meso-DMSA and DMPS in chelation therapy for mercury intoxication is well established, and while vicinal dithiols can form chelate complexes with Hg^{2+}, these are rare and there is only one report of such a complex in the Cambridge Structural Database [39, 40]. Work upon the solution chemistry of DMSA and DMPS with Hg^{2+} has shown these vicinal dithiols do not form bona fide chelation complexes with Hg^{2+} [41]. This study used a combination of XAS, element-specific chromatography, and DFT to characterize the species formed in solution, and showed that neither DMSA nor DMPS binds mercury as a chelator [41], instead acting as mono-dentate ligands, forming ring-shaped species involving more than one mercury atom, as shown in Figure 5.14. In addition, DMSA and DMPS differ in that, when DMPS is present in excess, it forms complexes in which four DMPS are coordinated to Hg via a single thiolate. An analogous complex does not form with DMSA because a four-coordinate DMSA would bear a large negative charge (14−) localized near to the central Hg which would make this mode of binding less preferential. DMPS bears only a single negative charge which is localized further from the dithiolate, and hence can form stable four-coordinate species [41].

Thus, despite much literature and many claims to the contrary, while DMSA and DMPS bind Hg^{2+}, they are not in fact mercury chelators in solution. The effectiveness of DMSA, DMPS and BAL in clearing Hg^{2+} from intoxicated victims has in all probability nothing to do with the chelate effect, and their use cannot therefore legitimately be called chelation therapy; rather they act as mono-dentate mercury *sequestration* agents. They may also be poorly optimized for this clinical role as they only act as mono-functional thiols in their interaction with mercuric ions. The same is true for all of the agents currently used in mercury chelation therapy – none of them are actually mercury chelators [42]. The properties of some water-soluble homologues of DMSA based on 1,3-propanedithiols have been explored [43]. Although the motivation for this study was to improve properties by developing new chelators, as expected from previous work [42] these still cannot bind Hg^{2+} in the linear geometry that would be preferred for a 1:1 dithiolate chelator. Moreover, the study showed that the 1,3-propanedithiol-based mercury sequestration agents were effective in reducing the renal burden of Hg^{2+} in rats, but less so than DMPS or DMSA [43]. As we have mentioned, one preferred mode of binding for Hg^{2+} is the linear digonal coordination, with an S–Hg–S bond-angle of ~180°. The solution binding properties of straight-chain alkane dithiols [$^-S(CH_2)_nS^-$] with n of 4, 6 and 8 have been investigated, and all of these do form 1:1 chelate complexes with Hg^{2+} in solution [42]. DFT has also been used to explore in silico the preferred geometry of chelates with a wider range of straight-

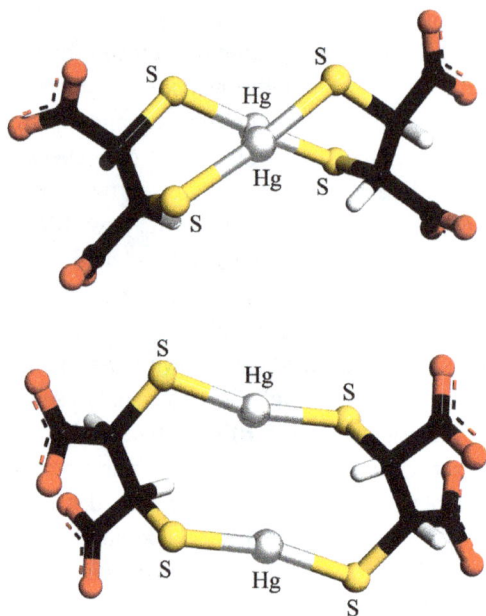

Figure 5.14: DFT geometry optimized structures for the two possible diasteromers of the smallest possible meso-DMSA:Hg^{2+} complex. The two disasteriomers are predicted to form in approximately equal amounts, and have Hg···Hg distances that differ (top 3.13 Å, bottom 3.04 Å) so that the Hg···Hg EXAFS almost cancels. Because DMPS is chiral with commercial material being a racemic mixture, DMPS will form four different conformers with analogous ring-shaped structures (not illustrated) and in excess it forms four-coordinate species, as discussed in the text.

chain alkane dithiolates [42], and this study concluded that a chain length of 8 or 9 allows mercury binding in a close to ideal metal geometry [42]. Of some substantial interest are the mercury chelators based on benzene-1,3-diamidoethanethiolate discussed by Atwood and co-workers [44, 45], which can bind mercury in an energetically close to optimal geometry [42].

Returning now to DMSA and DMPS, however, there is no doubt they are effective in promoting removal of mercury following intoxication [46–50], although not in cases where the mercury is organometallic in nature [51]. Of significant interest is the finding that the mechanism of elimination via the kidneys involves the multidrug resistance protein Mrp2 transporter [52], as Mrp2 deficient rats do not eliminate Hg^{2+} when treated with DMSA or DMPS [53]. Mrps are members of the C family of ATP-binding cassette (ABC) transporters, and share a mechanism in which ATP hydrolysis drives transport. Because Mrp2 tends to transport glutathione conjugates it seems plausible that the mechanism of Hg^{2+} sequestration by DMSA and DMPS and subsequent renal excretion may involve coordination of the metal by glutathione, or the formation of glutathione conjugates of some type.

5.6.3.3 Interaction of vicinal dithiol chelation drugs with Cd²⁺ and Zn²⁺

Cadmium, like its group-12 neighbor mercury, is considered a problematic toxic metal. All of arsenic, cadmium and mercury are what is known as chalcogenophilic, meaning they have a high affinity for chalcogenides such as sulfur and selenium. Cadmium is distinguished from mercury and arsenic in that it does find a specific use in nature, serving as a catalytic metal at the active site of an alternate carbonic anhydrase found in marine diatoms [54], a role normally fulfilled by its group 12 neighbor zinc. Major sources of human exposure can be linked to trace levels of cadmium in tobacco, through inhalation of cigarette smoke [55], and also, albeit at lower levels, in e-cigarette "vapes" [56]. Industrial activities also provide potential sources of human exposure, and the Centers for Disease Control and Prevention (CDC) of the United States have estimated that approximately 12 million adults in the USA may be exposed to problematic levels of cadmium. In particular, there are clear links between urinary cadmium levels and increased risk of diabetes [57, 58], and it has been suggested that Cd^{2+} exposure may be a cause of prediabetes and diabetes in humans. A number of Cd^{2+} complexes with vicinal dithiols have been structurally characterized [40, 59], and the binding of Cd^{2+} to DMSA and DMPS at physiological pH has been explored by using a combination of XAS, size exclusion chromatography and DFT [60]. The results show a complex chemical behavior [60], but that unlike Hg^{2+}, Cd^{2+} does form bona fide chelate complexes with both of DMSA and DMPS. Figure 5.15 shows two selected DFT-computed structures of chelate complexes with DMPS, a 1:1 chelate complex with water donors completing the Cd^{2+} coordination sphere, and a 2:1 chelate complex with two DMSA bound to the metal.

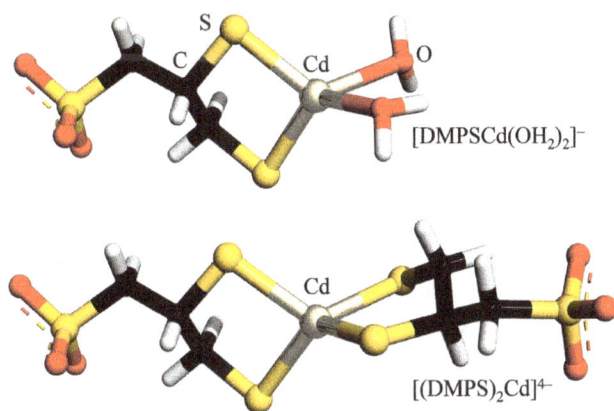

[DMPSCd(OH₂)₂]⁻

[(DMPS)₂Cd]⁴⁻

Figure 5.15: DFT geometry optimized Cd-DMPS chelate complexes. The upper structure shows the 1:1 complex with the coordination completed by two water molecules, while the lower structure shows the structure of the bis-chelate species. Since DMPS is chiral, and the commercial preparation is a racemate, in solution four different conformers are possible for the bis-chelate.

The lightest element in the group 12 of the periodic table is zinc. Unlike mercury and cadmium, zinc is an essential element in humans; indeed, if one excludes s-block elements such as Na, K and Ca, it is the most abundant metal in living organisms, with 11% of proteins containing Zn^{2+} as a structural or catalytic site [17]. It therefore will not be discussed in detail here, except to point out that vicinal dithiol chelators BAL and DMPS have recently been shown to be highly effective in the treatment of hemotoxic snakebite [61], via chelation of the Zn^{2+} of zinc-dependent protease constituents of snake venom [61]. An injection of BAL or DMPS could be administered in the field, and given that approximately 75% of snakebite fatalities occur because the victim is remote from medical care [62], this important finding could further validate the WHO's inclusion of vicinal dithiol chelators in its list of essential medicines.

5.6.4 Chelation therapy and Wilson's disease

As a final topic in this section we turn to discuss chelation therapy of Wilson's disease, a dyshomeostasis of copper which is due to a mutation in the ATP7B transporter. Wilson's disease is more prevalent in males than in females, and has a world-wide incidence of about 1 in 30,000. It causes accumulation of copper in tissues, and particularly in the liver and brain, which manifests with liver disease and neurological symptoms [63]. In healthy individuals copper is exported from cells by "packaging" in the Golgi via the ATP7B transporter (the Wilson's protein). Wilson's disease is a chelation therapy success story [64], and we have already mentioned the pioneering work using BAL [34]. Modern chelation therapy uses D-penicillamine (Figure 5.16) and trientine (N'-[2-(2-aminoethylamino)ethyl]ethane-1,2-diamine), with both having seen effective clinical use in combination with oral Zn^{2+} supplementation, which serves to antagonize copper uptake from the gut. Although successful in treatment of Wilson's disease, the detailed molecular basis of their action is unclear. The Cambridge Structural Database contains a very large number of structurally characterized copper complexes with trentine [39], all of which contain Cu^{2+}, and a number of structures with penicillamine, although most of these are polynuclear clusters. If D-penicillamine is mixed with Cu^{2+} salts in aqueous solution at physiological pH then an intensely purple colored stable solution forms. The complex and beautiful crystal structure for this purple species has been determined [65]; it contains 14 copper ions as a mixture of 6 Cu^{2+} and 8 Cu^+ (Figure 16). Whether or not this cluster is important biologically is unclear, but the beauty of the structure means that it merits brief description here. The six Cu^{2+} sites all contain the distorted square planar type geometry that is typical of Cu^{2+} with N_2S_2 coordination from two D-penicillamines each, and lie external to the Cu^+ sites which are bound by three μ_4 sulfur donors in a distorted trigonal planar geometry connected by a long axial bond (Cu–Cl = 2.82–2.92 Å) to a central μ_8 Cl^-. The crystallographic unit cell is large for a small molecule, showing cubic symmetry, containing 24 of the aforementioned 14-copper clusters. To date, there are few studies of the coordination chemistry of solution chelate complexes with Cu^{2+} or Cu^+, and the chemical nature of the chelate complex that is responsible for the successful copper chelation therapy of Wilson's disease remains uncertain. That said, there are some newly developed chelators based on D-penicillamine [66] and on N-substituted cysteine amides [67], which could provide superior performance in copper chelation therapy of Wilson's disease.

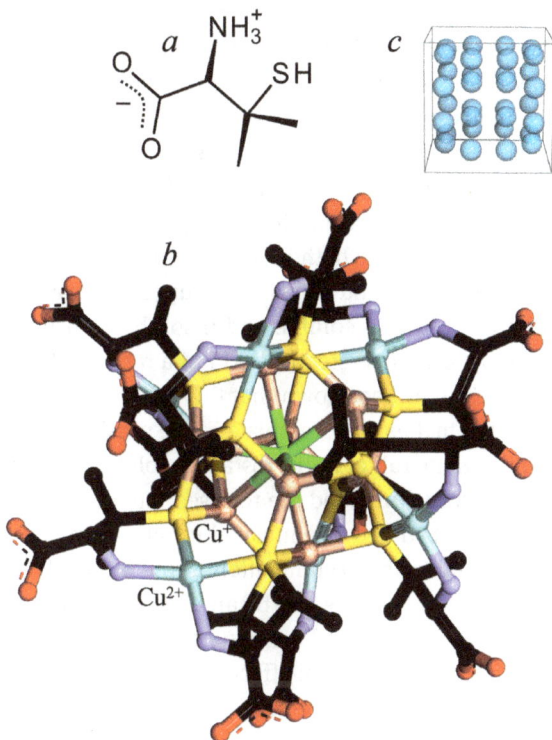

Figure 5.16: *a* shows a schematic structure of D-Penicillamine (D-Pen), *b* the crystal structure of the mixed-valent $[Cu^{2+}{}_6Cu^+{}_8(D\text{-}Pen)_{12}Cl]^{5-}$ complex. The large crystallographic unit cell (a =50.847 Å) gave limited crystallographic resolution, which (not surprisingly) precluded determination of hydrogen atom positions. *c* shows the crystallographic unit cell, with the positions of the 24 $[Cu^{2+}{}_6Cu^+{}_8(D\text{-}Pen)_{12}Cl]^{5-}$ clusters denoted as blue spheres.

We now turn to a novel and promising approach to Wilson's disease chelation therapy, which differs from other approaches in that the chelator itself contains a metal. Early work showed that dietary tetrathiomolybdate $[MoS_4]^{2-}$ could block copper uptake from gut in humans, but interperitoneal injection of tetrathiomolybdate in a rat model of Wilson's disease shows a considerably greater effect [68]. Tetrathiomolybdate was found to protect rats from the adverse effects of copper accumulation in early stages, before the onset of signs, and restores apparent good health within 24 h at later stages when advanced neurological problems have developed [68]. The molecular basis for these remarkably effective animal experiments has been shown to be formation of highly stable chelates between tetrathiomolybdate and cuprous ions, via two Mo =S groups, to form multi-metallic clusters [69, 70]. The $[MoS_4]^{2-}$ anion is a tetrahedral species (Figure 5.17), and each of the six edges of the tetrahedron can in principal provide two sulfur donors to a Cu^+ entity, so that a range of complexes $[(RCu)_xMoS_4]^{2-}$ with x =1–6 are possible; examples of all of these complexes have been structurally characterized [39]. In the rat models [69, 70], XAS (Figure 5.17 B, C) indicates that the predominant cluster is $[(RCu)_3MoS_4]^{2-}$, which is also the most prevalent of the many examples of such complexes in the Cambridge Structural Database [39]. These complexes tend to be highly stable, and may be nearly biochemically inert; it seems likely that chelation with tetrathiomolybdate effectively soaks up the excess copper, preventing or relieving the signs of disease, depending upon whether it is given early or late. While these complexes have been shown to form in the liver [69, 70], and accumulate both in liver and in kidney, they are mobilized and excreted at least to a degree, and it has been suggested that the $[(RCu)_xMoS_4]^{2-}$ complexes might be divided into soluble and insoluble and fractions, with the latter being mobile and slowly excreted [71]. XFI of a rat model following treatment with $[MoS_4]^{2-}$ (Figure 5.17A) has shown that Cu and Mo exhibit a cell-specific co-localization in kidney; the metals co-accumulate in the proximal convoluted tubules of kidney but not in the glomeruli or in the distal convoluted tubules [70]. This suggests that complexes transported in blood are released into the filtrate in the glomerulus cavity, and then re-absorbed and accumulated locally when passing through the proximal tubules, explaining the accumulation in the kidney [70]. Recent clinical trials conclude that tetrathiomolybdate can be an effective treatment for Wilson's disease [72, 73] especially if neurological symptoms are present [73]. Some caution may be appropriate as adverse long-term effects of $[MoS_4]^{2-}$ are still unclear, but its ability to restore animals with advanced neurological signs to apparent good health may suggest a similar role in treatment of humans.

The formation of structurally similar clusters is believed to be responsible for molybdenum-induced copper deficiency, well-known in veterinary medicine, which occurs in ruminants feeding on a molybdenum rich diet [74]. In this case molybdates $[MoO_4]^{2-}$ in the diet become converted into thiomolybdate $[MoS_4]^{2-}$ in the reducing sulfide-rich, essentially anaerobic ferment of the rumen [75] (the first of the four ruminant stomachs). Under these conditions Cu^+ will predominate, and will readily react with the $[MoS_4]^{2-}$. The resulting highly stable copper-molybdenum complexes effectively sequester the copper, preventing absorption and causing copper deficiency, which can be fatal. Finally, in this section, as we are considering ruminants, we turn briefly to the North Ronaldsay sheep. This intriguing breed of sheep originates from North Ronaldsay which is the most northerly of the Orkneys, off the north coast of Scotland. The sheep are confined to the shore by a wall that encircles the entire island (some 13 miles in circumference). The wall is actually a fairly recent

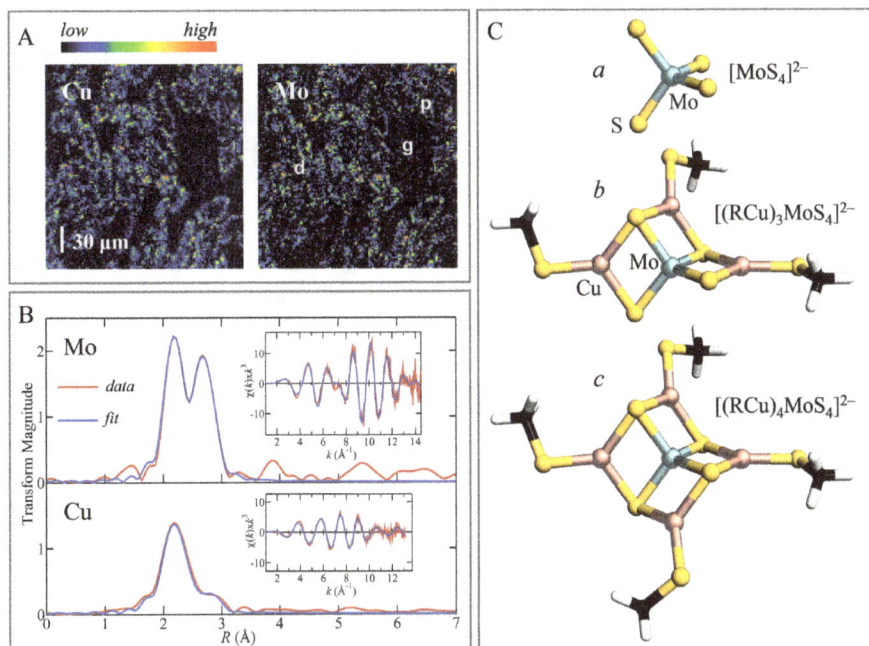

Figure 5.17: Panel A shows XFI for Cu and Mo for a section of Wilson's disease rat model treated with $[MoS_4]^{2-}$. Clear co-localization of the Cu and Mo is observed. **d** indicates the distal convoluted tubule, **p** the proximal convoluted tubule and **g** the glomerulus. Panel B shows the Mo and Cu K-edge EXAFS Fourier transforms (phase-corrected for sulfur backscatterers) of kidney tissues, showing the Mo–S and Mo⋯Cu interactions (Mo) and the Cu–S interaction and weaker Cu⋯Mo/Cu interaction (Cu) together with theoretical fits. The EXAFS oscillations are shown in the insets. For the Cu EXAFS the Cu⋯Mo and Cu⋯Cu partially cancel. C shows the DFT geometry optimized structures for *a* $[MoS_4]^{2-}$, *b* $[(RCu)_3MoS_4]^{2-}$ and *c* $[(RCu)_3MoS_4]^{2-}$, in which the external ligands are approximated as methane thiolates for computational purposes.

addition, built in 1831 to keep sheep from feeding on the fields and crofts of the islanders in the interior of the island. The North Ronaldsay breed is very much older, and perhaps the oldest known breed of sheep. A large part of the normal diet of the sheep is seaweed, which is rich in arsenic compounds such as arsenosugars [76] and poor in copper. As a consequence, the sheep are resistant to seaweed derived arsenic exposure [77], and are very effective at absorbing copper from the seaweed. Indeed, when these sheep are allowed to feed on terrestrial herbage, such as grass, they can die from copper poisoning [78, 79]. Whether or not these two intriguing phenomena are related to each other in any way remains to be seen. Before moving on, we note that seaweeds such as *Phaeophyceae* (brown kelp) eaten by the North Ronaldsay sheep contain very little selenium [80], so that the complex interrelationship between arsenic and selenium, addressed next, probably has little relevance to the remarkable resilience to arsenic of these sheep.

5.7 Arsenic and selenium

5.7.1 Arsenic and selenium antagonism

Trivalent arsenic compounds, such as arsenite, which is $As(OH)_3$ at physiological pH, can be highly poisonous. Pentavalent arsenate is also toxic, but much less so, and owes its toxicity to its similarity to phosphate, although conversion to the more toxic arsenite [81] by phosphatases that show adventitious arsenate reductase activity may also play a role [82]. Selenium compounds such as selenite, which with a pKa of 7.3 occurs as $[HSeO_3]^-$ and $[SeO_3]^{2-}$ at physiological pH, are also highly toxic, although selenium differs from arsenic in that it is an essential trace element [83]. Given independently these compounds are both toxic, but when given together a surprising thing happens – their toxicities cancel. This antagonism was first reported in 1938 in a short paper by Alvin Moxon [84]. Moxon was an agricultural chemist, who was interested in selenium intoxication of prairie livestock. The prairie of North America has selenium-rich soils [85], and a number of plant species accumulate it in their tissues, some, called selenium hyperaccumulators [86, 87] to levels that can be toxic. In 1938 there was much less awareness of the dangers of arsenic than there is today, but nonetheless Moxon was appropriately cautious about his finding, stating that "The feeding of arsenic to livestock to prevent selenium poisoning is not recommended on the basis of these results" [84]. Following this early report, a large number of studies confirmed and extended Moxon's findings of a curious toxicological cancellation, but the molecular basis was not revealed until XAS and other spectroscopic methods were applied [88]. This work showed that a novel arsenic-selenium compound – the seleno-bis(S-glutathionyl)arsinium ion (Figure 5.18) is formed in the blood and the liver and subsequently excreted in the bile [89–91]. The core of the species contains arsenic coordinated to two glutathione-derived thiolates and a single terminal selenide, as shown in Figure 5.18. Since the core of this complex bears a single negative charge, the species is thus normally abbreviated as $[(GS)_2AsSe]^-$, where GS denotes a glutathione thiolate donor to arsenic. The overall charge of $[(GS)_2AsSe]^-$ is pH-dependent; at physiological pH, values will actually be 3– because of deprotonated carboxylates $(-CO_2^-)$ and protonated amides $(-NH_3^+)$ of the two glutathione thiolates. The As–Se bond-length is 2.32 Å [88, 92], indicating a bond order of approximately 1.5, or As–Se [88, 92], with both Raman and ^{77}Se NMR being in agreement with this [88]. The formation of $[(GS)_2AsSe]^-$ is currently thought to be purely chemical, and not specifically a product of enzyme catalyzed processes [88, 93]. XAS [88, 92] in combination with DFT [92] and XFI [94] and biochemical studies have been essential in unravelling the some of the details of this mutual detoxification. Studies using membrane vesicles incorporating various different multidrug resistance protein transporters have shown that Mrp2 but not Mrp1, Mrp4 or Mrp5 transport $[(GS)_2AsSe]^-$ [95]. As $[(GS)_2AsSe]^-$ is formed in erythrocytes, which lack the Mrp2 transporter; it cannot readily leave the cell, and serves to slow the distribution of both arsenic and selenium to tissues, potentially reducing toxic effects. The importance of this finding may in part explain the situation with insects, which lack any cells that would correspond to vertebrate erythrocytes, which show no significant antagonistic response to co-administered arsenite and selenite, but instead a synergism resulting in 100% mortality when both are given [96].

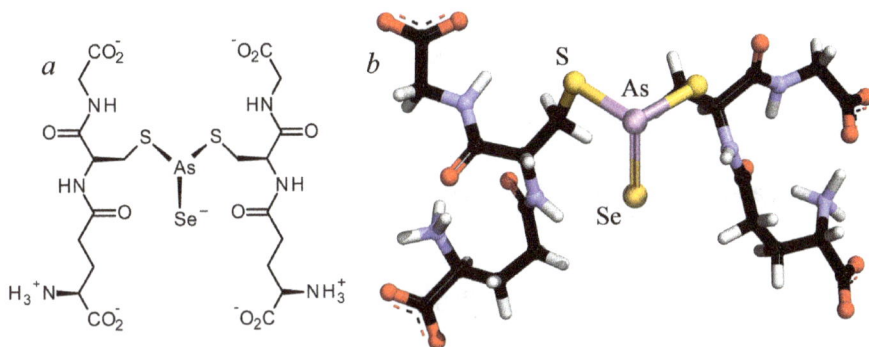

Figure 5.18: Molecular basis for the antagonism between arsenite and selenite – the seleno-bis (S-glutathionyl)arsinium ion; *a* shows the schematic structure and *b* shows the DFT geometry optimized structure.

As was pointed out in the very first report upon the formation of $[(GS)_2AsSe]^-$ [88], these findings may have profound consequences for populations whose water supply is contaminated with arsenic. An estimated 200 million people world-wide are exposed to arsenic in their drinking water at levels that are above the World Health Organization's limit of 10 $\mu g \cdot l^{-1}$ [97]. In many, but not all, affected areas this causes what is known as arsenicosis, a syndrome characterized by development of skin lesions (melanosis and keratosis), skin cancers (e.g., squamous cell carcinoma) and finally internal tumors, often leading to death. One of the worst hit areas is the Ganges River delta, in particular Bangladesh and West Bengal, where tens of millions of people primarily in rural populations are affected, which has been called the world's worst mass poisoning by the World Health Organization. In such populations, medical treatment of cancers, when it is available, frequently involves amputation of an affected limb, with devastating effects on affected individuals, whose livelihood depends upon rural subsistence farming. The issue is a complex one, as only some individuals in communities developing arsenicosis, despite all drinking the same arsenic-contaminated water. The mystery deepens when one considers that there are communities elsewhere in the world [98, 99], with similar or even higher arsenic in their drinking water, but no evidence of arsenicosis among the population. It seems very clear that arsenicosis is completely distinct from acute high-dose arsenic poisoning, and hence it seems likely that the underlying molecular mechanisms differ. As we have discussed, for every atom of arsenic that is bound via formation of

[(GS)₂AsSe]⁻ one atom of selenium is sequestered. Selenium is, however, essential for good heath, and we have suggested that chronic low-level arsenicosis might in fact be an arsenic-induced selenium deficiency [88], and it is now clear both that endogenous selenium can produce [(GS)₂AsSe]⁻ [100] and that selenium deficiency can mimic some of the metabolic effects of low-level arsenic exposure [101]. In support of this hypothesis the diets of affected communities in the vicinity of the Ganges river delta are naturally low in selenium, and high-arsenic communities with no arsenicosis have adequate dietary selenium [102]. Because of this, we have suggested that selenium supplements might be used as a treatment [88]. However, attention to the details of chemical speciation of the form of selenium given is very important, because, as we will shortly discuss, not all selenium forms act to cancel arsenic's toxic effects, and some act instead to magnify them.

5.7.2 Arsenic and selenium synergism

The antagonistic relationship between As^{3+} and some selenium compounds in which there is a mutual detoxification is now relatively well understood, but methylated selenium metabolites can produce the opposite effect. Thus, when arsenite is given to rats at a doses that would normally be non-toxic and then followed by methylated selenium derivatives, again at normally non-toxic doses, then the two together combine to have a lethal effect [103, 104]. A broad range of different selenium compounds show substantial synergistic toxicities with arsenite, for which the schematic structures are shown in Figure 5.19. Of these L-selenomethionine was the least toxic, perhaps because it acts as a molecular mimic of L-methionine and hence tends to be incorporated into proteins. While the mechanism for the synergistic response is not yet proven, it has been suggested to be due to arsenite-inhibition of components of detoxification pathways for methylated selenium species [103]. Interestingly, the same synergistic response with arsenite is observed with trimethyltelluronium species [(CH₃)₃Te]⁺, suggesting that whatever the pathways involved in the synergistic effect, they are shared between selenium and tellurium.

Plants grown on high-selenium soils naturally produce significant levels of many of the selenium compounds in Figure 5.19, depending upon the plant. Moreover, some of these have been claimed to have beneficial properties, such as the "nutraceutical" Se-methyl-L-selenocysteine. Thus, caution is warranted when using any plant-based treatment of existing arsenicosis, since further study of this interesting facet of the interrelationship between arsenic and selenium compounds is needed.

methylseleninic acid   selenobetaine   trimethylselenonium

dimethylselenoxide   selenobetaine-methyl ester   Se-methyl-L-selenocysteine

L-selenomethionine

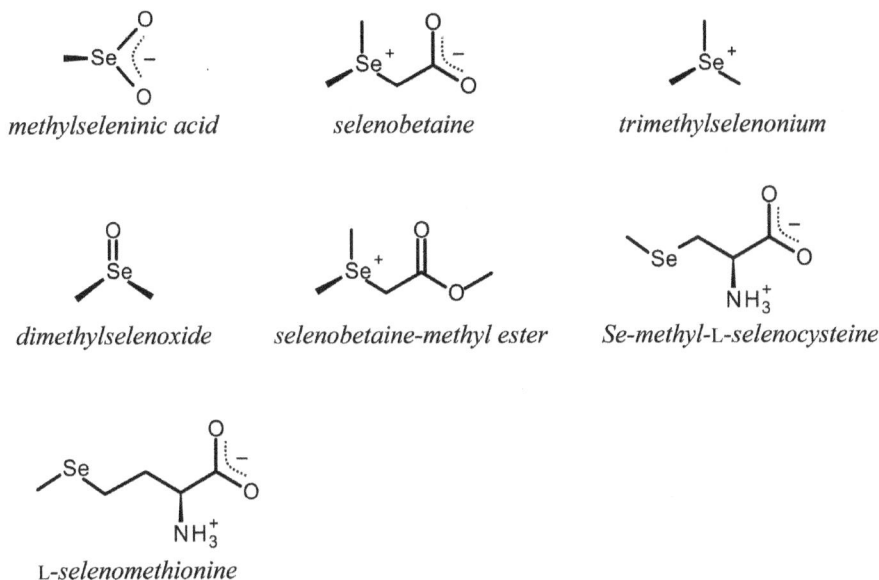

Figure 5.19: Structures of methylated selenium compounds for which synergistic toxic effects have been reported with arsenite.

5.8 Mercury and selenium antagonism and synergism

Similarly to arsenic, there is also a well-known toxicological antagonism between Hg^{2+} and selenium. Also similarly to arsenic, the situation is far from the simple picture view "selenium cancels mercury." The first report that described a molecular basis for the mutual detoxification of mercuric mercury and selenite was published more than two decades ago, and an early example of a combination of computational chemistry and XAS to toxicology [105]. In the case of blood-borne Hg^{2+}, it is likely that the metal will be associated with serum albumin. Selenite is thought to form selenide [HSe]⁻ by the reducing cellular environment within erythrocytes, which is mobile to cell membranes and may then react with the Hg^{2+}, forming an inorganic mercuric selenide nano-particle with approximate composition $Hg_{100}Se_{100}$. Like bulk HgSe this probably possesses a zincblende core structure, although the alternative wurtzite structure could not be excluded from the XAS data [105]. An effective model system with very similar spectroscopic properties can be made by reacting aqueous solutions of reduced glutathione, selenite and mercuric salts. This black-colored solution species has external glutathiones probably linked both to selenium and to mercury [105]. A computer-generated model is shown in Figure 5.20. Nano-HgSe has been observed in a range of organisms, including humans exposed to deadly levels of organomercury species.

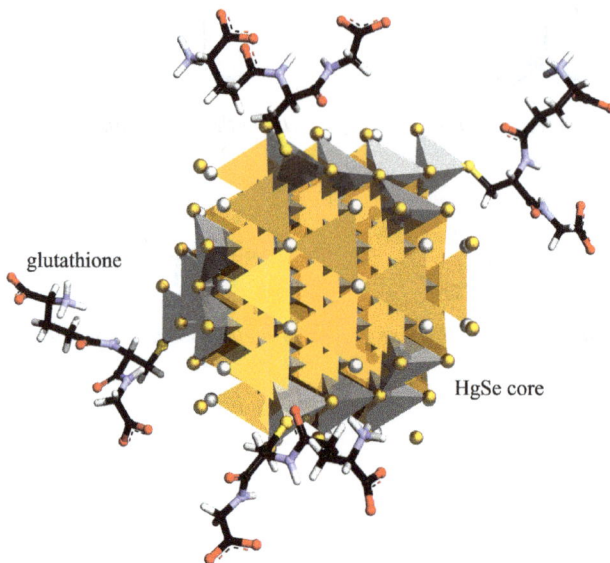

Figure 5.20: Model of a $Hg_{79}Se_{79}$ nanoparticle with four external glutathione attached.

These interactions between mercury and selenium have been studied using lar-val stage zebrafish as a vertebrate model with X-ray fluorescence imaging to probe the localizations of mercury, selenium and other elements [106]. Fish exposed to Hg^{2+} were observed to contain nano-scale structures comprising co-localized mer-cury and selenium, but no similar co-localization was seen with methylmercury expo-sure under similar conditions [106]. Quantitative elemental stoichiometries together with Hg L_{III} μ-XAS spectroscopy, showed the presence of mixed chalcogenide clus-ters, $HgS_xSe_{(1-x)}$, where x is 0.4–0.9, probably related to the nano-particulate HgSe ob-served in the mammalian system. In these experiments no selenium was given, and the resulting $HgS_xSe_{(1-x)}$ must therefore have come from endogenous selenium in the bodies of the fish. In agreement with this, control fish showed specific high-selenium regions, in particular pigment spots, that were depleted in the Hg^{2+} treated fish [106]. In a subsequent zebrafish study the response to fish treated with both selenium in the form of L-selenomethionine and methylmercury or inorganic mercury was exam-ined [107]. When selenomethionine was given before Hg^{2+} both mercury and selenium levels in the brain increased, but toxicity was decreased. Conversely, with selenium pre-treatment methylmercury levels were unchanged but toxicity was substantially in-creased above that from methylmercury alone. Importantly, both inorganic and organ-ometallic mercury effectively blocked selenium transport, observed in pre-treatments, and profoundly disrupted thyroid metabolism. The study suggested that this may be due to mercury-based inhibition [108] of selenocysteine-dependent deiodinases that are essential for thyroid function [108]. Similar nano-particulate HgSe has been ob-served in the brain tissues of humans fatally poisoned by exposure organometallic mercury [109], which must be a result of methylmercury demethylation to inorganic forms, a reaction known to occur relatively slowly [110]. Similar nano-HgSe contents have been observed in a range of tissues of other organisms exposed to methylmercury [111, 112]. Clues to a role for selenoprotein P have also been recently discussed, with evidence from Hg $L_{\alpha1}$ HERFD-XAS, and it has been suggested that this transporter plays and important role in the demethylation reaction [111, 112].

5.9 Arsenic, selenium and mercury

We now turn to a rather curious three-way antagonism which illustrates the extent to which these complex interrelationships can combine. A three-way cancellation of toxicities has been reported for arsenic, selenium and methylmercury [113]; this work showed that, as expected from previous studies, survival rates of Japanese quail fed methylmercuric chloride and selenite were poor when the compounds were given alone and improved when given together (5% to 76% survival after 16 weeks). Remarkably, survival increased to 100% when arsenite was given along with both methylmercury chloride and selenite [113]. This three-way antagonism has been rationalized as the formation of an excretable species involving all three toxic elements, since $[(GS)_2AsSe]^-$ can form in erythrocytes, methylmercury species can cross the erythrocyte membrane [114], and mercury and selenium donors have a strong affinity. A complex produced in vitro has been formulated as $[(GS)_2AsSeHgCH_3]$ with a structure (Figure 5.21) that has been confirmed by XAS at As and Se K-edges and the Hg L-edges, together with support from DFT calculations [115]. This species could form in erythrocytes and may be the molecular basis for this exotic triple-antagonism.

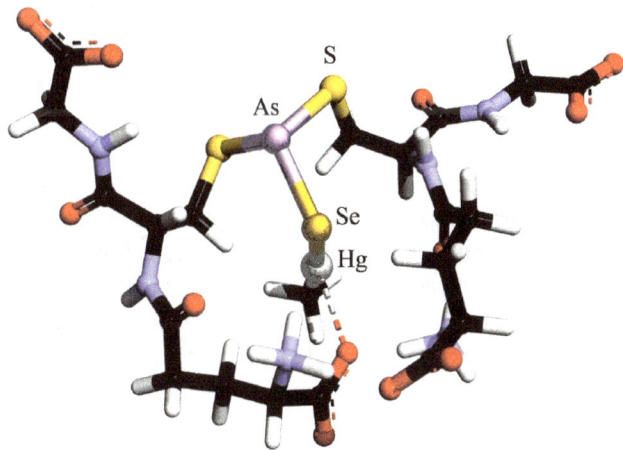

Figure 5.21: DFT geometry optimized structure of methylmercury(II) seleno bis(S-glutathionyl) arsenic (III).

While this triple antagonism might save an individual unfortunate enough to be simultaneously exposed to all three poisons, we note this would be a most unlikely and unlucky occurrence, and the molecular toxicology of this phenomenon is thus probably only of academic interest.

5.10 Closing remarks

Our goal in this chapter was to review and illustrate the in situ experimental and computational techniques that can give insights into the molecular toxicology of metals and metalloids. In this chapter we have reviewed methods that are normally thought of as the purview of the physical chemist or physicist, but only relatively recently have become part of the toolbox of the molecular toxicologist. Our examples deliberately have been drawn primarily from our own research, and this should not be taken as marginalizing the many elegant and outstanding research accomplishments of other groups.

In closing, we feel compelled to point out that metal and metalloid toxicology may be an under-appreciated threat both to humanity and to the environment. We have not discussed lead here, except to note that its toxicity is lower than some other important toxic elements. However, the World Health Organization estimates that globally the annual death rate from lead intoxication exceeds that from AIDS [116]. At the time of writing this chapter the U.S. National Institutes of Health spends approximately three billion dollars on AIDS research annually, but comparatively speaking almost nothing on toxic heavy metal research. Mercury, the most toxic of the heavy elements, is dramatically increasing in our environment due to pollution, much of which is due to burning of coal at unprecedented rates by China and India, with recent increases in atmospheric levels (three- to fivefold), ocean surface levels (two- to threefold) [117] and human consumption (up by 38%) [118]. There is evidence that mercury played a partial role in at least three geologically recorded mass-extinction events [119, 120]. One of these, the end-Permian extinction, also called the Great Dying, was the largest such event in earth's history, and is thought to have been associated with volcanic activity causing the burning of coal deposits [121], much as the modern exploitation of coal is now elevating both carbon dioxide and atmospheric mercury. However, unlike carbon dioxide, for which technologies exist for removal and sequestration, toxic heavy metals can be very challenging to remove from the environment and may be the most underestimated threat of our time. Research into heavy metal and metalloid toxicology is thus an important and timely topic, for which a detailed molecular-level understanding is an imperative.

Further reading

Our description of the experimental techniques reviewed in this chapter was necessarily (and deliberately) superficial. A more detailed description of these topics may be found in the following texts.

- *A general text on synchrotron-based X-ray spectroscopy; an excellent and comprehensive introductory text to many aspects of this field*: Cramer SP. X-Ray Spectroscopy with Synchrotron Radiation. Fundamentals and Applications. Springer Nature Switzerland AG 2020.
- *A text with focus on X-ray absorption spectroscopy of molybdenum and tungsten enzymes, but generally applicable*: George GN. X-ray absorption spectroscopy of molybdenum and tungsten enzymes. In: Kirk ML, Hille R, Schulzke C, ed. Molybdenum and Tungsten Enzymes: Spectroscopic and Theoretical Investigations. Royal Society of Chemistry, Series on Metallobiology, 2016, 121–67.
- *A comprehensive review of X-ray fluorescence imaging, including methods and applications*: Pushie, MJ, Pickering IJ, Korbas M, Hackett MJ, George GN. Elemental and chemically specific X-ray fluorescence imaging of biological systems. Chem Rev 2014, 114, 8499–541.
- *A review of many of the methods of computational chemistry, including the details deliberately omitted in this chapter*. Cramer CJ. Essentials of Computational Chemistry. Theories and Models. Second Edition, Wiley, 2002.

References

[1]   Das Buch Paragranum, 1529–30. Goodrick-Clarke N. Paracelsus Essential Readings. North Atlantic Books. Berkeley, California, 1999.
[2]   Cutler EG, Bradford EH. Action of iron, cod-liver oil, and arsenic on the globular richness of the blood. Am J Med. 1878, 75, 74–84.
[3]   Kwong YL, Todd D. Delicious poison: Arsenic trioxide for the treatment of leukemia. Blood. 1997, 89, 3487–3488.
[4]   Hoonjan M, Jadhav V, Bhatt P. Arsenic trioxide: Insights into its evolution to an anticancer agent. J Biol Inorg Chem. 2018, 23, 313–329.
[5]   Zheng J, Zhang K, Liu Y, Wang Y. Fatal acute arsenic poisoning by external use of realgar: Case report and 30 years literature retrospective study in China. Forensic Sci Int. 2019, 300, e24–30.
[6]   Nathwani AC, Down JF, Goldstone J, Yassin J, Dargan PI, Virchis A, Gent N, Lloyd D, Harrison JD. Polonium-210 poisoning: A first-hand account. Lancet. 2016, 388, P1075–80.
[7]   Pushie MJ, George GN. Spectroscopic studies of molybdenum and tungsten enzymes. Coord Chem Rev. 2011, 255, 1055–1084.
[8]   George GN. X-ray absorption spectroscopy of molybdenum and tungsten enzymes. Kirk ML, Hille R, Schulzke C. ed. Molybdenum and Tungsten Enzymes: Spectroscopic and Theoretical Investigations. Royal Society of Chemistry, Series on Metallobiology. Cambridge, UK, 2016, 121–167.

[9]   Nehzati S, Dolgova NV, James AK, Cotelesage JJH, Sokaras D, Kroll T, George GN, Pickering IJ. High energy resolution fluorescence detected X-ray absorption spectroscopy: An analytical method for selenium speciation. Anal Chem. 2021, 93, 9235–9243.

[10]   Bissardon C, Proux O, Bureau S, Suess E, Winkel LHE, Conlan RS, Francis LW, Khan IM, Charlet L, Hazemann JL, Bohic S. Sub-ppm level high energy resolution fluorescence detected X-ray absorption spectroscopy of selenium in articular cartilage. Analyst. 2019, 144, 3488–3493.

[11]   Le Pape P, Blanchard M, Juhin A, Rueff J-P, Ducher M, Morin G, Cabaret D. Local environment of arsenic in sulfide minerals: insights from high-resolution X-ray spectroscopies, and first-principles calculations at the As K-edge. J. Anal. At. Spectrom. 2018, 33, 2070–2082.

[12]   Burke K, Wagner LO. DFT in a nutshell. Int J Quant Chem. 2013, 113, 96–101.

[13]   Baseden KA, Tye JW. Introduction to density functional theory: Calculations by hand on the helium atom. J Chem Edu. 2014, 91, 2116–2123.

[14]   Kim C-Y, Parkin S, Bharara M, Atwood D. Linear coordination of Hg(II) by cysteamine. Polyhedron. 2002, 21, 225–228.

[15]   Calhorda MJ, Pregosin PS, Veiros LF. Geometry optimization of a Ru(IV) allyl dicationic complex: A DFT failure?. J Chem Theory Comput. 2007, 3, 665–670.

[16]   Jacobsen H. A failure of DFT is not necessarily a DFT failure – Performance dependencies on model system choices. J Chem Theory Comput. 2011, 7, 3019–3025.

[17]   Cotelesage JJH, Pushie MJ, Grochulski P, Pickering IJ, George GN. Metalloprotein active site structure determination: Synergy between X-ray absorption spectroscopy and X-ray crystallography. J Inorg Biochem. 2012, 115, 127–137.

[18]   Pushie MJ, Cotelesage JJH, Lyashenko G, Hille R, George GN. X-ray absorption spectroscopy of a quantitatively Mo(V) dimethyl sulfoxide reductase species. Inorg Chem. 2013, 52, 2830–2837.

[19]   Warelow TP, Pushie MJ, Cotelesage JJH, Santini JM, George GN. The active site structure and catalytic mechanism of arsenite oxidase. Sci Rep. 2017, 7, 1757/1–9.

[20]   Nehzati S, Summers AO, Dolgova NV, Zhu J, Sokaras D, Kroll T, Pickering IJ, George GN. Hg(II) binding to thymine bases in DNA. Inorg Chem. 2021, 60, 7442–7452.

[21]   Edmonds JS, Francesconi KA, Cannon JR, Raston CL, Skelton BW, White AH. Isolation, crystal structure and synthesis of arsenobetaine, the arsenical constituent of the western rock lobster *Panulirus longipes cygnus* George. Tetrahedron Lett. 1977, 18, 1543–1546.

[22]   Popowich A, Zhang Q, Le CX. Arsenobetaine: The ongoing mystery. Natl Sci Rev. 2016, 3, 451–458.

[23]   Marafante E, Vahter M, Denker L. Metabolism of arsenocholine in mice, rats and rabbits. Sci Tot Environ. 1984, 34, 223–240.

[24]   Styblo M, Del Razo LM, Vega L, Germolec DR, LeCluyse EL, Hamilton GA, Reed W, Wang C, Cullen WR, Thomas DJ. Comparative toxicity of trivalent and pentavalent inorganic and methylated arsenicals in rat and human cells. Arch Toxicol. 2000, 74, 289–299.

[25]   Casdorph HR, Walker M. Toxic Metal Syndrome: How Metal Poisonings Can Affect Your Brain. Avery Publishing Group, Garden City Park, New York, 1995.

[26]   Ord MG, Stocken LA. A contribution to chemical defense in World War II. Trends in Biochem Sci. 2000, 25, 253–256.

[27]   Antsyshkina AS, Sadikov G, Poznyak AL, Sergienko VS. Crystal structure of the (diaqua (ethylenediaminetetraacetato)calciate) Tris(1,10-phenanthroline)nickel crystal hydrate (Ni (Phen)$_3$)(Ca(Edta)(H$_2$O)$_2$)·10.5H$_2$O. Russ J Inorg Chem. 2002, 47, 43–54.

[28]   Lamas GA, Issa OM. Edetate disodium-based treatment for secondary prevention in post-myocardial infarction patients. Curr Cardiol Rep. 2016, 18, 20/1–8.

[29] Brown MJ, Willis T, Omalu B, Leiker R. Deaths resulting from hypocalcemia after administration of edetate disodium: 2003–2005. Pediatrics. 2006, 118, e534–36.

[30] Abergel RJ. Chelation of actinides. Crichton RR, Ward RJ, Hider RC. ed. Metal Chelation in Medicine. Royal Society of Chemistry, Series on Metallobiology. Cambridge, UK, 2017, 183–212.

[31] Vilensky JA, Redman K. British anti-Lewisite [dimercaprol]: An amazing history. Ann Emerg Med. 2003, 42, 378–383.

[32] World Health Organization https://list.essentialmeds.org/ October 1 2021 release

[33] Danielli JF, Danielli M, Fraser JB, Mitchell PD, Owen LN, Shaw G. BAL-INTRAV: A new non-toxic thiol for intravenous injection in arsenical poisoning. Biochem J. 1947, 41, 325–333.

[34] Denny-Brown D, Porter H. The effect of BAL (2,3-dimercaptopropanol) on hepatolenticular degeneration (Wilson's disease). N Engl J Med. 1951, 245, 917–925.

[35] Kim LHC, Abel SJC. Survival after a massive overdose of arsenic trioxide. Crit Care Resusc. 2009, 11, 42–45.

[36] Vantroyen B, Heilier JF, Meulemans A, Michels A, Buchet JP, Vanderschueren S, Haufroid V, Sabbe M. Survival after a lethal dose of arsenic trioxide. J Toxicol Clin Toxicol. 2004, 42, 889–895.

[37] Kamijo Y, Soma K, Asari Y, Ohwada T. Survival after massive arsenic poisoning self-treated by high fluid intake. J Toxicol Clin Toxicol. 1998, 36, 27–29.

[38] Yamauchi H, Takata A. Arsenic metabolism differs between child and adult patients during acute arsenic poisoning. Toxicol Appl Pharmacol. 2021, 410, 115352/1–9.

[39] Groom CR, Bruno IJ, Lightfoot MP, Ward SC. The Cambridge structural database. Acta Cryst. B 2016, 72, 171–179.

[40] Govindaswamy N, Moy J, Millar M, Koch SA. A distorted mercury [Hg(SR)$_4$]$^{2-}$ complex with alkanethiolate ligands: The fictile coordination sphere of monomeric [Hg(SR)$_x$] complexes. Inorg Chem. 1992, 31, 5343–5344.

[41] George GN, Prince RC, Gailer J, Buttigieg GA, Denton MB, Harris HH, Pickering IJ. Mercury binding to the chelation therapy agents DMSA and DMPS, and the rational design of custom chelators for mercury. Chem Res Toxicol. 2004, 17, 999–1006.

[42] Fu J, Hoffmeyer RE, Pushie MJ, Singh SP, Pickering IJ, George GN. Towards a custom chelator for mercury: Evaluation of coordination environments by molecular modeling. J Biol Inorg Chem. 2011, 16, 15–24.

[43] Satter W, Palmer JH, Bridges CC, Joshee L, Zalups RK, Parkin G. Structural characterization of 1,3-propanedithiols that feature carboxylic acids: Homologues of mercury chelating agents. Polyhedron. 2013, 64, 268–279.

[44] Zaman KM, Blue LY, Huggins FE, Atwood DA. Cd, Hg, and Pb compounds of benzene-1,3-diamidoethanethiol (BDETH$_2$). Inorg Chem. 2007, 46, 1975–1980.

[45] Hutchison A, Atwood DA, Santilliann-Jiminez QE. The removal of mercury from water by open chain ligands containing multiple sulfurs. J Hazard Mater. 2008, 156, 458–465.

[46] Aposhian HV, Maiorino RM, Gonzalez-Ramirez D, Zuniga-Charles M, Xu Z, Hurlbut KM, Junco-Munoz P, Dart RC, Aposhian MM. Mobilization of heavy metals by newer, therapeutically useful chelating agents. Toxicology. 1995, 97, 23–38.

[47] Campbell JR, Clarkson TW, Omar MD. The therapeutic use of 2,3-dimercaptopropane-1-sulfonate in two cases of inorganic mercury poisoning. J Am Med Assoc. 1986, 256, 3127–3130.

[48] Gonzalez-Ramirez D, Maiorino RM, Zuniga-Charles M, Xu Z, Hurlbut KM, Junco-Munoz P, Aposhian MM, Dart RC, Diaz Gama JH, Echeverra D, Woods JS, Aposhian HV. Sodium 2,3-dimercaptopropane-1-sulfonate challenge test for mercury in humans: II. Urinary mercury,

porphyrins and neurobehavioral changes of dental workers in Monterrey, Mexico. J Pharmacol Exp Ther. 1995, 272, 264–274.

[49] Garza-Ocanas L, Torres-Alanis O, Pineyro-Lopez A. Urinary mercury in twelve cases of cutaneous mercurous chloride (calomel) exposure: Effect of sodium 2,3-dimercaptopropane -1-sulfonate (DMPS) therapy. J Toxicol Clin Toxicol. 1997, 35, 653–655.

[50] Böse-O'Reilly S, Drasch S, Beinhoff C, Maydla S, Voskob MR, Roidera G, Dzajaa D. The Mt. Diwata study on the Philippines 2000 – Treatment of mercury intoxicated inhabitants of a gold mining area with DMPS (2,3-dimercapto-1-propane-sulfonic acid, Dimaval®). Sci Tot Env. 2003, 307, 71–82.

[51] Pfab R, Mueckter H, Roider G, Zilker T. Clinical course of severe poisoning with thiomersal. J Toxicol Clin Toxicol. 1996, 34, 453–460.

[52] Bridges CC, Joshee L, Zalups RK. Multidrug resistance proteins and the renal elimination of inorganic mercury mediated by 2,3-dimercaptopropane-1-sulfonic acid and meso-2,3-dimercaptosuccinic acid. J Pharmacol Exp Ther. 2008, 324, 383–390.

[53] Bridges CC, Joshee L, Zalups RK. MRP2 and the DMPS- and DMSA-mediated elimination of mercury in Tr– and control rats exposed to thiol S-conjugates of inorganic mercury. Toxicol Sci. 2008, 105, 211–220.

[54] Lane T, Saito MA, George GN, Pickering IJ, Prince RC, Morel FFM. A cadmium enzyme from a marine diatom. Nature. 2005, 435, 42.

[55] Pappas RS. Toxic elements in tobacco and in cigarette smoke: Inflammation and sensitization. Metallomics. 2011, 3, 1181–1198.

[56] Prokopowicz A, Sobczak A, Szuła-Chraplewska M, Ochota P, Kośmider L. Nicotine Tob Res. Exposure to cadmium and lead in cigarette smokers who switched to electronic cigarettes. 2018, 21, 1198–1205.

[57] Schwarz G, Il'yasova D, Ivanova A. Urinary cadmium, impaired fasting glucose, and diabetes in the NHANES III. Diabetes Care. 2003, 26, 468–470.

[58] Tinkov AA, Filippini T, Ajsuvakova OP, Aaseth J, Gluhcheva YG, Ivanova JM, Bjørklund G, Skalnaya MG, Gatiatulina ER, Popova EV, Nemereshina ON, Vinceti M, Skalny AV. The role of cadmium in obesity and diabetes. Sci Total Environ. 2017, 601–602, 741–755.

[59] Pulla Rao C, Dorfman JR, Holm RH. Synthesis and structural systematics of ethane-1,2-dithiolato complexes. Inorg Chem. 1986, 25, 428–439.

[60] Zeini Jahromi E, Gailer J, Pickering IJ, George GN. Structural characterization of Cd^{2+} complexes in solution with DMSA and DMPS. J Inorg Biochem. 2014, 136, 99–106.

[61] Abulesch L-O, Hale MS, Ainsworth S, Alsolaiss J, Crittenden E, Calvete JJ, Evans C, Wikinson MC, Harrison RA, Kool J, Casewell NR. Preclinical validation of a repurposed metal chelator as an early-intervention therapeutic for hemotoxic snakebite. Sci Transl Med. 2020, 12, eaay8314/1–13.

[62] Mohapatra B, Warrell DA, Suraweera W, Bhatia P, Dhingra N, Jotkar RM, Rodriguez PS, Mishra K, Whitaker R, Jha P. Million death study collaborators, snakebite mortality in India: A nationally representative mortality survey. PLOS Negl Trop Dis. 2011, 5, e1018/1–8.

[63] Shyamal DK, Ray K. Wilson's disease: An update. Nat Clin Pract Neurol. 2006, 2, 482–493.

[64] Mohr I, Weiss K-H. Current anti-copper therapies in management of Wilson disease. Ann Transl Med. 2019, 7, Suppl. 2(S69), 1–7.

[65] Birker PJMWL, Freeman HC. Structure, properties, and function of a copper(I)-copper(II) complex of D-penicillamine: Pentathallium(I) μ_8-chloro-dodeca(D-penicillaminato]octacuprate (I)hexacuprate(II) n-hydrate. J Am Chem Soc. 1977, 99, 6890–6899.

[66] Jullien A-S, Gateau C, Lebrun C, Kieffer I, Testemale D, Pascale Delangle P. D-penicillamine tripodal derivatives as efficient copper(I) chelators. Inorg Chem. 2014, 53, 5229–5239.

[67] Gauthier L, Charbonnier P, Chevallet M, Delangle P, Texier I, Gateau C, Deniaud
      A. Development, formulation, and cellular mechanism of a lipophilic copper chelator for the
      treatment of Wilson's disease. Int J Pharm. 2021, 609, 121193/1–8.
[68] Klein D, Arora U, Lichtmannegger J, Finckh M, Heinzmann U, Summer K-H.
      Tetrathiomolybdate in the treatment of acute hepatitis in an animal model for Wilson
      disease. J Hepatol. 2004, 40, 409–416.
[69] George GN, Pickering IJ, Harris HH, Gailer J, Klein D, Lichtmannegger J, Summer K-H.
      Tetrathiomolybdate causes formation of hepatic copper-molybdenum clusters in an animal
      model of Wilson's disease. J Am Chem Soc. 2003, 125, 1704–1705.
[70] Zhang L, Lichtmannegger J, Summer K-H, Webb S, Pickering IJ, George GN. Tracing copper-
      thiomolybdate complexes in a prospective treatment for Wilson's disease. Biochemistry.
      2009, 48, 891–897.
[71] Ogra Y, Chikusa H, Suzuki KT. Metabolic fate of the insoluble copper/tetrathiomolybdate
      complex formed in the liver of LEC rats with excess tetrathiomolybdate. J Inorg Biochem.
      2000, 78, 123–128.
[72] Weiss KH, Askari FK, Czlonkowska A, Ferenci P, Bronstein JM, Bega D, Ala A, Nicholl D, Flint
      S, Olsson L, Plitz T, Bjartmar C, Schilsky ML. Bis-choline tetrathiomolybdate in patients with
      Wilson's disease: An open-label, multicentre, phase 2 study. Lancet Gastroenterol Hepatol.
      2017, 2, 869–876.
[73] De Fabregues O, Viñas J, Palasí A, Quintana M, Cardona I, Auger C, Vargas V. Ammonium
      tetrathiomolybdate in the decoppering phase treatment of Wilson's disease with
      neurological symptoms: A case series. Brain Behav. 2020, 10, e01596/1–7.
[74] Sas B. Secondary copper deficiency in cattle caused by molybdenum contamination of
      fodder: A case history. Vet Hum Toxicol. 1989, 31, 29–33.
[75] Huang Y, Marden JP, Julien C, Bayourthe C. Redox potential: An intrinsic parameter of the
      rumen environment. J Anim Physiol Anim Nutr. 2018, 102, 393–402.
[76] George GN, Prince RC, Singh SP, Pickering IJ. Arsenic K-edge X-ray absorption spectroscopy
      of arsenic in seafood. Mol Nutr Food Res. 2009, 53, 552–557.
[77] Hansen HH, Raab A, Francesconi KA, Feldman J. Metabolism of arsenic by sheep chronically
      exposed to arsenosugars as a normal part of their diet. Environ Sci Technol. 2003, 37,
      845–851.
[78] Weiner G, Field AC, Smith C. Deaths from copper toxicity of sheep at pasture and the use of
      fresh seaweed. Vet Rec. 1977, 101, 424–425.
[79] McLachan GK, Johnston WS. Copper poisoning in sheep from North Ronaldsay maintained on
      a diet of terrestrial herbage. Vet Rec. 1982, 111, 299–301.
[80] Hou X, Yan X. Study on the concentration and seasonal variation of inorganic elements in 35
      species of marine algae. Sci Total Environ. 1998, 222, 141–156.
[81] Radabaugh TR. Aposhian HV enzymatic reduction of arsenic compounds in mammalian
      systems: Reduction of arsenate to arsenite by human liver arsenate reductase. Chem Res
      Toxicol. 2000, 13, 26–30.
[82] Bhattacharjee H, Sheng J, Ajees AA, Mukhopadhyay R, Rosen BP. Adventitious arsenate
      reductase activity of the catalytic domain of human Cdc25B and Cdc25C phosphatases.
      Biochemistry. 2010, 49, 802–809.
[83] Dolgova NV, Nehzati S, Choudhury S, MacDonald TC, Regnier NR, Crawford AM, Ponomarenko
      O, George GN, Pickering IJ. X-ray spectroscopy and imaging of selenium in living systems.
      Biochim Biophys Acta. 2018, 1862, 2383–2392.
[84] Moxon AL. The effect of arsenic on the toxicity of seleniferous grains. Science. 1938, 88, 81.

[85] Thavarajah D, Vandenberg A, George GN, Pickering IJ. Chemical form of selenium in naturally selenium-rich lentils (*Lens culinaris* L.) from Saskatchewan. J Ag Food Chem. 2007, 55, 7337–7341.

[86] Pickering IJ, Prince RC, Salt DE, George GN. Quantitative, chemically-specific imaging of selenium transformation in plants. Proc Natl Acad Sci USA. 2000, 97, 10717–10722.

[87] Pickering IJ, Wright C, Bubner B, Ellis D, Persans MW, Yu EY, George GN, Prince RC, Salt DE. Chemical form and distribution of selenium and sulfur in the selenium hyperaccumulator *Astragalus bisulcatus*. Plant Physiol. 2003, 131, 1460–1467.

[88] Gailer J, George GN, Pickering IJ, Prince RC, Ringwald SC, Pemberton J, Glass RS, Younis HS, DeYoung DW, Aposhian HV. A metabolic link between arsenite and selenite: The seleno-bis (S-glutathionyl) arsinium ion. J Am Chem Soc. 2000, 122, 4637–4639.

[89] Manley S, George GN, Pickering IJ, Prince RC, Glass RS, Prenner EJ, Yamdagni R, Wu Q, Gailer J. The seleno-bis(S-glutathionyl) arsinium ion is assembled in erythrocyte lysate. Chem Res Toxicol. 2006, 19, 601–607.

[90] Kaur G, Ponomarenko O, Zhou JR, Swanlund DP, Summers KL, Dolgova NV, Antipova O, Pickering IJ, George GN, Leslie EM. Studies of selenium and arsenic mutual protection in human HepG2 cells. Chem-Biol Interact. 2020, 327, 109162/1–10.

[91] Kaur G, Javed W, Ponomarenko O, Shekh K, Swanlund DP, Zhou JR, Summers KL, Casini A, Wenzel M, Casey JR, Pickering IJ, George GN, Leslie EM. Human red blood cell uptake and sequestration of arsenite and selenite: Evidence of seleno-bis(S-glutathionyl) arsinium ion formation in human cells. Biochem Pharmacol. 2020, 180, 114141/1–13.

[92] George GN, Gailer J, Ponomarenko O, La Porte PF, Strait K, Alauddin M, Ahsan H, Ahmed S, Spallholz J, Pickering IJ. Observation of the seleno bis-(S-glutathionyl) arsinium anion in rat bile. J Inorg Biochem. 2016, 158, 24–29.

[93] Galier J. Arsenic-selenium and mercury-selenium bonds in biology. Coord Chem Rev. 2007, 251, 234–254.

[94] Ponomarenko O, La Porte PF, Singh SP, Langan G, Fleming DEB, Spallholz JE, Alauddin M, Ahsan H, Ahmed S, Gailer J, George GN, Pickering IJ. Selenium-mediated arsenic excretion in mammals: A synchrotron-based study of whole-body distribution and tissue-specific chemistry. Metallomics. 2017, 9, 1585–1595.

[95] Carew MW, Leslie EM. Selenium-dependent and independent transport of arsenic by the human multidrug resistance protein 2 [MRP2/ABCC2]: Implications for the mutual detoxification of arsenic and selenium. Carcinogenesis. 2010, 31, 1450–1455.

[96] Andrahennadi R Biotransformation of selenium and arsenic in insects: environmental implications (Doctoral dissertation). University of Saskatchewan, Saskatoon, 2009.

[97] Naujokas MF, Anderson B, Ahsan H, Aposhian HV, Graziano JH, Thompson C, Suk WA. The broad scope of health effects from chronic arsenic exposure: Update on a worldwide public health problem. Environ Health Perspect. 2013, 121, 295–302.

[98] Sancha AM, Rodriguez D, Vega F, Fuentes S, Salazar AM, Hernan V, Moreno V, Baron AM. Exposure to Arsenic of the Atacameno Population in Northern Chile. Reichard EG, Zapponi GA ed. Assessing and Managing Health Risks From Drinking Water Contamination: Approaches and Applications. International Association of Hydrological Sciences (Publication Number 233), Oxfordshire, UK. 1995, 141–146.

[99] Gradecka D, Palus J, Wasowicz W. Selected mechanisms of genotoxic effects of inorganic arsenic compounds. Int J Occupat Med Env Health. 2001, 14, 317–328.

[100] Gailer JG, Ruprecht L, Benker B, Schramel P. Mobilization of exogenous and endogenous selenium to bile after the intravenous administration of environmentally relevant doses of arsenite to rabbits. Appl Organomet Chem. 2004, 18, 670–675.

[101] Carmean CM, Mimoto M, Landeche M, Ruiz D, Chellan B, Zhao L, Schulz MC, Dumitrescu AM, Sargis RM. Dietary selenium deficiency partially mimics the metabolic effects of arsenic. Nutrients. 2021, 13, 2894/1–11.

[102] Spallholz JE, Boylan LM, Ruaman MM. Environmental hypothesis: Is poor dietary selenium intake an underlying factor for arsenicosis and cancer in Bangladesh and West Bengal, India?. Sci Tot Environ. 2004, 323, 21–32.

[103] Kraus RJ, Ganther HE. Synergistic toxicity between arsenic and methylated selenium compounds. Biol Trace Elem Res. 1989, 20, 105–113.

[104] Obermeyer BD, Palmer IS, Olson OE, Halvarson AW. Toxicity of trimethylselenonium chloride in the rat with and without arsenite. Toxicol Appl Pharmacol. 1971, 20, 135–146.

[105] Gailer J, George GN, Pickering IJ, Madden S, Prince RC, Yu EY, Denton MB, Younis HS, Aposhian HV. Structural basis of the antagonism between inorganic mercury and selenium in mammals. Chem Res Toxicol. 2000, 13, 1135–1142.

[106] MacDonald TC, Korbas M, James AK, Sylvain NJ, Hackett MJ, Nehzati S, Krone PH, George GN, Pickering IJ. Interaction of mercury and selenium in the larval stage zebrafish vertebrate model. Metallomics. 2015, 7, 1247–1255.

[107] Dolgova NV, Nehzati S, MacDonald TC, Summers KL, Crawford AM, Krone PH, George GN, Pickering IJ. Disruption of selenium transport and function is a major contributor to mercury toxicity in zebrafish larvae. Metallomics. 2019, 11, 621–631.

[108] Pickering IJ, Cheng Q, Mendoza Rengifo E, Dolgova NV, Kroll T, Sokaras D, George GN, Arnér ESJ. Direct observation of methylmercury and auranofin binding to selenocysteine in thioredoxin reductase. Inorg Chem. 2020, 59, 2711–2718.

[109] Korbas M, O'Donoghue JL, Watson GE, Pickering IJ, Singh SP, Myers GJ, Clarkson TW, George GN. The chemical nature of mercury in human brain following poisoning or environmental exposure. ACS Chem Neurosci. 2010, 1, 810–818.

[110] Lind B, Friberg L, Nylander M. Preliminary studies on methylmercury biotransformation and clearance in the brain of primates: II. Demethylation of mercury in brain. J Trace Elem Exp Med. 1988, 1, 49–56.

[111] Manceau A, Bourdineaud J-P, Oliveira RB, Sarrazin SLF, Krabbenhoft DP, Eagles-Smith CA, Ackerman JT, Stewart AR, Ward-Deitrich C, Del Castillo Busto ME, Goenaga-Infante H, Wack A, Retegan M, Detlefs B, Glatzel P, Bustamante P, Nagy KL, Poulin BA. Demethylation of methylmercury in bird, fish, and earthworm. Environ Sci Technol. 2021, 55, 1527–1534.

[112] Manceau A, Gaillot A-C, Glatzel P, Cherel Y, Bustamante P. In vivo formation of HgSe nanoparticles and Hg–tetraselenolate complex from methylmercury in seabirds–implications for the Hg–Se antagonism. Environ Sci Technol. 2021, 55, 1515–1526.

[113] El-Begearmi MM, Ganther HE, Sunde ML. Dietary interaction between methylmercury, selenium, arsenic, and sulfur amino acids in Japanese quail. Poult Sci. 1982, 61, 272–279.

[114] Naganuma A, Imura N. Methylmercury binds to a low molecular weight substance in rabbit and human erythrocytes. Toxicol Appl Pharmacol. 1979, 47, 613–616.

[115] Korbas M, Percy AJ, Gailer J, George GN. A possible molecular link between the toxicologies of arsenic, selenium and methyl-mercury: Methyl mercury(II) seleno bis(S-glutathionyl) arsenic (III). J Biol Inorg Chem. 2008, 13, 461–470.

[116] World Health Organization. Lead Poisoning and Health. https://www.who.int/news-room /fact-sheets/detail/lead-poisoning-and-health (accessed December 16, 2021)

[117] Lamborg CH, Hammerschmidt CR, Bowman KL, Swarr GJ, Munson KM, Ohnemus DC, Lam PJ, Heimbürger L-E, Rijkenberg MJA, Saito MA. A global ocean inventory of anthropogenic mercury based on water column measurements. Nature. 2014, 512, 65–68.

[118] Lavoie RA, Bouffard A, Maranger R, Amyot M. Mercury transport and human exposure from global marine fisheries. Sci Rep. 2018, 8, 6705/1–9.

[119] Grasby SE, Liu X, Yin R, Ernst RE, Chen Z. Toxic mercury pulses into late Permian terrestrial and marine environments. Geology. 2020, 48, 830–833.

[120] Grasby SE, Beauchamp B, Bond DPG, Wignall PB, Sanei H. Mercury anomalies associated with three extinction events (Capitanian crisis, latest Permian extinction and the Smithian/ Spathian extinction) in NW Pangea. Geol Mag. 2016, 153, 285–297.

[121] Sun Y, Joachimski MM, Wignall PB, Yan C, Chen Y, Jiang H, Wang L, Lai X. Lethally hot temperatures during the early Triassic greenhouse. Science. 2012, 338, 366–370.

A. L. Norman, J. Xie, C. Kruschel, and M. Wieser
Chapter 6
Using isotopic abundances to follow anthropogenic emissions: an example of sulfur, oxygen and boron isotopes in a Canadian watershed

6.1 Introduction

Tracing the fate of emissions in the environment using a multi-elemental isotope approach has gained traction in recent years since it provides significantly more information about the transformations and/or the source of emissions than either concentration data or concentration and isotope data for a single element (e.g., [1, 2]).

Elemental concentration data are a critical tool to assess and characterize the source and presence of pollutants in the natural environment. Careful measurement and integration of the data for specific chemical elements can help to reveal potential sources of chemicals to environmental receptors as well as monitor changes over time. The challenge is, however, to resolve among several different sources of the same chemical element since concentration data alone cannot distinguish, for example, sulfur emissions from anthropogenic activities and sulfur released by natural biogeochemical processes in soils. The incorporation of stable isotope abundance data with concentration data for an element can add an entirely new dimension because the relative abundances of the stable isotopes of many elements are not constant. This is because the masses and nuclear structure of isotopes of a given element are not the same because of differing numbers of neutrons in the nuclei of the atoms. Atoms of a particular element will have the same number of protons (the number of protons effectively defines the element) and, in a neutral atom, the same number of electrons. Therefore, isotopes of an element participate in the same chemical processes. However, the different masses of the isotopes will change the reaction rates and chemical equilibria such that "lighter" isotopes may be preferentially incorporated into different phases or products of a chemical/physical process. In

Acknowledgments: The authors thank Courtney Kruschel and Kerri Miller for their contributions of $\delta^{11}B$ and boron concentrations. Funding for this research was provided by Norman and Wieser's NSERC discovery grants

A. L. Norman, J. Xie, C. Kruschel, M. Wieser, Department of Physics and Astronomy, 2500 University Drive NW, University of Calgary, Calgary, AB, T2N 1N4, Canada, e-mail: alnorman@ucalgary.ca, e-mail: mwieser@ucalgary.ca

https://doi.org/10.1515/9783110626285-006

this manner, the isotopic composition of a sample of the element is an archive of the source of the element and processes that the element may have undergone along its journey from source to the sampling location.

To enable meaningful comparison of results between laboratories, all isotope ratios (R) are expressed in delta notation (equation 1) to eliminate possible instrumental bias

$$\delta = \left(\left(R_{sample}/R_{standard} \right) - 1 \right) \times 1000 \tag{6.1}$$

Where R_{sample} is the ratio of the numbers of heavy to light atoms in the sample and $R_{standard}$ is the isotope amount ratio of the standard. In the case of boron, the delta zero isotope reference material is a boric acid supplied by the National Institute of Standards and Technology (NIST), SRM 951. In the case of oxygen, the zero delta is defined by a water sample distributed by the International Atomic Agency (IAEA), VSMOW (Vienna Standard Mean Ocean Water). For sulfur, the reference material is a silver sulfide available from IAEA, S-1, and has a value of – 0.3 ‰ VCDT (Vienna Canyon Diablo Troilite).

In this case study, concentration and stable isotopic composition of boron, oxygen and sulfur are integrated to determine the origins of these elements where industrial activities impact the watershed. In Alberta, sulfur is an excellent example of the opportunity to employ sulfur isotope abundance data to elucidate industrial sources to pristine regions. Sulfur from hydrocarbon reserves is typically enriched in the heavier ^{34}S isotope by 3% relative to sulfur found in soils and plants and can be resolved easily using analytical methods. Krouse and Grinenko, [3] for example, showed that foliage, litter and bark for coniferous trees in a study of the rural West Whitecourt in Alberta had $\delta^{34}S$ values up to + 24 ‰ close to sulfur emissions from sour gas processing, while soil sulfur, 43 km distant, had $\delta^{34}S$ values that were close to 0 ‰. Norman et al. [4] applied similar techniques in an urban environment. That study demonstrated sour gas processing and flaring emissions moved into the city of Calgary, Alberta, from upwind rural locations and made up approximately 70% of airborne sulfur (SO_2 and aerosol sulfate) as well as precipitation sulfate in the city.

Plots of δ values versus concentration are a powerful tool that has been used in such studies. Mixing of two isotopically distinct sources (A and B) of an element leading to concentration C, for example, can be used along with a mass balance approach for the isotopes (δX where X = A, B, C) to find the proportion and/or amount of the element derived from each source.

$$[C] = [A] + [B] \tag{6.2}$$

$$\delta_C[C] = \delta_A[A] + \delta_B[B] \tag{6.3}$$

Studies where the isotopic composition of several elements is combined have proven useful to track the origin of agricultural nitrate and sources affecting boron within the Oldman River basin [5]. Kruk and colleagues examined the nitrogen and oxygen isotopes of nitrate along with $\delta^{34}S$ values for sulfate [5]. The isotope compositions of nitrate nitrogen and oxygen helped to distinguish the proportion of fertilizer and manure-derived nitrate affecting the Oldman River downstream of the Oldman Reservoir (first map enlargement in Figure 6.1). However, as pointed out in Kruk's study, less is known about activities affecting the headwaters of the Castle River which feeds into the Oldman River, and whether an influence from atmospheric sulfur emissions is detectable. This study sets out to address how sulfate and boron geochemistry in the Castle River changes downstream during baseflow (fall) conditions and whether atmospheric pollutant deposition to the Castle Watershed can be deduced based on a multi-isotope approach using B, and S and O in sulfate.

6.1.1 Context for the study site

The Castle River is an ideal study area to examine the effects of long-term atmospheric deposition on river geochemistry because the watershed is located in a sparsely populated region supporting tourism and recreation with grazing leases and industrial activities. Two large sour gas (natural gas, consisting of methane containing more than 1% hydrogen sulfide or H_2S) processing facilities that operated from the mid-1950s until the 2010s, were located approximately 25 km to the east (717960mE, 5466090mN, 1520masl) and 35 km to the northwest (675350mE, 550009mN, 1250masl) of the Castle River. The northwest facility was closed in 2012 but was operational prior to and during this study. Natural gas produced by gas wells dotted throughout the landscape typically has >4% H_2S, reaching as high as 34%. Hydrogen sulfide is toxic to adult humans at hundreds of ppm in air and can be detected by the human nose at levels a million times lower [6]. Therefore, the H_2S needs to be stripped from the methane gas stream prior to sale to consumers. Gas produced at wellheads within the Castle watershed is typically sent through pipelines to one of these two facilities.

A previous study examined the geology and sources of sulfate affecting the Oldman River between 2002 and 2003 [7]. In that study, sedimentary sulfate at upstream tributaries was interpreted to mix with agricultural and/or pyrite oxidation eastward of the foothills. This study differs from Rock and Mayer's work, as it is focused on baseline flow conditions (during fall) within the Castle River watershed to assess whether atmospheric deposition of sulfur and boron is evident [7]. It is worthwhile to note that atmospheric deposition may be retained within the pedosphere and groundwater for long time periods, integrating atmospheric deposition inputs to watersheds. Considerable lag times between watershed deposition and riverine inputs have been noted in steep river catchments worldwide by Jasechko and colleagues [8]. In another example, manure nitrate inputs to river catchments in New Zealand were later found in downstream groundwaters, suggesting a lag time of several decades [9].

The headwaters of the Oldman River are an important source of drinking water and irrigation affecting the livelihoods of approximately 161,000 residents in southwest Alberta [10]. The Castle River is a tributary of the Oldman River that originates high in the Rocky Mountains just west of the eastern slopes. The watershed is utilized for forestry, grazing, as well as recreation and industrial activities. Emissions from motorized vehicles (on road and off) as well as oil and gas facilities in particular, have the potential to impact atmospheric deposition that may be detectable within this otherwise pristine region. Understanding whether and how specific human activities are discernable within the headwaters using the geochemistry of the Castle River can help decision makers mitigate detrimental outcomes and better manage water resources to retain the river's pristine condition. Knowing what factors affect ion geochemistry in watersheds is particularly relevant as river water levels decrease. Kruk and colleagues [5] point out that hydrometric flows in the Oldman River have declined by more than 30% since 1913. An increase in atmospheric deposition of pollutants in the headwaters as higher levels of emissions from motorized recreation and industrial activities grow, adds to potential challenges in maintaining a pristine state for the Castle River and the and the Oldman River downstream.

The objective of this study is to use a multi-isotope approach ($\delta^{34}S$, $\delta^{18}O_{sulfate}$, $\delta^{11}B$) to investigate whether the cumulative deposition of atmospheric emissions over the past half century is evident under baseflow conditions upstream to downstream along the Castle River.

6.1.2 Study area

The Castle River watershed is a part of the Oldman River Basin that lies along the eastern slopes of the Rocky Mountains in southwestern Alberta. The Castle River flows from south to north where it joins the Oldman River (Figure 6.1). The geology for the entire area is dominated by Mesozoic and Cretaceous limestone and dolomites. Paleozoic carbonates and evaporites occur toward the northern end of the valley, while metasedimentary rocks (Purcell Supergroup) occur at higher elevations in the south of the valley [5, 7, 11]. Bare rock is evident at the highest elevations (up to ~ 2,500 masl, Figure 6.2), with forests dominated by Lodgepole Pine (*P. Contorta*), White Spruce (*Picea glauca*), Trembling Aspen (*Populus tremuloides*) and Douglas Fir (*Pseudotsuga menziezzi*) in the Montain Cordillera ecoregion along the slopes lower down. Soils range from poorly developed luvisols (standard soil type) in the alpine reaches to Brunisols and Black and Grey Chernozems lower in the valleys. Soils are typically underlain by glacial till containing pyrite (FeS_2) as the terrain levels out in the valleys. This area is characterized by strong prevailing southwest and westerly winds and receives 650 mm precipitation annually: including 380 mm as rain and 270 mm as snow. Maximum precipitation occurs in May and June, whereas the minimum amount occurs in January and February. The mean annual temperature is 4.6 °C, and the mean monthly temperature varies between 15.5 °C in the late summer and – 6.6 °C in winter [12] based on data reported over a 30-year time frame (1971–2000) at Beaver Mines Station. The Castle River is characterized by peak stream flows during spring and minimum stream flows are observed in late winter [13].

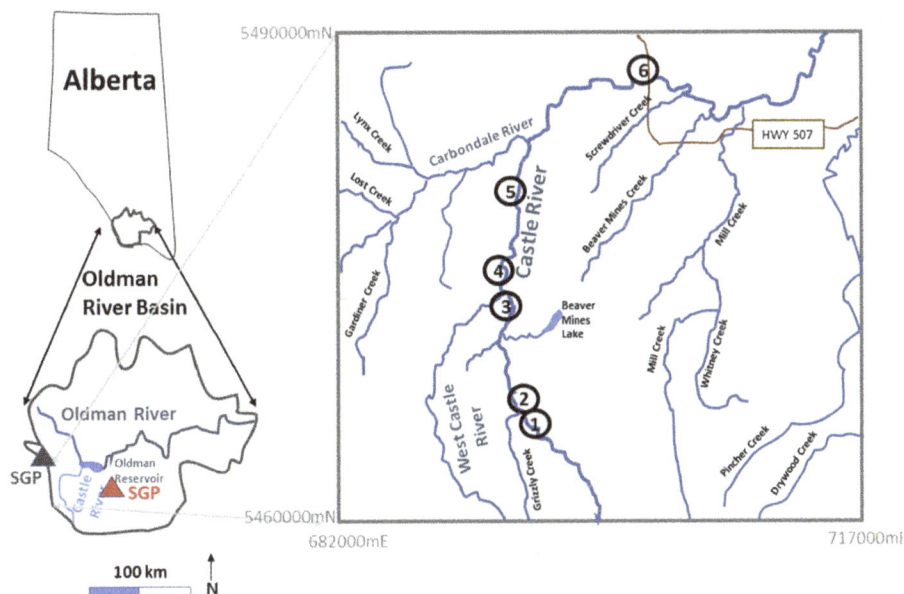

Figure 6.1: Map of the study area – the Castle River watershed which belongs to the Oldman River Basin of southern Alberta, Canada. The map shows the Castle River and its tributaries as well as sampling sites 1 through 6, UTM coordinates and roads.

6.1.3 Sulfate sources and their $\delta^{34}S$ values

Sulfate in the Castle River may be derived from weathering of bedrock geology, which is dominated by carbonate and dolomitic limestone, and dissolution of sulfate from oxidation of pyrite within the glacial tills that are abundant at the surface in the region. Other potential sources of sulfate may result from long- or short-term atmospheric deposition from on- and off-road vehicles, nearby volcano Mt. St. Helens, and the natural gas industry. Both lithospheric and atmospheric sources of sulfate may have an influence on both the groundwater and surface water for the area as sulfate is potentially retained, cycled and remobilized within the pedosphere before being flushed into the headwaters.

An earlier study used the CALPuff (CALifornia Puff) Lagrangian air dispersion model to track the mass of sulfur emissions from the source stack to follow how such as plume spreads and disperses over time across specific terrain. The model uses three-dimensional grid cells and downscaled weather forecasting model outputs for specific terrain as well as emissions (g/s) and the temperature at which they are released (°C). The term "Puff" reflects that the model computes the concentration, chemistry, and flux that occur downwind of the emissions within each grid box before moving to the next time step. This modeling showed the annual atmospheric deposition patterns from the two sour gas processing plants, identifiable at the center of the highest emission contour plots. Contour plots for emissions are similar to geographical contours on a map that show meters above sea level: instead of height above sea level these contour plots show, using lines across the map, the spatial extent where pollutant concentrations are a constant value. Figure 6.2 demonstrates how the sulfur emissions affect the headwaters region of the Castle River [14]. A key point to note is although he dominant wind directions are southwest and westerly, deposition also occurs to the west as well as the east of the Rocky Mountains' easternmost ranges. This is likely the effect of downwash – strong winds aloft over the mountains descend along the eastern slopes creating turbulent downdraughts and turbulence that funnel emissions back into the nearby Rocky Mountains. The modeling clearly shows this effect in Figure 6.2. However, the impact of this atmospheric deposition relative to background sources in the Castle River has not been measured previously.

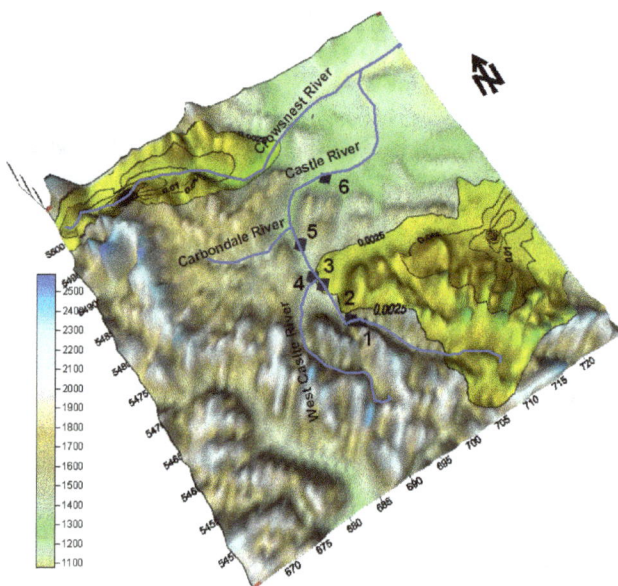

Figure 6.2: Map of the sampling locations (squares), terrain features (masl) and modeled pattern of annual sulfur deposition (contour labels are given in units of µgS/m²/s) based on results from Chandrasekaran et al., 2007 (9). Coordinates are reported in km along the x and y axes in UTM (Zone 11 U).

In order to distinguish atmospheric from geologic sources of sulfate, $\delta^{18}O_{sulfate}$ and $\delta^{11}B$ values need to be included in this investigation. The inclusion of $\delta^{18}O_{sulfate}$ combined with $\delta^{34}S$ values provides a means to distinguish between atmospheric primary (formation at high temperature prior to emission) and secondary (low temperature formation after emission) sulfate deposition and sulfate from weathering of glacial till within the watershed. The sulfur isotope composition of sulfate for a variety of atmospheric emissions has been previously assessed. A volcano that periodically emits large amounts of SO_2, sulfate and ash, Mount St. Helens, lies approximately 700 km to the southwest in the state of Washington (USA). Its emissions have produced significant ash within southern Alberta, most recently in the late 1980s and early 1990s. The $\delta^{34}S$ of Mt. St. Helen's ash has a $\delta^{34}S$ near + 9 ‰ [15]. Vehicle exhaust, that represents a mixture of sulfur from fuel, oil and anti-knock agents, has $\delta^{34}S$ values near + 5 ‰ [4]. Sulfate in aerosols and SO_2 downwind of the sour gas processing plants within this region of Alberta is + 20 ± 2 ‰. Sulfate from pyrite weathering has $\delta^{34}S$ values that are isotopically lighter ranging from – 18 to 0 ‰ with lower values in regions further from roadways and vehicle emissions [3–5, 16, 17].

6.1.4 Primary versus Secondary sulfate and $\delta^{18}O_{sulfate}$

Atmospheric pollutant sulfate can either be formed by high temperature (>350 °C) combustion before emission into the atmosphere (primary sulfate), or it can be formed from oxidation of emitted gaseous sulfur compounds (such as SO_2) after they are released to the atmosphere (secondary sulfate) [3, 18]. Although the $\delta^{34}S$ values for primary and secondary sulfate from the same emission source might be identical, $\delta^{18}O_{sulfate}$ values will differ depending on formation conditions. Primary sulfate incorporates oxygen from O_2 ($\delta^{18}O$ = + 24 ‰) and fractionated isotopes in oxygenated emissions contribute to sulfate formation with $\delta^{18}O$ values up to + 41 ‰ [18]. Secondary sulfate, typically formed at ambient temperature during SO_2 transport from the emission source, has $\delta^{18}O_{sulfate}$ values in the range of – 18 to 0 ‰ in southwestern Alberta [4, 16, 19].

6.1.5 Boron sources

Boron is an essential nutrient for living organic material at low concentration. Decayed organic material that was deposited and accumulated in sedimentary layers was eventually converted into fossil fuels because of increased pressures and temperatures as well as chemical and bacterial processes. Therefore, boron along with sulfur is an component of fossil fuels such as coal, oil and natural gas [20]. During the processing of fossil fuel, boron is released to the atmosphere, most likely in the form of boric acid $B(OH)_3$ [21], which is eventually removed by dry deposition or precipitation. In aqueous solution it is mobilized in its most common forms of B $(OH)_3$ and $B(OH)_4^-$. Boron concentrations in atmospheric emissions for seven gas plants located in central Alberta were found to have a value 79.5 µg/m^3 relative to sites distant from gas plant emissions where boron concentrations were one to two orders of magnitude lower (0.7 to 2.1 µg/m^3), The $\delta^{11}B$ values for the gas plant emissions ranged from − 10 to + 3 ‰ [20]. Boron from marine sedimentary matter is expected to have $\delta^{11}B$ values near + 40 ‰ [22].

The $\delta^{11}B$ in pristine western mountain headwaters across the Mackenzie Basin including the Rocky Mountains were assessed by Lamarchand and Gaillardet [23] and were found to range from + 3.6 to + 12.1 ‰. Within the Oldman River tributaries in the eastern flanks of the Rocky Mountains, Kruk and colleagues found near $\delta^{11}B$ values near + 10 ‰ in the most pristine locations rising to + 19 ± 3.2 ‰ when manure inputs were deemed large [5].

6.2 Methods

The Oldman River region is a good example of a complex ecosystem where the impact of industrial sources of sulfur and boron can be assessed using an integrated approach that combines stable isotope abundance data collected from environmental samples. In the following section, the technical methods to measure and interpret isotopic compositions in the watershed and understand anthropogenic deposition are described. Careful attention to where, when and how field samples are collected, and some of the limitations of conducting such work in remote and wintery environments, are described. Then we consider how the samples are treated in order to measure the isotope composition for sulfate ($\delta^{34}S$, $\delta^{18}O$) and boron ($\delta^{11}B$) values.

6.2.1 Fieldwork

Surface water grab samples were collected from six locations downstream of the headwaters (assumed to be at UTM Zone 11 U 708,152 mE, 5,452,538 mN) along the Castle River once per month from October to December in 2008. Water samples were collected several cm below the surface of the river within a fetch where water reached between 10 and 50 cm depth. Care was taken to find open water over a shallower fetch when collecting water samples after ice had formed. One-liter wide-mouthed polyethylene bottles were submerged upstream, rinsed at least three times with river water before sample collection. Sample bottles were then capped and placed in a cooler (~ < 4 °C) during transport to the lab where they were stored at 4 °C prior to analysis.

Other water parameters that were measured in the field included river flow rate (m³/s), width (m) and depth at 1 m intervals which were used to calculate discharge using the point and quarter method. Flow rate was not possible to obtain at all locations throughout the sampling period due to ice formation or lack of light.

Late fall and early winter were chosen for the sampling times rather than spring and summer to ensure baseflow conditions. Baseflow conditions (1) minimized the contributions of surface and soil derived ions to the river, and (2) bacterial activity should be lower at lower temperatures experienced at this time of year, minimizing the effects of bacterial activity on the geochemistry impacting the river. Based on these field measurements and the ion concentration measured by Ion Chromatography (IC), ion fluxes were calculated using eq. (6.4).

$$\text{Ion Flux (mg/s)} = \text{Discharge (m}^3/\text{s)} * 1000 \ (\text{L/m}^3) * \text{Ion Concentration (mg/L)} \quad (6.4)$$

6.2.2 Laboratory analysis

Sulfate from river water samples was recovered as $BaSO_4$ precipitate by the addition of $BaCl_2$ and then HCl to the water samples [4]. $\delta^{34}S$ and $\delta^{18}O_{sulfate}$ values were analyzed by a gas chromatography combustion mass spectrometry at the Isotope Science Laboratory of University of Calgary. Raw δ values were corrected using sample-standard bracketing with known standards: for $\delta^{34}S$ values ($1\sigma \pm 0.3$ ‰) to Vienna Canon Diablo Troilite (V-CDT), while $\delta^{18}O_{sulfate}$ values ($1\sigma(\delta^{18}O) = \pm 0.5$ ‰) is expressed relative to Vienna Standard Mean Ocean Water (VSMOW). The prefix Vienna occurs where the International Atomic Energy Association (IAEA) has supplied uniform materials (e.g., $BaSO_4$) with agreed-upon isotope ratios ($\delta^{34}S$, $\delta^{18}O_{sulfate}$) for distribution to the isotope community to use as primary standards that day-to-day laboratory standards that individual labs around the world affix to.

Boron was separated from the water samples based on the ion exchange method described by Kruk et al. [5]. A volume of water containing ~ 300 nanograms (ng) of boron was adjusted to a minimum pH of 8 by the addition of 0.5 NaOH and pumped through 100 uL of Amberlite IRA-743 boron-specific ion exchange resin. At this high pH, boron in solution is coordinated as the tetrahedral $B(OH)_4^-$ ion, which is retained by the ion exchange resin material. The boron adsorbed on the resin can then be recovered in 100 uL of water such that the boron is $B(OH)_3$ and no longer retained by the resin. The eluted boron was brought to a final volume of 1.0 mL. The prepared samples were aspirated by a PFA nebulizer into a PFA spray chamber (Elemental Scientific Inc.) and the ^{11}B and ^{10}B isotope abundance ratios were measured by a Neptune multiple collector inductively coupled plasma mass spectrometer. Isotope abundance ratios were calibrated by sample-standard bracketing with SRM-951 boric acid as the zero-delta reference. Seawater reference material from the International Association for the Physics

Sciences of the Oceans (IAPSO) seawater was run as a check on the isotope abundance data with each analytical session and returned a value of +39 ($1\sigma(\delta^{11}B) = 1$). Boron concentrations were measured on a Perkin Elmer Elan DRC II quadrupole Mass Spectrometer (MS).

6.3 Results

The six sampling sites are shown by sampling month along with elevation, discharge (Q), distance from headwaters (HD, based on Google Earth) and drainage area (DA from ArcGIS) in Table 6.1. River water sampling was performed along a 64 km stretch from the headwaters along the valley bottom where there was access to the river and it was possible to cross the river to perform depth and flow determinations. The change in elevation between the first site (at 27.6 km) and the last (at 64 km) was small (~200 m). Discharge ranged from 0.48 m^3/s at site 1 in November, to five times that amount, 2.7 m^3/s at the Water Survey of Canada hydrometric flow station by highway (HWY) 507 (6 km downstream of site 6). Baseflow conditions are evident in this dataset since discharge remained relatively constant between sampling periods except where the river was braided (at site 4A) or in December when it was not possible to determine the full flow since significant ice covered the edge of the river with clearly flowing water beneath.

Concentrations for boron, sulfate, sodium as well as the isotope composition for boron and sulfate are reported in Table 6.2. A clear decrease in concentrations downstream is observed for both boron (0.0167 to 0.0103 mg/L) and sulfate (19.4 to 11.4 mg/L). In contrast sodium concentrations remained relatively constant (average = 1.6 ± 0.1 mg/L, n = 14) except at site 6 downstream of the confluence with the Crowsnest River, where they rose (average = 2.4 ± 0.1, n = 3).

$\delta^{11}B$ values did not exhibit a large change downstream for the first five sites (average = +4.2 ± 0.6 ‰) despite the clear decrease in boron concentrations downstream. However, $\delta^{11}B$ for site 6 (including $\delta^{11}B$ = +6.9 ‰ for a second site sampled near site 6 in November) was somewhat higher (average = +6.0 ± 1.6 ‰). $\delta^{34}S$ values ranged from +23.5 ‰ upstream to +14.2 ‰ downstream as sulfate concentrations decreased. $\delta^{18}O_{sulfate}$ values were highest in December at site 3C (+10.4 ‰) and lowest at site 6B (+1.7 ‰) in November.

It is worthwhile noting that sites 2, 3 and 4 are within the contour lines (see Section 6.1.2 for a description of what contour lines represent) for average sulfur deposition from the eastern sour gas plant in Figure 6.2 while sites 1, 5 and 6 are further away.

A comparison of ion concentrations to distance from headwaters and drainage area can give insight whether surface runoff contributes to ground- and surfacewater ion concentrations: a decrease in concentration with increased distance downstream and/or drainage area is expected if dilution takes place, an increase in

Table 6.1: UTM Coordinates (Zone 11 U), discharge (Q), distance from headwaters (HD), and drainage area (DA) for the six sampling locations, site 1 being closest to the headwaters, along the Castle River in southern Alberta. (n.d. = not measured and/or detected).

Sample ID	Easting (m)	Northing (m)	Elev (masl)	Q (m³/s)	HD (km)	DA (km²)
Headwaters	708,152	5,465,902				
October						
1A	695,144	5,465,902	1,414	0.683	27.6	179.1
2A	694,031	5,467,163	1,411	1.318	30.0	205.9
3A	693,569	5,473,499	1,356	1.523	39.4	238.6
4A	693,065	5,475,360	1,347	0.737*	42.2	373.4
5A	694,026	5,480,240	1,294	n.d.	48.8	406.7
6A	702,631	5,486,879	1,214	2.679**	64.0	800.9
November						
1B	695,144	5,465,902	1,414	0.477	27.6	179.1
2B	694,031	5,467,163	1,411	1.261	30.0	205.9
3B	693,569	5,473,499	1,356	0.851	39.4	238.6
4B	693,065	5,475,360	1,347	1.448	42.2	373.4
5B	694,026	5,480,240	1,294	n.d.	48.8	406.7
6B	702,631	5,486,879	1,214	n.d.	64.0	800.9
December						
1C	695,144	55,465,902	1,414	0.527	27.6	179.1
2C	694,031	5,467,163	1,411	0.624	30.0	205.9
3C	693,569	5,473,499	1,356	1.713	39.4	238.6
4C	693,065	5,475,360	1,347	1.463	42.2	373.4
5C	694,026	5,480,240	1,294	1.429	48.8	406.7
6C	702,631	5,486,879	1,214	n.d.	64.0	800.9

*may be an underestimate as the river was braided and only the largest part of the river was sampled

**based on Castle River discharge measurements by Water Survey of Canada hydrometric gauge for the Castle River, by HWY 507.

Table 6.2: Anion, cation concentrations and fluxes (mg/s) (*italicized text below concentration data*) and B, S and O$_{sulfate}$ δ values (‰).

Sample ID	B (mg/L)	SO$_4$ (mg/L)	Na (mg/L)	δ^{11}B	δ^{34}S	δ^{18}O$_{sulfate}$
October						
1A	0.0163	19.6	1.56	+4.0	+23.0	+7.0
	11.1	*13,400*	*1,070*			
2A	n.d.	16.3	1.55	+4.2	+22.8	+7.4
	n.d.	*21,400*	*2,050*			
3A	n.d.	14.2	1.63	+3.9	+22.3	+9.3
	n.d.	*21,500*	*2,490*			
4A	0.0139	13.7	1.47	+3.8	+20.4	+4.4
	10.3	*10,100*	*1,090*			
5A	0.0143	13.6	1.60	+3.3	+20.4	+4.6
	n.d.	*n.d.*	*n.d.*			
6A	0.0117	11.5	2.23	+3.7	+15.4	+6.2
	31.4	*30,900*	*5,960*			
November						
1B	0.0167	17.8	1.60	+4.5	+22.3	+6.9
	8.0	*8,490*	*762*			
2B	0.0164	17.2	1.56	+4.6	+23.2	+9.1
	20.7	*21,700*	*1,970*			
3B	n.d.	14.7	1.66	+5.0	+21.7	+7.9
	n.d.	*12,500*	*1,411*			
4B	0.0132	12.3	1.52	+3.9	+20.8	+3.6
	18.8	*17,800*	*2,200*			
5B*	n.d.	n.d.	n.d.	n.d.	n.d.	n.d.
	n.d.	*n.d.*	*n.d.*			
6B	0.0103	12.2	2.36	+6.2	+15.1	+1.7
	n.d.	*n.d.*	*n.d.*			
6Bb**	0.0103	n.d.	n.d.	+6.9	+14.8	+3.0
	n.d.	*n.d.*	*n.d.*			
December						
1 C	n.d.	12.0	1.61	+4.6	+23.5	+7.5
	n.d.	*6,340*	*852*			
2 C	0.0159	19.4	1.53	+3.4	+22.6	+7.2
	10.2	*12,400*	*980*			
3 C	0.0138	18.7	1.63	+4.4	+22.2	+10.3
	22.1	*30,000*	*2,620*			
4 C	n.d.	15.4	1.57	+5.4	+21.3	+7.7
	n.d.	*22,600*	*2,290*			
5 C	n.d.	13.1	1.89	+4.4	+20.5	+8.8
	n.d.	*18,700*	*2,700*			
6 C	0.0108	12.6	2.47	+7.3	+14.2	+7.3
	n.d.	*n.d.*	*n.d.*			

*site 5B was not sampled in November. **site 6Bb, at an unspecified location between sites 5&6 was sampled in place of site 5 in November only concentration and discharge (Q) suggests there is either more mineral weathering or atmospheric deposition from long-range transport (which would be expected to be spatially uniform: [24]). A lack of change in ion concentration as Q increases downstream suggests groundwater and mineral weathering are dominant processes and there is chemostatic equilibrium for the ion along the river [25]. Figure 6.3 demonstrates that both sulfate and boron concentrations decline as the distance downstream increases (a) and drainage area increases (b) downstream. In contrast, sodium (and calcium) concentrations increased downstream with higher drainage area (data not shown, r^2 = +0.71 (+0.95) and +0.80 (+0.86), respectively). Higher variability is observed for sulfate than for boron closer to the headwaters and suggests more of both ions in the headwaters area that is diluted downstream.

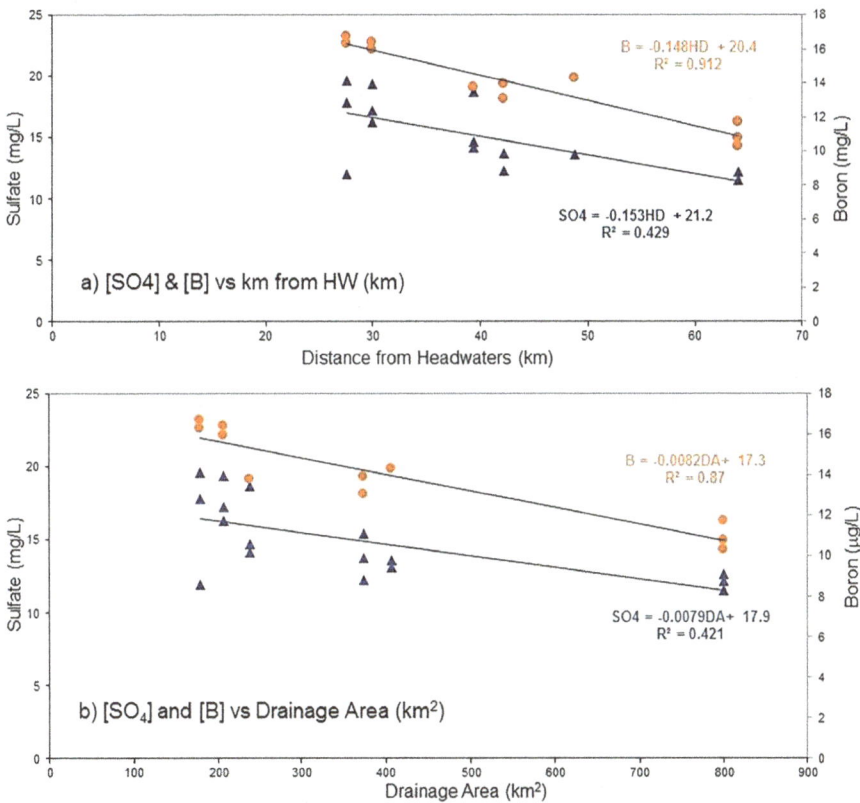

Figure 6.3: Castle River sulfate and boron concentrations decrease downstream as distance from headwaters (HW) (a) and as drainage area (DA) increased (b). This is consistent with concentrated source(s) of sulfate and boron in headwaters that is(are) diluted downstream.

Similarly, river ion fluxes that normalize for variations in discharge (Q) help discern whether ions are derived from mineral weathering (Table 6.2, *italicized text*). A strong increase in ion concentration with Q is indicative of mineral weathering. Fluxes for all ions measured increased downstream, though sodium (r^2 = +0.75) and calcium (r^2 = +0.73) were more strongly correlated with discharge than sulfate (r^2 = +0.27) or boron (r^2 = +0.55). Therefore, sodium is the best indicator for mineral dissolution in this study.

6.4 Discussion

The concentration data presented in Tables 6.1 and 6.2 as well as the plots in Figure 6.3 suggests that sulfate and boron may have a common source high in the headwaters region. In contrast, concentration data for sodium and calcium show an *increase* in concentration with distance from headwaters and drainage area (r^2 = 0.80, 0.86 respectively, plots not shown). This is an interesting result and demonstrates a decoupling between sodium that is dominated by carbonate mineral dissolution [17] from sulfate and boron in the headwaters.

To determine the source of sulfur and boron, $\delta^{34}S$ values for sulfate, as well as sulfate and $\delta^{11}B$ and boron concentrations were compared. The results are shown in Figure 6.4. Boron and sulfate are reasonably correlated (Figure 6.4a) but boron correlates better with $\delta^{34}S$ values (r^2 = +0.67 and +0.88, respectively), as shown in Figure 6.4b. The isotopic composition for boron and sulfate are potentially better indicators of the element's source(s) than are concentrations: concentrations are affected on spatiotemporal scales by processes such as dilution and evaporation for example, while δ values for S and B are not [3, 22]. If, for example, a point source high in sulfur and boron affected the headwaters, then concentrations for both elements would decline downstream with larger inputs of precipitation-derived (meteoric) and groundwater discharge. This would be an example of dilution by water low in sulfur and boron compared to the point source as catchment area increased. If evaporation significantly increased the concentration of the elements downstream, then δ values would remain constant while concentrations increased. In contrast to what is observed in this study (e.g., other regions of the globe may have considerably different hydrologic conditions and the isotope compositions for B and S may be indistinguishable), the isotope composition of the boron and sulfur in the headwaters might be similar to that in precipitation and groundwater. In that case no change in the δ values would be evident as dilution or evaporation took place.

Figure 6.4c shows where the data on $\delta^{11}B$ versus $\delta^{34}S$ is expected to fall for deposition of emissions from sour gas activities. Mixing between deposition from sour gas emissions with background sulfate from pyrite in tills, and background boron associated with more pristine locations in the Oldman Watershed [5], would follow the white arrow to the upper left in Figure 6.4c. Modification of sulfur in the pedosphere over decades of atmospheric deposition by sulfur mineralization would shift the $\delta^{34}S$ values toward the right (arrow to the right of the grey box in Figure 6.4c). This is because as sulfur mineralization in soils occurs, there is a loss of lighter sulfate ($^{32}SO_4$) from the soil column over time resulting in the preferential retention of ^{34}S within the pedosphere [26]. The data from this study are shifted toward more positive $\delta^{11}B$ values than expected for simple mixing of emissions deposition and geogenic sources of sulfate and boron. The very positive $\delta^{11}B$ values described by Kruk et al. [5] for boron from manure ($\delta^{11}B = +19.2 \pm 3.2$ ‰) applied relatively uniformly throughout the valley of the Castle River [11] are expected to drive the mixing line upward.

Added confirmation for this interpretation is obtained by comparing boron and sulfate data to the $\delta^{18}O_{sulfate}$ as shown in Figure 6.5. When boron concentrations are low (downstream), the oxygen isotopes in sulfate ($\delta^{18}O_{sulfate}$) values tend toward meteoric water, suggesting more secondary sulfate is present as shown in

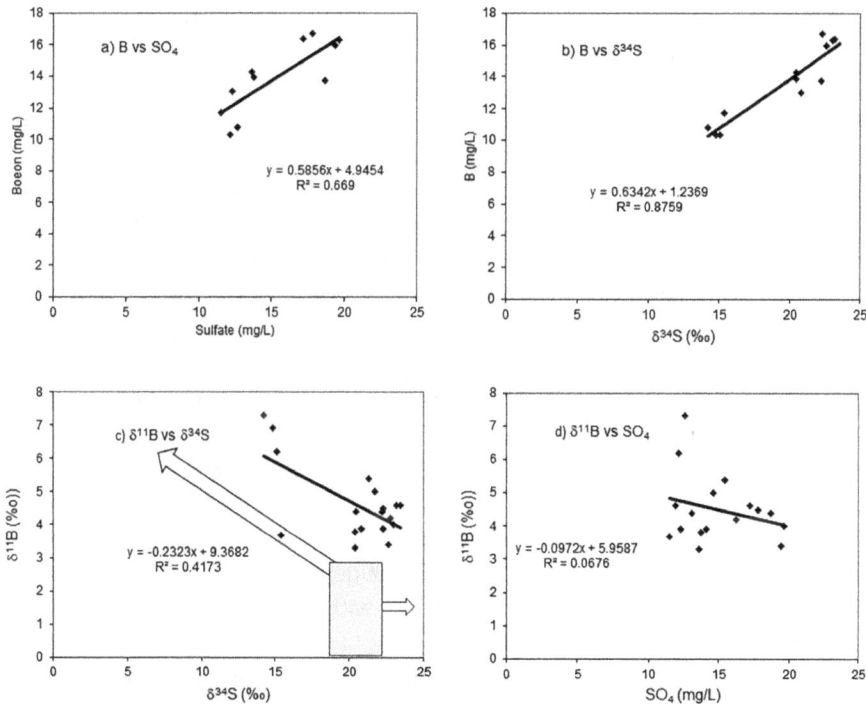

Figure 6.4: Relationships between sulfate and boron ions under baseline conditions in the Castle River. Correlations between boron and sulfate concentrations (a), boron and $\delta^{34}S$ values (b), $\delta^{11}B$ and $\delta^{34}S$ (c) and $\delta^{11}B$ and sulfate (d) are plotted. In Figure 4c, the grey shaded region shows where sour gas emissions would be expected. The arrow in the lower right of Figure 4c pointing right shows how soil sulfur mineralization is expected to affect remaining organic S present in the pedosphere. The white arrow up and to the left represents a mixing line if sour gas emissions mixed with boron under pristine conditions.

Figure 6.5a. This is in reasonable agreement with the results from Rock and Mayer [7] who found $\delta^{18}O_{sulfate}$ values near +1‰ in the Castle River at a point ~ 15 km further downstream of site 6 over a two-year sampling period. As boron concentrations increase going upstream toward the headwaters of the Castle River, more sulfate is enriched in $\delta^{18}O$ which is consistent with a primary (high temperature) sulfate source at high elevation in the headwaters region (see Section 6.1.3). An alternate source of sulfate that might produce these results is bedrock that displays positive $\delta^{18}O$ sulfate values and high boron and sulfate concentrations such as evaporites or sedimentary sulfates (Figures 6.5a,b): $\delta^{18}O_{sulfate}$ values found at high boron and concentrations are consistent with seawater values from the Paleozoic that are generally more than +10 ‰ [3, 7]. However, if weathering of seawater evaporites or sedimentary rocks were responsible for these data, sodium should positively correlate with boron and sulfate concentrations. Instead the opposite was found: sodium increased even as boron and sulfate decreased downstream (Figure 6.3a,b). Mineral weathering is associated with high Na and Ca in this region, so it is more likely that the high boron and sulfate and high $\delta_{18}O_{sulfate}$ at high elevations toward the headwaters is from atmospheric deposition.

Plots of $\delta^{18}O_{sulfate}$ versus $\delta^{11}B$ and $\delta^{34}S$ can help resolve the source of boron and sulfate in the headwaters region. Boron with an isotope composition ranging between +3 and +12 ‰ from the Oldman tributaries was interpreted to represent background, or relatively pristine conditions (orange oval in Figure 6.5c) by Kruk et al. [5]. If this boron were from an organic matrix and was associated with organic sulfur that was oxidized to sulfate, then the data should lie within the orange oval in Figure 6.5c as meteoric water with negative $\delta^{18}O$ is incorporated into sulfate. The data from the bulk of the samples in this study, however, lies above the orange shaded oval. A theoretical two-source mixing between a source of atmospheric deposition that is dominantly primary sulfate (upper left of the blue square in Figure 6.5c) and manure (lower right) is shown as dashed line in Figure 6.5c. The data from this study do overlap with such an interpretation. Note that the trend shown in Figure 6.4c, is not consistent with boron from a seawater source (seawater $\delta^{11}B$ ~ +40 ‰) since the trend in $\delta^{11}B$ with $\delta^{34}S$ should be positive rather than negative. There is the potential for a non-industrial source of sulfate from atmospheric deposition that contains a mix of primary and secondary sulfate that then mixes with a pedogenic and/or geologic source of sulfate (orange oval in Figure 6.5c).

Finally, Figure 6.5d demonstrates the utility of the multi-isotope approach in defining atmospheric versus terragenic sources of sulfate. The bulk of the sulfate $\delta^{34}S$ values are within the range expected from atmospheric deposition from industrial emissions that represent a mix of dominantly secondary with a smaller proportion of primary sulfate (top box Figure 6.5d). The data show that this source of sulfate with quite positive $\delta^{34}S$ and low $\delta^{11}B$ values (with high boron concentrations at higher elevations) mixes with another sulfate source at low elevations that has the following characteristics:

(1) Lower boron concentration (Figure 6.4b) and higher Na and Ca downstream (Figure 6.3)
(2) Is associated with atmospheric secondary rather than primary sulfate (Figure 6.5a) and/or the incorporation of meteoric water as sulfides are oxidized in the pedosphere or lithosphere downstream (Figure 6.5c)

These characteristics are consistent with background boron in the Castle tributaries as described by Kruk et al. [5]. Contributions from manure/soil sulfur oxidation or pyrite dissolution are potential explanations for Figure 6.5c: mixing of industrial deposition with these sources would be displayed as a trend toward the red box in Figure 6.5d and a single data point from the most downstream site in October does suggest this process may occur. Low $\delta^{34}S$ values in manure as grass eaten by herbivores is isotopically close to the $\delta^{34}S$ signature of glacial till, or sulfate from the oxidation of FeS_2, should correspond to low $\delta^{18}O_{sulfate}$ from meteoric water input during sulfur oxidation. Turnover of organic matter associated with manure and/or pyrite oxidation would be expected to trend toward the lower left (clear arrow pointing down to the left) in Figure 6.5d . Another three data points in Figure 6.5d with $\delta^{34}S <$

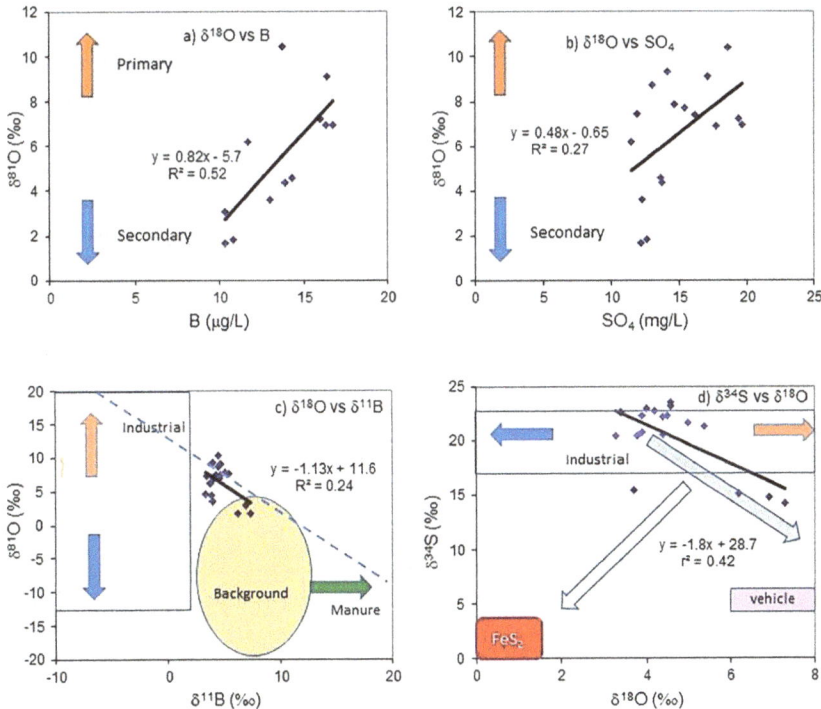

Figure 6.5: $\delta^{18}O_{sulfate}$ is shown relative to boron (a) and sulfate (b) concentrations as well as boron $\delta^{11}B$ (c) and sulfate $\delta^{34}S$ (d) values.

+15 ‰ do not follow this trend: they have higher $\delta^{18}O_{sulfate}$ than industrial emissions (transparent green arrow pointing down to the right) in the lower right of Figure 6.5d . Vehicle exhaust or volcanic emissions from Mt. St. Helens ($\delta^{34}S \sim +9$ ‰) both could plausibly have the signature of sulfate formed at high temperature that is then deposited across the Castle watershed. An area source for this sulfate is consistent with higher Na and Ca downstream and lower sulfate as mineral weathering by meteoric water contributes to watershed runoff.

6.5 Conclusion of case study

Atmospheric deposition of a mix of primary and secondary emissions from the sour gas industry is apparent at baseline flows toward the headwaters in the fall in the Castle River in southern Alberta. Further downstream, this industrial source mixes with boron from the pedosphere (possibly manure) and sulfate that is consistent with high temperature emissions (primary): possibly long-term deposition of volcanic emissions from Mt. St. Helens and/or vehicle exhaust ($\delta^{34}S \sim +5$ ‰: [16]). Rock and Mayer [7] suggested sulfate from organic matter turnover was significant within the tributaries feeding the Oldman River, and a single data point from October for the most downstream site in this study does agree with that interpretation, but three other data points during baseflow conditions did not. Instead, these downstream data suggest long-term atmospheric deposition should be considered as a potential source of sulfate in the watershed.

Further research would help identify if manure, soil organic matter turnover, vehicle exhaust and/or Mt. St. Helen's sulfate contribute to this background boron and sulfate that dilutes the signature of industrial deposition downstream in within the lower reaches of the Castle watershed.

6.6 Summary

The consideration of stable isotopic composition in addition to elemental concentrations when searching for sources of pollutants to the natural environment offers significant advantages. This is because the stable isotopic composition of an element associated with a particular source may be distinct from occurrences of that element in the pristine environment. Therefore, one not only characterizes "how much" of the element is present, but also "what source" can be attributed to the measurement. This isotopic archive is recorded in the sample itself and can be revealed through a careful measurement using modern analytical methods. When the isotopic information from specific elements is brought together, a comprehensive picture of the sources and pathways of the elements is achieved.

References

[1]   Widory D, Kloppmann W, Chery L, Bonnin J, Rochdi H, Guinamant J-L. Nitrate in groundwater: An isotopic multi-tracer approach. Contam Hydrol. 2004, 72, 165–188.

[2]   Widory D, Petelet-Giraud E, Negrel P, Ladouche B. Tracking sources of nitrate in groundwater using coupled nitrogen and boron isotopes: A synthesis. Environ Sci Technol. 2005, 39, 539–548.

[3]   Krouse HR, Grinenko VA. Stable Isotopes: Natural and Anthropogenic Sulfur in the Environment Scope 43. Chichester. John Wiley and Sons, New York. Brisbane. Toronto. Singapore, 1992, 440.

[4]   Norman AL, Krouse HR, MacLeod JM. Appointment of pollutants in an urban airshed: Calgary, Canada, A case study. In: Air Pollution Modeling and Its Application XVI. Borrego C, Incecik S, eds., Kluwer Acadmic / Plenum Publishers, New York, 2004, 2004, 107–125.

[5]   Kruk MK, Mayer B, Nightingale M, Laceby JP. Tracing nitrate sources with a combined isotope approach in a large mixed-use watershed in southern Alberta, Canada. Sci Total Environ. 2020, 703, 135043 15. pgs.

[6]   ATSDR (2006) Agency for Toxic Substances and Disease Registry (https://www.atsdr.cdc. govlastcheckedNovember302021).

[7]   Rock L, Mayer B. Identifying the influence of geology, land use, and anthropogenic activities on riverine sulfate on a watershed scale by combining hydrometric, chemical and isotopic approaches. Chem Geol. 2009, 262, 121–130.

[8]   Jasechko S, Kirchner JW, Welker JM, McDonnell JJ. Substantial proportion of global streamflow less than three months old. Nat Geosci. 2016, 9(2), 126–129. https://doi.org/ 10.1038/ngeo2636.

[9]   Stewart MK, Stevens G, Thomas JT, van der Raaij R, Trompeetter V. Nitrate sources and residence times of groundwater in the Waimea Plains, Nelson. J Hydrol (NZ). 2011, 50(2), 313–338.

[10] OMBW (2021) Old Man Basin Watershed https://caringforourwatersheds.com/canada/ alberta/watershed-information/ (last accessed October 18, 2021).

[11] Baker H, DeMaere C, France T, Willoughby M, Alexander M (2020) Alberta Environment and Parks Land Division Rangeland Conservation and Stewardship Section Pincher Creek, Alberta Range plant communities and range health assessment guidelines for the south ecosystem of the montane natural subregion of Alberta, 8th approximation. Publication No. T/136,236 pgs. ISBN No.: 978-1-4601-4835-8 https://open.alberta.ca/publications/9781460148358 (last accessed October 27 2021)

[12] CCN (2021) Canadian Climate Normals & Averages https://climat.meteo.gc.ca/climate_ normals/index_e.html. (last accessed Oct 28, 2021)

[13] Mayhood DW (1995). Some effects of natural gas operations on fishes & their habitats on Canada's Rocky Mountains East slopes. Technical report Rocky Mountain ecosystem coalition. p.26.

[14] Chandrasekaran R, Jevne A, Li S, Morrice B, Sappier D (2007) Air Quality Modeling: In Cumulative Impacts from Sour Gas Emissions near Pincher Creek, Alberta. Submitted to the Alberta Energy and Utilities Board (EUB) for the hearing of the well and pipeline applications Nos. 1498479 and 1483571 by Shell. p11.

[15] Martin E, Bekki S, Ninin C, Bindeman I. Volcanic sulfate aerosol formation in the troposphere. J Geophys Res. 2014, 119(2), 12,660-12 673.

[16] Norman AL, Anlauf K, Hayden K, Thomson B, Brook. J, Li S-M, Bottenheim J. Aerosol sulphate and its oxidation on the Pacific NW coast: S and O isotopes in PM2.5 from Pacific 2001. Atmos Environ. 2006, 40, 2676–2689.

[17] Hendry MJ, Cherry JA, Wallick EI. Origin and distribution of sulfate in a fractured till in southern Alberta. Water Resour Res. 1986, 22, 45–61.

[18] Holt BD, Kumar R, Cunningham PT. Primary sulfates in atmospheric sulfates; estimation by oxygen isotope ratio measurements. Science. 1982, 217(4554), 51–53.

[19] Rock L, Mayer B. Isotope hydrology of the Oldman River basin, southern Alberta, Canada. Hydrol Process. 2007, 21(24), 3301–3315.

[20] Bradley CE. A Multiple-Isotope Approach to Tracing Natural Gas Processing Emissions. (MSc. Thesis) Department of Civil Engineering, The University of Calgary, 2005, 157.

[21] Fogg TR, Duce RA, Fasching JL. Sampling and determination of boron in the atmosphere. Anal Chem. 1983, 55(13), 2179–2184.

[22] Barth S. Utilization of boron as a critical parameter in water quality evaluation: Implications for thermal and mineral water resources in Germany and N Switzerland. Environ Geol. 2000, 40, 73–89.

[23] Lemarchand D, Gaillardet J, Göpel C, Manhès G. An optimized procedure for boron separation and mass spectrometry analysis for river samples. Chem Geol. 2002, 182(2–4), 323–334.

[24] Evans J, Norman AL, Reid M. Evidence of smoke from wildland fire in surface water of an unburned watershed. Water Resour Res. 2021, 57, e2021WR030069 14, pgs, https://doi.org/10.1029/2021WR030069.

[25] Basu NB, Destouni G, Jawitz JW, Thompson SE, Lukinova NV, Darracq A, Zanardo S, Yaeger M, Sivapalan M, Rinaldo A, Rao PSC. Nutrient loads exported from managed catchments reveal emergent biogeochemical stationarity. Geophys Res Lett. 2010, 37, 1–5.

[26] Norman AL, Giesemann A, Krouse HR, Jager HJ. Sulphur isotope fractionation during sulphur mineralization: Results of an incubation-extraction experiment with two Black Forest soils. Soil Biol and Biogeochem. 2002, 34, 1425–1438.

Daniel Sánchez-Rodas, Ana M. Sánchez de la Campa,
and María Millán Martínez

Chapter 7
The role of metalloids (As, Sb) in airborne particulate matter related to air pollution

7.1 Airborne particulate matter

7.1.1 Definition and classification

Airborne particulate matter (PM), also called atmospheric aerosol, has been defined as material that is suspended in the air in form of minute solid particles or liquid droplets. The term "aerosol" refers to the solid or liquid particles suspended in air or another gas [1, 2]. PM is one of the major air pollutants affecting both human health and the environment, in addition to other well-known pollutants, such as gases (e.g., CO_2, SO_2, NO_2), radioisotopes (e.g., uranium, caesium, radon), inorganic (mercury, ammonia) and/or volatile organic compounds (e.g., hydrocarbons, organic chlorinated compounds).

PM can be classified according to the aerodynamic properties of the particles because they govern the transport and removal of particles from the air, as well as the deposition within the respiratory system. Aerodynamic properties are associated with the chemical composition and sources of the particles [3]. In this sense, PM is commonly classified in three categories according to the diameter of the particles, which is in the range of microns (μm):

i) Total suspended particles (TSP) refers to particles of diameters up to about 50 μm,
ii) PM_{10}, also called coarse particles, includes particles with diameter equal or less than 10 μm, and
iii) $PM_{2.5}$, which consists of fine particles with diameters ≤ 2.5 μm. A comparison between relative sizes of PM_{10} and $PM_{2.5}$ particles and a human hair or a sand grain is shown in Figure 7.1. More recently, the term "ultra fine particles" (UFP) has been introduced, which corresponds to particles with a diameter <0.1 μm [4].

In addition to size, PM can also be classified based on their anthropogenic or a natural source of the particles. Among the anthropogenic sources, the most import ones are traffic, power generation, industry, and manufacturing. Furthermore, the main natural sources can be volcanic eruptions, geothermic activity, wildfires, atmospheric resuspension due to strong winds, transport of natural particles from arid regions,

Daniel Sánchez-Rodas, Ana M. Sánchez de la Campa, María Millán Martínez, Center for Research in Sustainable Chemistry-CIQSO. University of Huelva, Spain, e-mail: rodas@uhu.es

https://doi.org/10.1515/9783110626285-007

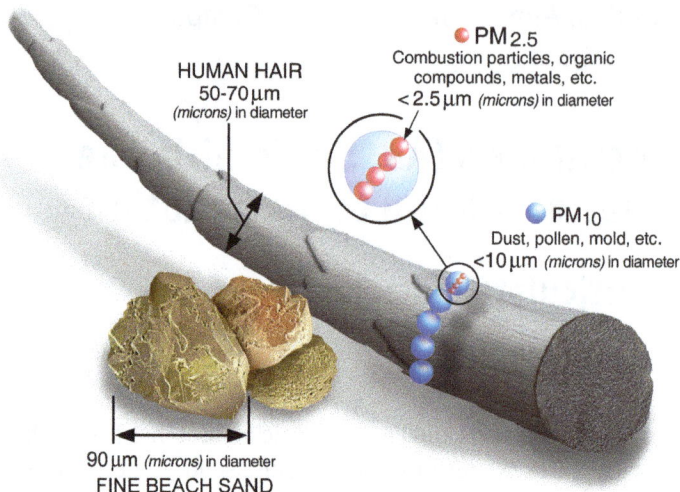

Figure 7.1: Size comparison of PM [5].

and marine aerosol [6]. Due to the worldwide industrialization, the largest fraction of PM in the atmosphere comes from anthropogenic sources. Anthropogenic PM can have harmful effects on human health, the climate, and ecosystems [7].

7.1.2 Effects on human health

Within the last decade the World Health Organization (WHO) has estimated that air pollution involving PM has contributed to approximately 800,000 premature deaths each year, making it the 13th leading cause of mortality worldwide [8]. More recently air pollution has been attributed to cause >6 million deaths based on data from 2015 [9]. It is inferred that the smaller the particle size, the greater the capacity to enter the alveoli of the lungs, thus increasing its potential toxic effect (Figure 7.2). Depending on the capacity to penetrate into the respiratory system, PM can be classified into inhalable particles (<30 µm), the extra thoracic fraction (>10 µm), the thoracic fraction (<10 µm), the tracheobronchial fraction (10–2.5 µm) and the alveolar fraction (<2.5 µm).

Ambient PM is a recognized threat to public health on a global scale and not only in highly polluted environments. Epidemiological and human exposure studies show that both short- and long-term exposure to PM correlate with cardiovascular and respiratory morbidity and mortality [11]. Lung problems such as chronic bronchitis, reduced function, and even premature death have been associated with long-term exposure to aerosols. Lower relative risks were reported for PM short-term exposure [12].

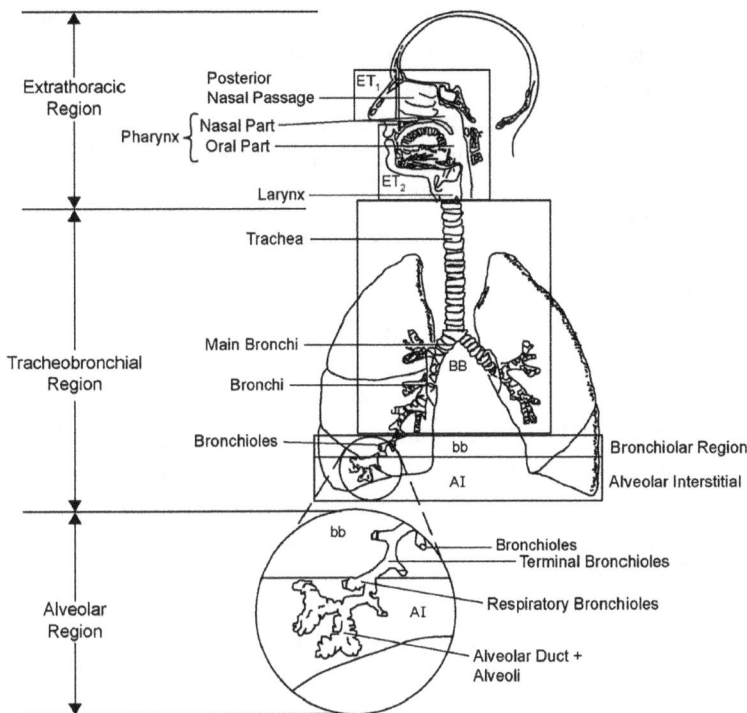

Figure 7.2: Scheme of the human respiratory system [10].

The relationship between PM and human health has been studied since the 1970s. Early studies performed in urban areas of the United States showed that particulate air pollution was associated with cardiopulmonary and lung cancer mortality but not with mortality due to other causes. Increased mortality is associated with fine particulate air pollution at levels commonly found in American cities [13, 14]. Other studies performed in Europe, such as the APHEA project (Air Pollution and Health: a European Approach) have established a similar relation between PM exposure and daily mortality data [15]. This project, encompassing several European cities, revealed that mortality was associated with PM exposure: when $PM_{2.5}$ concentrations increased by 10 µg m^{-3}, the daily mortality increased by 0.52% and that of deaths from cardiovascular and respiratory causes by 0.76% and 0.71%, respectively. PM alone was responsible for a loss in statistical life expectancy of up to 12–36 months.

The adverse effects of PM on human health have resulted in legislation to regulate its concentration in air. For the European Union, according to the Directive 2008/50/EC, the limit of daily average concentration of PM_{10} is 50 µg m^{-3}, and the allowed annual average is 40 µg m^{-3}. In the United States, the National Air Quality Standard (Final Rule/Decision 78 FR 3085) establishes a daily average of 150 µg m^{-3}, and an annual average of 50 μg m^{-3}. The WHO recommendations for PM$_{10}$ are more restrictive: a daily average of 50 μg m^{-3} and an annual average of 20 μg m^{-3}.

Regarding PM$_{2.5}$, the European Union has established a target annual average value of 25 μg m^{-3}. In the United States, the National Air Quality Standard for PM$_{2.5}$ is 35 μg m^{-3} (daily average), and the annual average is 12 μg m^{-3}. Again, the WHO recommends concentrations for PM$_{2.5}$ that are more restrictive than current legislations: a daily average of 25 μg m^{-3} and an annual average of 10 μg m^{-3}.

In addition to the size of the particles, the chemical composition of PM is another cause for its potential toxic effect. PM can contain a great variety of metals and metalloids associated with different natural and anthropogenic sources. A particular metal or metalloid can originate from more than one source. There are metals that are typical of the Earth's crust, such as Al, Mn, Rb, K, Fe, Ca, Sr, Ba, or Ti, whereas Na and Mg often originate from marine aerosols. On the other hand, the main anthropogenic sources are traffic and industry. The origin of metals and metalloids in PM due to traffic is related to combustion processes (Pb, Zn, Sb, V), brake-linings (Sb, Cu) and tyre abrasion (Ba, Mn). Others are related to combustion processes (Pb, Zn, Sb, Cd, and V). Also, the industrial activity present in a certain area can be responsible for comparatively large concentrations of some metals and metalloids. This is the case of copper smelting (As, Cu, Pb, Zn), the pigment industry (As, Cr, Mo, Ni, Cu, Co) and the steel industry (Pb, Zn, Mn, Cd, and As) [16].

7.2 Arsenic and antimony in PM

7.2.1 Toxicity, origin, and legislation of arsenic and antimony in PM

The metalloid arsenic (As) is an element that has no known biological function. The chronic and acute exposure of humans to arsenic oxyanions has been related to cancer, cardiovascular disease, neurological disorders, gastrointestinal disturbances, liver and renal diseases, reproductive health effects and dermal changes [17]. To this end, the International Agency for Research on Cancer (IARC) has classified inorganic arsenic within Group 1A, corresponding to substances that are carcinogenic to humans. On the other hand, antimony (Sb) is a less studied metalloid, which is not essential for life either. It is classified by IARC in Group 2B, as possibly carcinogenic to humans [18].

Arsenic is a constituent that can be found in PM. Natural emission sources of arsenic (e.g., volcanoes) are minor compared to anthropogenic ones, which account for about 97% of the arsenic that is emitted to the atmosphere. The main human activities that release arsenic to the atmosphere are metallurgy (copper, zinc, and lead smelters), coal-fired power plants, and waste incineration. These three sources together account for about 60% of the anthropogenic emissions [19].

Antimony is also found in the air associated with PM mainly due to anthropogenic emissions. The most import sources are fuel combustion related to traffic, metallurgy (iron, steel, and non-ferrous metal production) and waste incineration [20]. Regarding traffic, antimony is also a constituent of the brake pads in vehicles in which antimony compounds (i.e., Sb_2S_3) are used as a lubricant in friction material that is released in the form of particles during braking. This emission produces road dust that contains Sb [21, 22].

A health risk evaluation of arsenic in the air was performed by the WHO in 2000, where the relation between arsenic and lung cancer was evaluated. It was concluded that a safety level for the inhalation of arsenic cannot be recommended [23]. One of the first regulations about arsenic in air was established by the European Union (Directive 2004/107/EC), and mandated a target value of 6 ng m^{-3} of As in PM_{10} as the annual average, starting in 2013 [24]. China has adopted a National Ambient Quality Standard (NAAQS) (GB 3095–2012) with the same value of 6 ng m^{-3} [25], whereas the United States has not adopted a national regulation for arsenic in PM_{10}. Regarding antimony, there is not yet any national legislation about its concentration in air.

7.2.2 Sampling and chemical analysis of PM

The PM sampling procedures employ high-volume air samplers which are equipped with a pump that forces ambient air to pass through a size selective (e.g., PM_{10} or $PM_{2.5}$) inlet (Figure 7.3). Particles with larger diameters are trapped inside the inlet as the smaller particles migrate through the inlet, which are then retained on the surface of a filter. The volume of air sampled depends on the flow rate of the pump. The pump is programmable, allowing to select the sampling time. In most studies, the sampling time corresponds to 24 h, due to established norms (e.g., EN 12341) that regulate the gravimetric determination of PM_{10} or $PM_{2.5}$ [26]. Regarding the filters that retain the PM, they are commonly made of quartz, although other materials are also employed, such as Teflon, polycarbonate, or glass.

After sampling, the filters are submitted to a chemical treatment for the posterior analysis. The filters with PM are treated with strong acids, such as hydrofluoric (HF), nitric (HNO_3) and/or perchloric ($HClO_4$) acids. In this way, the chemical constituents of the sample, such as arsenic and antimony are dissolved from the solid PM into the acid. The acid treatment is often used in combination with heating or other forms of energy (e.g., microwave radiation) to accelerate the dissolution process [27].

The resulting acid solution can be analysed for total arsenic and antimony by several analytical techniques. The most common ones for the determination of both elements are based on atomic spectroscopy, in which a flame converts the elements into an atomic vapour that is capable of absorbing or emitting radiation at wavelengths

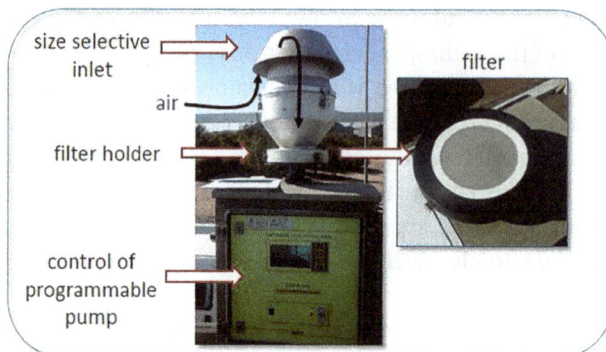

Figure 7.3: Left : device for airborne PM sampling. Right: quartz filter with retained PM (grey shade circle) after 24 of air sampling.

that are characteristic for each element. The main atomic spectroscopic techniques for arsenic and antimony determination are atomic absorption spectroscopy (AAS) and atomic fluorescence spectroscopy (AFS). Another technique which is widely employed is inductively coupled plasma-mass spectrometry (ICP-MS), where the acid solution is introduced into an argon plasma at a temperature of several thousand degrees, producing the vaporization and ionization of the elements. The formed ions are then separated and quantified according to their characteristic mass (m) to electric charge (z) ratio in a mass spectrometer.

7.3 Case study: Evolution of arsenic in PM related to metallurgical emissions

7.3.1 Study area

The city of Huelva is located on the southwest of mainland Spain in Andalusia and has about 145,000 inhabitants. Huelva lies at the confluence of the Odiel and Tinto Rivers on the Atlantic coast (Figure 7.4). The surrounding area of Huelva has been involved in mining and metallurgical activities over centuries as it is close to the so-called Iberian Pyrite Belt. The latter is an important mining region that has predominantly focused on copper, but also gold and silver since prehistoric times until today [28]. Moreover, several industrial estates were established around the city since the 1960s. They encompass a chemical and petrochemical complex, a phosphate-based fertilizer plant, and a metallurgical complex for copper production that involves copper smelting which is one of the largest in Europe [29].

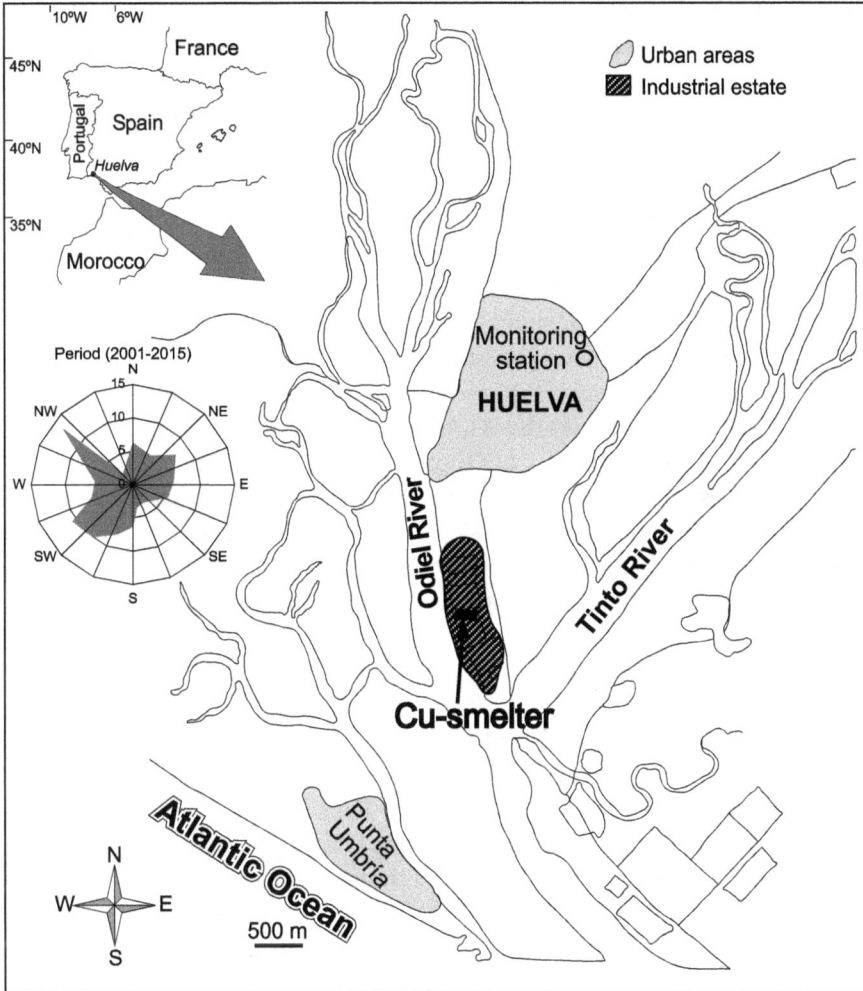

Figure 7.4: Location of the copper smelter and monitoring station in Huelva.

7.3.2 Copper metallurgy and air emissions

At the copper metallurgical complex, copper is obtained by a pyrometallurgical process (pyro, from the Greek word for fire), meaning that high temperatures are employed. It involves the crushing and drying of the raw material, which is a copper concentrate obtained from different sulphide minerals, such as chalcopyrite ($CuFeS_2$), bornite (Cu_5FeS_4), coveline (CuS), or tenantite ($Cu_{11}FeAs_4S_{13}$), among others. Copper concentrates typically contain about 20–30% of copper, 30–40% of sulphur, and 25–35% of iron. Arsenic is a frequent impurity that is present in copper concentrates, usually at concentrations less than 1% (i.e., several grams of As per t). Other impurities that are commonly found in copper concentrates are lead (Pb), cadmium (Cd), selenium (Se), bismuth (Bi), and nickel (Ni) [30–32].

The copper concentrate is submitted to a high temperature treatment in different types of furnaces. In the flash furnace the ore is smelted to form a matte enriched in copper and a waste called slag. After further treatment of the matte (e.g., in refining and moulding furnaces), the melted copper is cast into anodes with a purity of copper >99%, which are further purified by electrorefining. Simultaneously, part of the copper that is lost in the slag is recovered in an electric furnace [33].

In the different furnaces, sulphur (S) is converted into gaseous SO_2, a gas responsible for acid rain which is nowadays strictly regulated by air pollution legislation. Since the mid-twentieth century, the gaseous streams of copper smelters are mainly treated in acid plants to convert the SO_2 into H_2SO_4, a by-product that can be sold [33, 34]. During the whole process of Cu production, stack (chimney) and fugitive sources emit gases and particles with a fraction of the impurities that are originally present in the copper concentrate. Emission control of these contaminants is mandatory to prevent air pollution. The reduction of SO_2 and particles that are emitted from in the metallurgical industry is accomplished by implementing and enhancing various abatement technologies [35, 36].

7.3.3 Trend of arsenic in PM in Huelva

The main industrial air emissions stem from to the southern part of Huelva. This means that due to the predominant wind direction from the southwest, the air quality of this city can be impacted. This fact has motivated detailed studies on air quality during the last two decades, considering TSP, PM_{10} and $PM_{2.5}$ [29, 37, 38]. These studies have identified arsenic and other sulphide-related elements (e.g., Pb, Se, Bi, and Cd) as geochemical anomalies in the chemical composition of the PM of Huelva, meaning that their concentrations were above the background level [39]. Additional results that were obtained at the regional level indicated that the highest As concentrations in PM_{10} in Andalusia corresponded to the Huelva area (5.1–6.5 ng m^{-3}) with lower concentrations (0.7–1.8 ng m^{-3}) found in other cities in Andalusia [40].

The content of arsenic in the air of Huelva has been monitored during the last 20 years in both coarse particles (PM_{10}) and fine particles ($PM_{2.5}$) in a monitoring station located within the city limits (Figure 7.4). During this period, the annual average target value of 6 ng m^{-3} in PM_{10} was exceeded prior to the introduction of the European Directive that came into force in 2013. Therefore, the operators of the copper smelter addressed this problem by introducing abatement technologies to reduce its emissions of PM. These abatement technologies include the treatment of the gaseous streams with different types of filters (e.g., bag filters, wet electrofilters), scrubbers, or the injection of lime [$Ca(OH)_2$] into the stack exhaust.

The PM_{10} samples, each corresponding to a 24 h sampling period, were submitted to chemical analysis. A portion of the of quartz filter containing particulate matter was digested with acid and the concentration of metals and metalloids was determined by ICP-MS. The results of the chemical composition of PM_{10} samples collected between 2001 and 2020 (Table 7.1) in the nearby monitoring station of Huelva (distance of around 3 km) were classified into three periods, corresponding to before, during, and after the implementation of the abatement technologies.

Table 7.1: Mean and maximum concentrations of arsenic and other sulphide-related elements (ng m^{-3}) in PM_{10} samples from Huelva, prior (2001–2008), during (2009–2013) and after (2014–2020) emission abatement technology implementation at the copper smelter. PM_{10} units are µg m^{-3}. N indicates number of samples.

time period N	2001–2008 444		2009–2013 322		2014–2020 463	
	mean	max.	mean	max.	mean	max.
PM_{10}	37	171	29	72	27	135
As	6.3	62.1	5.2	71.2	2.9	26.8
Se	1.7	26.6	1.0	34.3	0.3	9.61
Cd	0.7	14.1	0.6	15.6	0.3	3.4
Sn	2.2	13.5	1.9	13.2	1.4	8.5
Sb	2.0	78.9	1.1	31.0	0.8	7.5
Pb	21.0	231	11.8	130	8.0	108
Bi	1.0	15.7	0.8	19.0	0.4	4.4

For the first period (2001–2008), the average arsenic concentration was comparatively high (6.3 ng m^{-3}), indicating that for some years the annual mean was likely above the annual mean target value of 6 ng m^{-3} stipulated by the European Directive. During this period, some daily PM_{10} samples showed extremely high arsenic concentrations of up to 62 ng m^{-3}. The industrial origin of the arsenic is in accord with the high concentration of this and other elements present in the raw material that was processed by the copper smelter. In this sense, Pb was also found in PM_{10} at a high mean concentration of 21 ng m^{-3}. Other elements, such as Se, Cd, Sn, Sb, and Bi showed mean concentrations ranging from 0.7–2.2 ng m^{-3}.

During the period of implementation of abatement technologies, the mean concentration of all pollutants in PM_{10} tended to decline. Afterwards, in the 2014–2020 period, the mean arsenic concentration dropped to 2.8 ng m^{-3}, which represents a 55% of reduction compared to the initial value. Similarly, the concentration of Pb dropped to 6.7 ng m^{-3}, and the remaining pollutants mean concentrations ranged 0.3–1.4 ng m^{-3}. Overall, the reduction in the concentration of the different pollutants was in the 69–75% range, while lower results were observed for Sb (49%) and Sn (28%). This indicates that for both of the latter elements other emission sources have to be considered. A recent study has pointed out that these elements may originate also from traffic emissions [41]. This means that further reduction of these two pollutants requires the implementation of alternative approaches in addition to the existing industrial abatement technologies.

In this context one may want to know if there is a correlation between arsenic and the other pollutants with the size of the PM particles. The results of a study that investigated $PM_{2.5}$ (Table 7.2) indicated the same trend considering the three time periods, with a reduction from an initial value of 5.2 ng m^{-3} for arsenic to 2.4 ng m^{-3} for the last period. The relatively high concentration of As in $PM_{2.5}$ compared to PM_{10} shows that it concentrates preferentially (up an 82%) in the finer fraction. Most of the other pollutants (Se, Cd, Pb, and Bi) also accumulated preferentially in the finer fraction (ca. 60–80%), thus representing an additional health risk. The main exception was Sb, which accumulated preferentially in the coarse fraction and indicates another major source for this element, such as traffic.

Table 7.2: Mean and maximum concentrations of arsenic and other sulphide-related elements (ng m^{-3}) in $PM_{2.5}$ samples from Huelva, prior (2001–2008), during (2009–2013) and after (2014–2020) emission abatement technology implementation at the copper smelter. $PM_{2.5}$ units are µg m^{-3}. N indicates number of samples.

time period	2001–2008		2009–2013		2014–2020	
N	377		317		373	
	mean	max.	mean	max.	mean	max.
$PM_{2.5}$	20	104	21	78	19	107
As	5.23	60.3	4.6	48.9	2.4	19.3
Se	1.15	16.9	1.0	76.7	0.2	5.9
Cd	0.62	12.7	0.5	18.9	0.2	2.2
Sn	1.3	11.0	1.2	10.3	0.8	3.0
Sb	0.7	6.3	0.7	27.6	0.5	7.5
Pb	15.8	164	10.8	87.7	6.1	51.2
Bi	0.8	16.3	0.6	11.7	0.4	3.4

The detailed trend for arsenic in each year between 2001 and 2020 indicated that the maximum mean annual concentrations of ≥ 10 ng m^{-3} in PM_{10} were reached in 2005 and 2006 (Figure 7.5). In the following years the mean annual concentration decreased to values above the target value of 6 ng m^{-3}. Since 2014 the concentration of arsenic in PM_{10} has 'steadily decreased and remained constant around 3 ng m^{-3}. The similarity in the shape of the arsenic in PM$_{2.5}$ and PM$_{10}$ graphs in Figure 7.5 reveals that the emission of fine particles from the copper smelter is the main arsenic emission source in Huelva. The fact that PM$_{2.5}$ represents about 70% of the arsenic emission indicates that this fraction represents an important human exposure source to this pollutant, as finer particles are more amenable to enter the respiratory system than coarse particles.

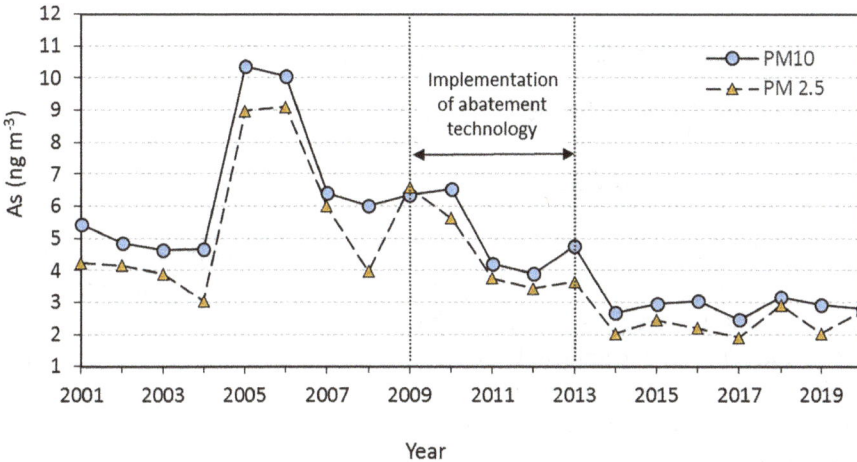

Figure 7.5: Trend of arsenic mean annual concentration in PM$_{10}$ and PM$_{2.5}$ at the monitoring station in Huelva.

These results highlight the importance of long-term monitoring of air pollutants to assess if legislatory means to prevent air pollution are met. Thus, the suitability of abatement technologies in industrial processes to mandate health and environmental protection was practically achieved. In this particular study on the emissions from a copper smelter it is clear that there is room for improvement. For instance, the arsenic concentration is still above the target value on some days, although the annual average is about 3 ng m^{-3}. This means that most of the time the concentration of arsenic found in the particulate matter of Huelva is low, but that there are still some days when 13–21 ng m^{-3} in PM$_{10}$ or 12–19 in PM$_{2.5}$ are encountered.

Another question that arises is if the change of the chemical composition of the airborne particles has been associated with a similar change in the levels of PM$_{10}$ or PM$_{2.5}$. In relation to PM$_{10}$ (Table 7.1), its concentration has diminished only to a certain extent, from 37 µg m^{-3} to 27 µg m^{-3}. This represents a 29% reduction and is lower than the overall reduction of the As and sulphide-related elements. Also, the concentration of PM$_{2.5}$ has remained constant around 20–21 µg m^{-3} (Table 7.2). This is an indication that the industrial source of both PM fractions represents a minority compared to other sources, such as particles that originated from traffic, the resuspension of dust from the earth crust's, marine aerosols, or secondary inorganic compounds (SIC). It has been estimated that the emissions of the industrial estate where the copper smelter is located accounts for 5–12% of the PM_{10} found in Huelva during the study period [37, 42].

The results of this study reveal that although industrial emissions are not the main source of PM, they are responsible for significant levels of arsenic and other sulphide-related elements in the airborne particulate matter of Huelva.

7.4 Case study: Antimony sources in PM due to traffic and smelting emissions

7.4.1 Study area and PM sampling

Cordoba and Granada are also located in Andalusia, with populations of about 200,000 and 300,000 inhabitants, respectively. As in most cities, the number of cars, buses, and motorcycles is considerable, as 220,000 vehicles are registered in Cordoba and 170,000 in Granada. In addition to traffic, Cordoba also has an industrial park which includes brass manufacturing (Cu and Zn alloy) and metallurgy located at the west of the city. The main products are brass ingots, as well as Cu bars, Cu rods, and the processing of Zn/Pb scrap.

Considering the Sb emissions sources related to PM, traffic should be the main one in Granada, whereas in Cordoba, in addition to traffic, there is a potential second source, namely the brass industry, as Sb is an impurity present in the raw material for brass production. Therefore, PM_{10} samples were collected at some monitoring stations, three in Cordoba and one in Granada (Figure 7.6). The PM_{10} samples of the station in Granada were collected between 2007 and 2013 and are mainly affected by local traffic. In Cordoba, the three stations selected were classified as traffic, urban background, and industrial, depending on their location in relation to the possible origin of particulate matter. The sampling periods in Cordoba were 2013 (traffic station), 2007–2013 (urban background), and 2012–2013 (industrial estate). In addition to Sb, some metals (Zn, Cu, Pb, and Cd) that are related to anthropogenic activities were also determined in the PM_{10} samples [43].

7.4.2 Chemical composition of PM_{10} collected in Cordoba and Granada

The concentration of the studied metals in PM_{10} of Cordoba highlights the importance of industrial emissions of the brass industry in this city. Mean concentrations of Zn (2743 ng m^{-3}), Cu (115 ng m^{-3}), Pb (66.8 ng m^{-3}), and Cd (4.09 ng m^{-3}) were found in the PM_{10} samples of the industrial monitoring station of Cordoba (Table 7.3). The mean concentration of these metals in the other two monitoring stations (traffic and

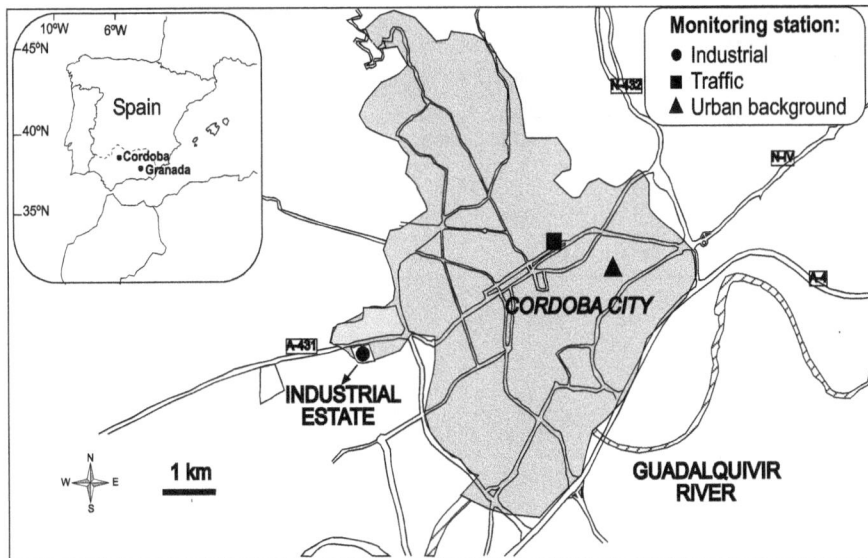

Figure 7.6: Location of the industrial estate and monitoring stations in Cordoba.

urban) of Cordoba were also high, indicating that the industrial activity also affects these areas. This is highlighted if we compare the data of the traffic station of Cordoba with the one in Granada. The results from these two traffic stations showed that higher mean concentrations of Zn, Cu, Pb and Cd were consistently present in Cordoba (1.5–10 times higher) than in Granada.

Regarding Sb, the analysis of the samples (Table 7.3) showed its presence in the PM_{10} samples collected at the four monitoring stations. Moreover, the mean concentrations were not very different, as they ranged 2.41–4.58 ng m^{-3} for the Cordoba stations, and a similar mean concentration (4.02 ng m^{-3}) was also registered in Granada, indicating the importance of traffic as an important Sb emission source. However, just the mean concentration of Sb does not help to distinguish between both possible emission sources (traffic and industry) of particulate matter containing Sb. Maximum daily concentrations of Sb did not help either to discriminate the traffic or industrial origin of the Sb, as similar maximum values were found at the industrial monitoring station in Cordoba (35.1 ng m^{-3}) and the traffic monitoring station in Granada (36.8 ng m^{-3}).

7.4.3 Speciation of Sb in PM_{10} of Cordoba and Granada

A different approach was sought to elucidate the origin of Sb by performing a speciation analysis to differentiate the Sb species according to their oxidation state, namely Sb(III) and Sb(V). The presence of both Sb species in PM is expected, as

Table 7.3: Mean and maximum concentrations of antimony and metals (ng m^{-3}) in PM$_{10}$ in three monitoring stations of Cordoba (industrial, traffic and urban) and one in Granada (traffic). PM$_{10}$ units are µg m^{-3}. N indicates number of PM$_{10}$ samples.

time period N	Industrial Cordoba		Traffic Cordoba		Urban Cordoba		Traffic Granada	
	2012–2013 21		2013 20		2007–2013 346		2007–2013 352	
	mean	max.	mean	max.	mean	max.	mean	max.
PM$_{10}$	28	93	31	48	33	242	42	249
Sb	2.43	35.1	4.58	13.4	2.41	109	4.02	36.8
Zn	2743	23,236	346	2486	211	3173	34.9	557
Cu	115	996	92.6	151	111	1051	61.3	365
Pb	66.9	660	14.8	76.9	12.0	79.6	6.3	72.5
Cd	4.09	56	0.53	2.99	0.70	14.2	0.14	1.28

there are studies that demonstrate their presence in dust particles that originate from the automotive brake abrasion [22, 44]. In this case, a different chemical treatment of the samples is necessary. We have to bear in mind that a strong acid treatment (addition of concentrated HF, HNO$_3$, and HClO$_4$) and heating is employed to determine the total Sb content of the PM samples. This strong acid treatment of the quartz filters dissolves the particulate matter that it contains, but also produces changes in the proportion of Sb(III) and Sb(V) due to the oxidizing character of the employed acids. Therefore, the information regarding the original distribution of Sb (III) and Sb(V) in the samples is lost.

Speciation analysis refers to the analytical activities of identifying and measuring species of an element, as defined by IUPAC [45]. In the present study, these activities involve an extraction procedure to quantitatively extract the Sb species from the particulate matter. At the same time the procedure has to guarantee that the molecular form of the Sb species does not change and that the extracting media is compatible with the posterior analytical methodology. In this sense, the procedure developed in this study consisted in cutting a portion of the filter containing the PM sample and adding an appropriated solvent, namely a hydroxyl amine solution NH$_2$OH · HCl 0.05 mol L^{-1}. The extraction was accelerated by heating (6 min) in a microwave oven operated at a low power setting (90 W).

Afterwards, the extract was submitted to analysis. In the case of speciation analysis it is required to combine a separation technique that separated the individual Sb species with an atomic detection technique to quantify the Sb species. In this sense, a liquid chromatohgraphic separation system (HPLC, high-performance liquid chromatography) was combined with hydride generation followed by AFS [46]. A small portion of the extract was injected onto an ion exchange column of the chromatographic system. Using an ammonium tartrate solution (0.2 M, pH 5) as the mobile phase the Sb species migrated through the column at different speeds. At the outlet of the column, solutions of HCl and a reductant ($NaBH_4$) were added, so the Sb species are converted into a gaseous molecule (SbH_3). The latter molecules are then transported with an inert gas (argon) to a hydrogen flame, where atomization occurs and where emission light is then recorded as function of time to obtain the Sb-specific chromatogram.

Figure 7.7 shows a typical chromatogram of a calibration solution which contains both Sb species. Each individual Sb species can be identified according to its retention time, which is the time that elapses between the injection of the sample and the maximum of the signal is recorded. The retention times for Sb(III) and Sb(V) using this particular analytical setup were 1.6 min and 4.9 min, respectively. On the other hand, the quantification of the concentration of Sb species is performed by measuring the area of the signals. Extracts with high concentrations will yield high areas. Therefore, a calibration curve is obtained by plotting the areas of the signals versus the concentration of known standards of Sb. Thereafter, PM_{10} samples were extracted and analysed with the described procedure. The speciation analysis of the PM samples indicated that Sb(V) was the main Sb species in all samples. A typical chromatogram of sample from Cordoba is shown in Figure 7.8.

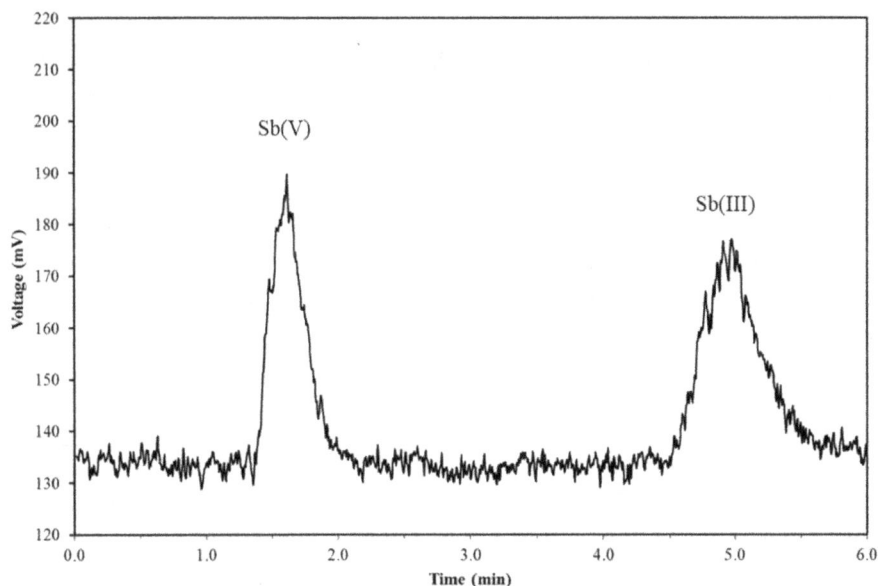

Figure 7.7: Chromatogram corresponding to a calibration solution of Sb(III) and Sb(V), obtained by HPLC coupled to atomic fluorescence spectroscopy.

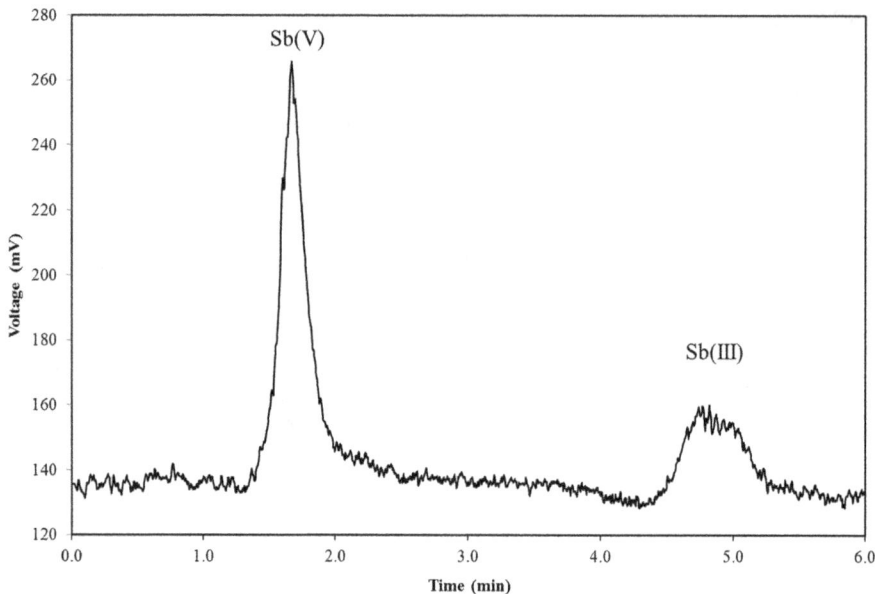

Figure 7.8: Chromatogram corresponding to an extract of PM_{10} sample from Cordoba, obtained by HPLC coupled to atomic fluorescence spectroscopy.

The speciation analysis of the PM_{10} samples showed that the relative percentage of Sb(III) and Sb(V) varied among the different monitoring stations, thus indicating that this parameter can be used as a fingerprint to distinguish the importance of both emission sources of Sb. The results of the industrial station of Cordoba corresponded to a higher percentage of Sb (85–85%) that was extracted. This percentage of Sb(V) decreased at the urban (74–77%) and traffic (73%) of this city. The lowest percentage of Sb(V) was found in the traffic station of Granada (64–69%). The fact that at the urban and traffic stations of Cordoba the percentages of Sb(V) were higher than in Granada indicates that the industrial emission is affecting the overall air quality of Cordoba.

The results indicate that traffic is associated with a smaller fraction of Sb(V) than the industrial source of the brass industry. This may be due to the different material composition of the cars brakes and the raw material employed during smelting. Also, higher temperatures are reached in smelting than in car braking, which favours the oxidation and prevalence of Sb(V) in the industrial emission. Similarly, the percentage of Sb(III) increased from a minimum of 14–15% at the industrial monitoring station of Cordoba to a maximum 31–36% at the traffic station of Granada, as is depicted in Figure 7.9.

To confirm this hypothesis, particulate matter was sampled at the industrial site where brass was manufactured. Ten total suspended particles (TSP) samples were collected in order to obtain a chemical profile of all the range of particles that

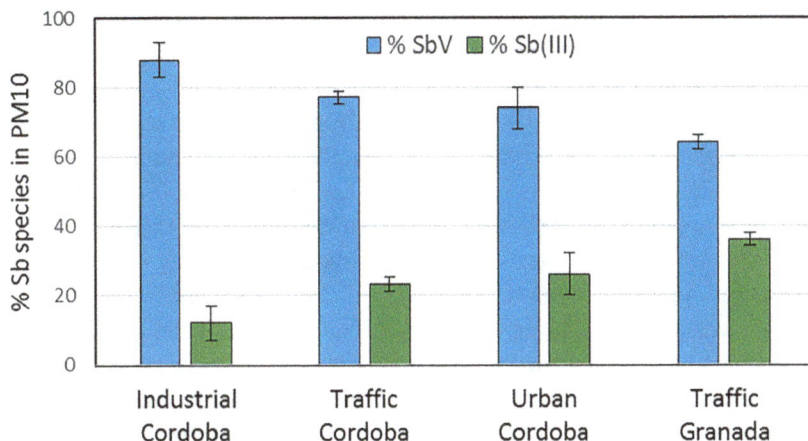

Figure 7.9: Percentages of Sb(III) and Sb(V) in the extracts of PM$_{10}$ samples from air monitoring stations in Cordoba and Granada.

Table 7.4: Total Sb and Sb speciation (Sb(III), Sb(V)) concentration (ng m^{-3}) of PM$_{10}$ samples at different monitoring stations and periods: industrial (2012–2013), urban (2011–2013) and traffic (2013) in Cordoba and traffic (2011–2013) in Granada. Concentrations correspond to mean ± standard deviation. N indicates number of samples.

Monitoring station	N	Sb$_{total}$	Sb(III)	Sb(V)	Sb extracted	% Sb(III)	% Sb(V)
Industrial Cordoba	7	6 ± 1	0.7 ± 0.4	4.8 ± 0.9	5.4 ± 0.9	14 ± 5	86 ± 5
	13	11 ± 10	2 ± 1	9 ± 8	11 ± 8	15 ± 4	85 ± 4
Urban Cordoba	8	9 ± 5	2 ± 1	6 ± 3	8 ± 3	23 ± 6	77 ± 6
	4	4.7 ± 0.8	1.2 ± 0.3	3.7 ± 0.7	4.9 ± 0.8	25 ± 2	75 ± 2
	3	5 ± 2	1.3 ± 0.2	4 ± 2	5 ± 2	26 ± 5	74 ± 5
Traffic Cordoba	5	6 ± 1	1.6 ± 0.3	4.3 ± 0.6	5.9 ± 0.7	27 ± 2	73 ± 2
Traffic Granada	8	5.9 ± 0.7	1.8 ± 0.2	4.0 ± 0.6	5.8 ± 0.6	31 ± 2	69 ± 2
	7	5.2 ± 0.7	1.8 ± 0.2	3.3 ± 0.5	5.1 ± 0.5	36 ± 2	64 ± 2
	9	6.2 ± 0.9	2.0 ± 0.3	4.0 ± 0.7	6.0 ± 0.8	33 ± 2	67 ± 2

are emitted during the smelting process. The analysis of the total Sb content of the samples was high, with a range of 27.3–120 ng m^{-3}. The subsequent speciation analysis of these samples provided a percentage of Sb(V) of 84–88%, which was very similar to that found at the industrial monitoring stations.

The results of this case study regarding Sb in PM indicate that the total element content is not enough to differentiate between to emission sources. In this case, speciation analysis of the oxidation states of Sb can be used as a fingerprint to highlight the presence and importance of two different emissions sources affecting the air quality.

In summary, PM is an important issue that directly impacts human health. Factors that play an important role in assessing the adverse effect of PM in humans include the concentration of particles in the air, their size, and their chemical composition. Studies into the latter factor have shown that PM is a considerable source for human exposure to toxic metals and metalloids, such as arsenic and antimony, which mainly originate from anthropogenic sources. This insight highlights the importance to perform studies to evaluate the exposure of the population to local sources of these contaminants, such as industry and/or traffic. State-of-the-art analytical methodologies allow to determine the concentrations of As and Sb in PM, which thus provides important information to estimate the human health risk. Speciation analysis can identify the oxidation state of these pollutants, which can help to identify their sources.

There are also some issues that need to be considered in the future that are related to the concentration of As and Sb in PM. For As, for example, contemporary policies only consider mean annual target concentration in PM_{10}. Future studies should address the possible importance of daily or hourly high concentration of As in air, which can represent a health issue that has not yet been fully evaluated. Another issue is the size of the particles because at present only PM_{10} are legislated, so scientists and policy makers should consider the health risk of As in fine or ultrafine particles. In the case of Sb at present there is not even a mean annual target concentration. Therefore, a scientific consensus about the urgency to solve these problems should be reached and studies for their execution promoted.

Suggested readings

Doug B. Particles in the Air. The Deadliest Pollutant Is One You Breathe Every Day. Springer International Publishing AG, Cham, Switzerland, 2018.

Harrison RM, Hester RE, Querol X. Airborne Particulate Matter: Sources, Atmospheric Processes and Health. Volume 42. Royal Society of Chemistry, Cambridge, UK, 2006.

Holgate ST, Samet JM, Koren HS, Maynard RL. Air Pollution and Health. Academic Press, London, UK, 1999.

Seigneur C. Air Pollution: Concepts, Theory, and Applications. Cambridge University Press, Cambridge, UJ. 2019.

Straif K, Cohen A, Samet J. Air Pollution and Cancer. IARC Scientific Publication No. 161. International Agency for Research on Cancer. World Health Organization. Lyon, France, 2013.

References

[1]   Finlayson-Pitts BJ, Pitts J. Chemistry of Upper and Lower Atmosphere: Theory, Experiments and Applications. 1st ed. Academic press, San Diego, CA, USA, 2000.

[2]   Seinfeld JH, Pandis SN. Atmospheric Chemistry and Physics. 3rd ed. Wiley, Hoboken, NJ, USA, 2016.

[3]   World Health Organization (WHO). Review of Evidence on Health Aspects of Air Pollution – REVIHAAP Project. WHO Regional Office for Europe, Copenhagen, Denmark, 2013.

[4]   Kumar P, Morawska L, Birmili W, Paasonen P, Hu M, Kulmala M, Harrison RM, Norford L, Britter R. Ultrafine particles in cities. Environ Int. 2014, 66, 1–10.

[5]   United States Environmental Protection Agency (US EPA). Particulate Matter (PM) Pollution. (2021. Accessed July 21, 2021, at https://www.epa.gov/pm-pollution/particulate-matter-pm-basics)

[6]   European Council (EU). Directive relating to limit values for sulphur dioxide, nitrogen dioxide and oxides of nitrogen, particulate matter and lead in ambient air. O J L 163, (1999/30/EC) 0041–0060.

[7]   Brimblecombe P, Maynard E. The Urban Atmosphere and Its Effects. Air Pollution Reviews. Imperial College Press, London, UK. Vol. 1, 2002.

[8]   Anderson JO, Thundiyil JG, Stolbach A. Clearing the air: A review of the effects of particulate matter air pollution on human health. J Med Toxicol 2012, 8(2),166–175.

[9]   Landrigan PJ, Fuller R, Acosta NJR, et al. The Lancet Commission on pollution and health. The Lancet Commissions. 2018, 391, 10119, 462–512.

[10] United States Environmental Protection Agency (US EPA). Air Quality Criteria for Particulate Matter. (Final report, Oct 2004) United States environmental protection agency, Washington, DC, EPA 600/P-99/002aF-bF, 2004. (Accessed July, 21, 2021, at https://cfpub.epa.gov/ncea/risk/recordisplay.cfm?deid=87903)

[11] Brook R, Rajagopalan S, Pope C, et al. Particulate matter air pollution and cardiovascular disease an update to the scientific statement from the American heart association. Circulation. 2010, 121, 2331–2378.

[12] United States Environmental Protection Agency (US EPA). Particle pollution and your health. Document EPA-452/F-03-001, 2003.

[13] Dockery DW, Pope C, Xu X, et al. An association between air pollution and mortality in six U.S. cities. New Eng J Med. 1993, 24, 1753–1759.

[14] Pope CA, Thun MJ, Namboodiri MM, et al. Particulate air pollution as a predictor of mortality in a prospective study of U.S. adults. Am J Respir Crit Care Med. 1995, 151, 669–674.

[15] CAFE: Baseline scenarios for the clean air for Europe (CAFE) programme: CAFE scenario analysis report Nr.1. International Institute for Applied Systems Analysis, Laxenburg, Austria, 2005. (Accessed July 7, 2021, at http://ec.europa.eu/environment/archives/cafe/activities/pdf/cafe_scenario_report_1.pdf)

[16] Moreno T, Querol X, Alastuey A, et al. Variations in atmospheric PM trace metal content in Spanish towns: Illustrating the chemical complexity of the inorganic urban aerosol cocktail. Atmos Environ. 2006, 40, 6791–6803.

[17] Jomova K, Jenisova Z, Feszterova M, et al. Arsenic: Toxicity, oxidative stress and human disease. J Applied Toxicol. 2010, 31, 95–107.

[18] International Agency for Research on Cancer (IARC). IARC Monographs on the identification of carcinogenic hazards to humans. (Accessed July, 21. 2021, at https://monographs.iarc.who.int/agents-classified-by-the-iarc)

[19] Wai KM, Wu S, Li X, Jaffe DA, Perry KD. Global atmospheric transport and source-receptor relationships for arsenic. Environ Sci Technol. 2016, 50, 3714–3720.

[20] Tian H, Zhou J, Zhu C, et al. A comprehensive global inventory of atmospheric antimony emissions from anthropogenic activities, 1995–2010. Environ Sci Technol. 2014, 48, 10235–10241.

[21] Von Uexküll O, Skerfving S, Doyle R, Braungart M. Antimony in brake pads a carcinogenic component? J Clean Prod. 2005, 13, 19–31.

[22] Iijima A, Sato K, Yano K, et al. Particle size and composition distribution analysis of automotive abrasion dust for evaluation of antimony sources of airborne particulate matter. Atmos Environ. 2007, 41, 4908–4919.

[23] World Health Organization (WHO). Air Quality Guidelines for Europe: Second Edition. WHO Regional Publications, European Series, No. 91. Copenhagen, Denmark, 2000.

[24] European Council (EU). Directive 2004/107/EC, As, cadmium, mercury, nickel, and polycyclic aromatic hydrocarbons in ambient air, O J L 2005, 23, 26.1.2005.

[25] Duan J, Tan J. Atmospheric heavy metals and arsenic in China: Situation, sources and control policies. Atmos Environ 2013, 74, 93–101.

[26] European Standard EN 12341. Standard Gravimetric Measurement Method for Determination of PM_{10} or $PM_{2.5}$ Mass Concentration of Suspended Particulate Matter. European Committee for Standarization (CEN), Brussels, Belgium, 2015.

[27] Sánchez-Rodas D, Sánchez de La Campa SM, Alsioufi L. Analytical approaches for arsenic determination in air: Critical review. Anal Chim Acta. 2015, 808, 1–18.

[28] Sáez R, Pascual E, Toscano M, Almodóvar GR. The Iberian type of volcano-sedimentary massive sulphide deposit. Miner Depos. 1999, 34, 549–570.

[29] Querol X, Alastuey A, de La Rosa J, Sánchez de La Campa A, Plana F, Ruiz RC. Source apportionment analysis of atmospheric particles in an industrialized urban site in southwestern Spain. Atmos Environ. 1999, 35, 3113–3125.

[30] European Copper Institute. 2014. Copper Concentrates. Environmental and human health hazard classification. (Accessed July, 21. 2021, at https://copperalliance.es/resources/copper-concentrates/)

[31] Muthumariappann S, David SA. A study of the chemical composition of copper concentrate and granulated slag. Int J Sci Res Rev. 2019, 07(03), 539–548.

[32] Pérez-Tello M, Parra-Sánchez VR, Sánchez-Corrales VM, et al. Evolution of size and chemical composition of copper concentrate particles oxidized under simulated smelting conditions. Metalll Mater Trans B. 2018, 49B, 627–643.

[33] Schlesinger ME, King MJ, Sole KC, Davenport WG. Extractive Metallurgy of Copper. 5th ed. Elsevier, Oxford, UK, 2011.

[34] Semrau KT. Control of sulfur oxide emissions from primary copper, lead and zinc smelters–A critical review. J Air Pollut Control Assoc. 1971, 21(4), 185–194.

[35] Liu Q, Wang Q. How China achieved its 11th Five-Year emission reduction target: A structural decomposition analysis of industrial SO_2 and chemical oxygen demand. Sci Tot Environ. 2017, 574, 1104–1116.

[36] Beavington F, Cawse PA, Wakenshaw A. Comparative studies of atmospheric trace elements: Improvement in ail quality near a copper smelter. Sci Total Environ. 2004, 332, 39–49.

[37] Alastuey A, Querol X, Plana F, et al. Identification and chemical characterization of industrial particulate matter sources in southwest Spain. 2008. J Air Waste Manage Assoc. 2008, 56, 993–1006.

[38] González-Castanedo Y, Sánchez-Rodas D, Sánchez de La Campa AM, et al. Arsenic species in atmospheric particulate matter as tracer of the air quality of Doñana Natural Park (SW Spain). Chemosphere. 2015, 119, 1296–1309.

[39] Sánchez de La Campa AM, Sánchez-Rodas D, González Castanedo Y, de La Rosa JD. Geochemical anomalies of toxic elements and arsenic speciation in airborne particles from Cu mining and smelting activities: Influence on air quality. J Hazard Mater 2015; 201: 18–27.

[40] de La Rosa JD, Sánchez de La Campa AM, Alastuey A, et al. Using PM_{10} geochemical maps for defining the origin of atmospheric pollution in Andalusia (Southern Spain). Atmos Environ. 2010, 44, 4595–4605.

[41] Millán-Martínez M, Sánchez-Rodas D, Sánchez de La Campa AM, Alastuey A, Querol X, de La Rosa JD. Source contribution and origin of PM_{10} and arsenic in a complex industrial region (Huelva, SW Spain). Environ Pollut. 2021, 274, 116268.

[42] Sánchez de La Campa AM, de La Rosa J, Querol X, Alastuey A, Mantilla. Geochemistry and origin of PM_{10} in the Huelva region, Southwestern Spain. Environ Res. 2007, 103, 305–316.

[43] Sanchez-Rodas D, Alsoufi L, Sánchez de La Campa AM, González-Castanedo Y. Antimony speciaction a geochemical tracer for anthropogenic emissions of atmospheric particulate matter. J Hazard Mater. 2017, 324, 213–220.

[44] Iijima A, Sato K, Ikeda T, Sato H, Kozawa K, Furuta N. Concentration distributions of dissolved Sb(III) and Sb(V) species in size-classified inhalable airborne particulate matter. J Anal At Spectrom. 2010, 25, 356–363.

[45] Templeton DM, Ariese F, Cornelis R, et al. Guidelines for terms related to chemical speciation and fractionation of elements. Definitions, structural aspects, and methodological approaches. Pure Appl Chem. 2000, 72(8), 1453–1470.

[46] AlSioufi L, Sánchez de La Campa AM, Sánchez-Rodas D. Microwave extraction as an alternative to ultrasound probe for antimony speciation in airborne particulate matter. Microchem J. 2016, 124, 256–260.

Eve Kroukamp and Victor Wepener

Chapter 8
Toxic trace metals in the environment, a study of water pollution

Objectives

This chapter will explore
- The sources of metals in surface waters
- physico-chemical interactions of metals with the receiving environment and how these affect the mobility, bioavailability, and ultimate fate of these contaminants
- The ecological impact of metals as illustrated by case studies of the effects of metal pollution on single organisms and whole ecosystems
- strategies toward treating industrial waste, the mitigation of the entry of these pollutants into the environment and the remediation of impacted environments.

By the end of the chapter the reader shall be able to:
- Define what are essential and non-essential metals
- Identify sources of metals in the aquatic environment
- Define a physico-chemical interaction and how this impact affects the mobility, bioavailability, and ultimate fate of metals
- Identify ecological impacts of metals in the environment and how this impacts single organisms and/or whole ecosystems
- Identify strategies for treating industrial waste
- Understand how the entry of metals into the environment can be mitigated
- Identify a few means by which impacted environments can be remediated

8.1 Introduction

Since metals and metalloids cannot be broken down and are persistent in nature, pollution involving these elements is one of the most significant challenges facing the environment and society today. For many years, "dilution is the solution to pollution" has been a strategy employed by industry and other polluters alike to

Eve Kroukamp, Victor Wepener, PerkinElmer Inc., Canada, North-West University, South Africa,
e-mail: Eve.Kroukamp@PerkinElmer.com and e-mail: evekroukamp@gmail.com

https://doi.org/10.1515/9783110626285-008

address the issue of waste management. This strategy assumes that if you dilute a contaminant to low concentrations then it no longer poses as a threat. We know now that although single element species under controlled conditions may have relatively predictable toxicity, the problem is that natural waterways are extremely complex and dynamic systems. Consequently, even low concentrations of metal/loids may be toxic to some organisms and may be transformed into more toxic forms that bioaccumulate and biomagnify, two terms which will be described later in this chapter.

Often when people refer to contamination of water by toxic metal/loids, they refer to this type of contamination as being "heavy metal contamination", a term which is generally not recommended. The reason being is that when this term was first coined it was used to describe metals such as mercury and lead which had a high atomic mass (200.59 g/mol and 207.2 g/mol respectively) and therefore the naming was appropriate. In more recent years, however, the term heavy metal contamination has been inappropriately extended to include metals such as Cu (63.55 g/mol) and metalloids, such as arsenic (74.92 g/mol), which have a comparatively lower low atomic mass and in the case of the latter not being a metal at all.

Currently, metal/loids such as arsenic, cadmium, chromium, lead, mercury, selenium, tin, and radionuclides are identified as being the elements of the greatest ecological and toxicological concern because even extremely low concentrations (µg/L or ng/L) can elicit a toxic response in organisms. Moreover, elements such as elemental Hg^0 and others can undergo biotransformation through enzymatic reactions into a more toxic and mobile form,which may undergo biomagnification in the food chain.

8.2 What are metals and metalloids

Metals form the largest group of chemical elements in the biosphere ("biosphere" being defined as all parts of the Earth that forms of life inhabit, i.e., all ecosystems). The periodic table provides a framework in which all elements are arranged according to their electron configurations (Figure 8.1). It is these electron configurations that are mainly responsible for the chemical behavior of the elements. The elements on the left of the periodic table are referred to as metals and display similar properties; i.e., they are good conductors of heat and electricity, they readily lose electrons, they oxidize in air and seawater and with the exception of Hg, they are solid at room temperature. There are several elements that exhibit one or more of the properties mentioned above (e.g., As and Te), which are called metalloids and represent the border to the non-metals (elements in blue in Figure 8.1), which includes selenium. Metals are further classified as alkali metals (highly reactive, e.g., Na and K), alkaline earth metals (less reactive, e.g., Mg and Ca) and transition metals (that form colored complexes, e.g., Fe, Au, Pt). For the purposes of this chapter both metals and metalloids are referred to as metals.

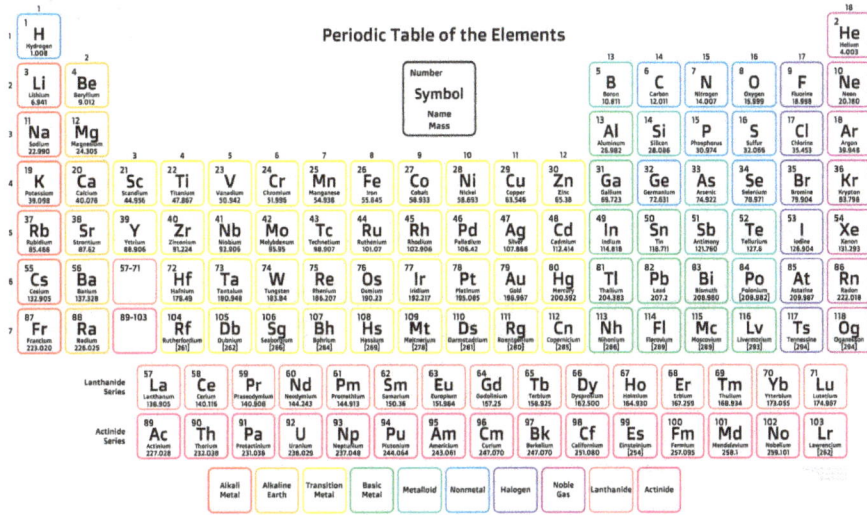

Figure 8.1: Periodic table of elements indicating metals on the left and metalloids (highlighted in blue) that effectively represent a border from the non-metals [1].

8.2.1 Essential and non-essential metals

Due to the fact that 90 chemical elements make up the earth's crust, metals are readily available for uptake by plants, animals, and humans via direct uptake from the environment and food. Metals that can be taken up by organisms can be categorized as either essential metals (i.e., important for normal physiological functioning) or non-essential metals (there is no known physiological need for the metal). Certain alkali and alkaline earth metals are regarded as macro-nutrients and are essential for critical physiological processes such as nerve impulse generation (e.g., Na and K), muscle contractions (e.g., Ca) and oxygen transport in vertebrates (e.g., Fe). These metals are usually under some form of physiological homeostatic control that regulates tissue levels within narrow limits by complex mechanisms that increase uptake or facilitate excretion. In humans, some essential metals are required in doses of mg/day (e.g., Cu and Zn for metalloenzyme assembly), whereas others are required in doses of µg/day (e.g., Se, Mo, Co).

Many non-essential metals have no physiological function (e.g., Cd, Hg, Pb), but when they increase in the environment due to anthropogenic activities, they may cause acute or chronic toxicity to the biota that are exposed to them. Even essential trace elements do become toxic if the levels exceed the physiological requirements of biota. Thus, the response of biota is dependent on the nature of the metal (i.e., its molecular form that is required for normal physiological functioning) and the actual concentration or the dose in mg/kg that is taken up. This relationship between the nature of the chemical and the dose administered has long been recognized and notably forms the basis of the observation by Paracelsus, a physician from the sixteenth century: "All substances are poisons. There is none which is not a poison. The right dose differentiates a poison from a remedy."

8.3 Sources and fate of metals

Metals enter the aquatic environment either by natural processes or by human activities, which are often referred to as "anthropogenic sources." Let's discuss these sources in a bit more detail.

8.3.1 Natural sources

Metals are present throughout the continental crusts in form of minerals and in geological formations (i.e., compounds and complexes with other elements) from which they can leach into the water catchment by processes involving physical and chemical weathering (i.e., erosion, run-off from land surrounding the catchment area and oxidation or reduction of both the land surface and sub-surface). Moreover, metals can enter waterways through less obvious means, such as the atmospheric deposition of wind-blown particles from shales and rocks, volcanic emissions, and forest fires, just to name a few.

Generally speaking, naturally occurring (otherwise referred to as background) concentrations of toxic metals in the hydrosphere ("hydrosphere" defined as being the total amount of water on the planet) tend to be at ultra-trace levels. Although all metals are present naturally in the hydrosphere to some degree, they are not always present in chemical forms which are readily available for biological uptake by organisms; i.e., some metals are not bioaccessible such as those present in colloids. Many industrial and mining processes (also referred to as being anthropogenic activities) can break down stable compounds found in the natural environment, releasing metal ions into solution at higher concentrations and/or at higher rates than those which are released via natural processes. Moreover, such activities can change the physical and chemical properties of the receiving environment (such as changing the pH) such that the chemical species of the metal ion may change to a more bioaccessible form.

8.3.2 Anthropogenic sources

As previously discussed, although some toxic metals are naturally released into water bodies, human activities can dramatically increase their concentration in these to levels up to a thousand-fold higher than what would be found due to natural processes alone. Anthropogenic activities tend to result in the mobilization of toxic metals from the earth's crust to the hydrosphere thereby increasing concentrations and the rate at which metals can enter the aquatic environment. These human activities include, but are not limited to, industry, sewage/municipal water production, vehicle emissions, agricultural activities, aquaculture, the production and use of algicides, paints/coatings and textiles, nuclear warfare (See Section 8.3.2 -Case study on Lake Karachay), nuclear power, the production of consumer goods (e.g., electronics), and mining activities. The pollution of natural waters by anthropogenic activities is especially concerning in countries with limited water resources, and therefore cannot afford water resources to be contaminated even slightly.

Case study of Pit lakes

Pit lakes are lakes resulting from abandoned open cast mines which have filled with ground, surface, and rainwater. These lakes are often designed to have a profile which reduces the cost of resource extraction. Consequently, the side banks typically have a steep gradient and the pits themselves are deep, often penetrating below the water table [2]. Pit lakes can be of environmental concern since they typically contain waters with a low pH which helps to mobilize metals into groundwater sources. This can result in contamination spreading far beyond the pit lake itself and may potentially have an impact upon

Figure 8.2: Berkley Pit Lake and Yankee Doodle Tailings Pond, Montana, USA.

nearby communities which rely on groundwater as a primary source of water. In the USA alone, more than 102 million people get their drinking water from public water systems that use groundwater and a further *ca.* 43 million Americans get their water from private groundwater reserves [3], bringing to light why such contamination is of particular concern.

The Berkeley Pit Lake in Montana, USA (Figure 8.2), for example, is known for having some of the largest accumulations of toxic mining water in the world. In this lake the acidic pH of the water (pH 2.6) greatly accelerates the mobilization of metals such as Fe, Al, Zn, Mn, Cu, Cd, and As into the lake water [2, 4] and has been a major source of water and soil contamination since the 1980s.

Case Study of Lake Karachay

In 1991, the Worldwatch organization defined the "Most Polluted Spot on Earth" as being Lake Karachay (also called Lake Karachai) in the southern Ural mountains in eastern Russia, located near the town of Ozyorsk [5]. Lake Karachay was used as a dumping site for radioactive waste from a nearby nuclear production, waste storage and reprocessing facility (Mayak Production Association) which was one of the largest nuclear facilities in the Russian Federation in 1951. The lake and surrounding rivers have been a major source of radiation sickness to people living in the region and for many years, the area where the radioactive effluent was discharged into the lake was said to deliver enough radiation to administer a lethal dose of radiation to the average human within an hour.

The lake had accumulated *ca.* 4.44 exabecquerels (EBq) of radioactivity, including 3.6 EBq of Cesium-137 and 0.74 EBq of Strontium-90. To put this into perspective these levels equate to between 37–89% of the estimated radioactivity released by the Chernobyl disaster with the key difference being that the radiation from the Chernobyl disaster was not limited to a single location. Moreover, between 1949 and 1956 the Mayak Production Association also dumped radioactive waste, comprised mainly of Sr-90, Cs-137, Pu-239, and Pu-240, directly into the Techa River, which flows into the Arctic Ocean. This activity delivered over 10^{17} Bq of waste into the river over this time. In 1957 a chemical explosion released a further 7.4×10^{16} Bq into the Techa River [6, 7].

After a drought had caused water levels in the lake to drop, contaminated silts were exposed and were dispersed by wind, further polluting the surrounding area. Due to concerns over dropping water levels and the increasing windblown radioactive dusts, the lake was backfilled with sand followed by being infilled with hollowed concrete blocks, rock, and dirt as recently as December 2016, some 60+ years after the contamination events. To this day, the infilled Lake Karachay and the surrounding rivers are still major contributors to groundwater contamination by radionuclides in the area, posing a significant risk to the surrounding communities. Moreover, should this load of radionuclides reach the Arctic Ocean, this could lead to one of the greatest ecological disasters the world would ever see.

8.3.3 Solubility and mobility

Water is ubiquitous on earth, being found throughout the biosphere. Consequently, water solubility or the ability of a chemical to dissolve in water (also known as hydrophilicity) is one of the most important factors affecting the mobility and the eventual uptake of elemental species by an organism or its deposition (e.g., in a sediment). Since water-soluble metal ions or organometallic compounds (e.g., humic acid-toxic metal complexes) move with water, they can be more widely dispersed throughout the environment than non-soluble ions/compounds. In the environment, water-soluble compounds and ions are often readily taken up by biological cells provided that specific uptake mechanisms are available at the cell membrane.

In contrast, lipophilic or fat-soluble contaminants (chemicals/molecules/compounds that can dissolve in oil or fat) require a biological carrier (e.g., an organism/algae/colloid) to move throughout the environment. Once in the human body, fat-soluble compounds can accumulate in fat deposits and be stored for many years. These contaminants can then be released during activities which access these fat reserves, such as breastmilk production by mothers and/or periods of weight loss or starvation.

Case study: Story of Mercury in the Amazon

Methylmercury (MeHg$^+$) is an example of a lipid-loving (lipophilic) compound and is well-known as being a Central Nervous System (CNS) disruptor. Since MeHg adversely affects the development of the CNS, exposure of children to this neurotoxin is of particular concern. The most common source of MeHg exposure in humans is via the ingestion of piscivorous fish (e.g., tuna, swordfish). The discharge of inorganic mercury (Hg^{2+}) and elemental mercury (Hg^0) from mining and industrial activities eventually finds its way into sediments where anaerobic bacteria can methylate these ions to MeHg$^+$.

In gold mining specifically, liquid mercury is often used to trap elemental gold particles as an amalgam, separating it from sediments in informal gold panning activities. Following this, the mercury is typically boiled off with a blow torch or other heat source, leaving behind a gold nugget. Although this is an effective means of extracting gold for prospectors, gold-panners in areas such as Brazil, Ecuador, and Bolivia and their families have been found to suffer from nerve damage as a consequence. In 2019, the World Health Organization (WHO) estimated that the number of people impacted by MeHg in the Amazon exceeded 1.5 million. The extent of mercury pollution in Amazon is not limited to the gold prospectors, but actually has had a significant impact upon the entire Amazon Basin and surrounding areas. For example, a recent study has found concentrations of MeHg in piscivorous fish exceeding the WHO limit by four times [8]. Moreover, a 2018 report by the WWF indicated that over 25% of the Amazon river dolphins and other iconic species were found to have MeHg concentrations which were higher than the WHO limit [9]. Since mercury is persistent in nature, the long-term impacts of this ongoing pollution are of particular concern.

8.3.4 Point and non-point sources

In defining the source of pollutants, it is necessary to distinguish between a point and a non-point source.

Point sources discharge pollutants from a specific and identifiable location where concentrations of the contaminant are typically higher closer to the source. Some examples of point sources include sewage treatment plants, factory effluent pipes, power plants, oil wells, and underground coal mines. To meet the definition of a point source, the source needs to be discrete and easily identified. Once a point source has been identified, it can be easily monitored and the implementation of regulatory stipulations verified to ameliorate the release of the pollutant into the environment. Point sources are often considered to be predictable and the release of the pollutant can be uniform throughout the year.

In contrast, a non-point source refers to the release of a pollutant from a variety of sources into lakes, rivers, wetlands, and groundwaters. These sources are diffuse; usually being found throughout a large area or scattered and have no specific location for discharge into the water body. The determination of the exact origin of these pollutants is often difficult because of the large number of contributing sources. Examples of non-point sources include agricultural and residential run-off, acid mine drainage, groundwater contamination, atmospheric deposition, and snowmelt.

8.4 Physicochemical interactions between metals and ligands

There are a number of physical and chemical factors which affect the availability of metals for uptake by organisms and the associated toxicity. Metal reactivity can be influenced by complexation with natural ligands. This in turn influences the mobility, catalytic activity, and toxicity of metals. Interactions between metals and ligands can include sorption processes, which in turn may affect a metals' 'speciation' (i.e., all molecular forms of a metal in solution). Sorption refers to the partitioning of a metal to a sorbent, e.g., a mineral or organic surface. Adsorption is a specific form of sorption that does not result in the precipitation of the sorbed species, which can then undergo desorption (e.g., due to changes in environmental conditions). Fine grained soils such as clay minerals and iron and manganese hydroxides have a higher tendency for higher metal adsorption than course-grained soils. This is due to the larger surface area and reactivities of the finer minerals. Similarly, organic matter displays the same properties for metal adsorption. Humic and fulvic acids are natural ligands that are break down products of decaying plant material. Depending upon the pH conditions of the environment, metal adsorption to humic and/or fulvic acids may be enhanced or reduced. For example, at neutral to slightly acidic pH (pH 5.5–7) values humic acids form strong complexes with metals and the formation of such metal-humic acid complexes decreases the potential metal uptake by organisms and increases metal motility using the humic acid as a vehicle [10].

The bioavailability of a metal is also impacted by its chemical species in solution. A chemical species is defined by IUPAC as the "specific form of an element defined as to isotopic composition, electronic or oxidation state, and/or complex or molecular structure" (Figure 8.3) where the latter two are used most commonly [11]. As an example of the impact of speciation on toxicity, let's take a quick look at the two species of chromium, namely trivalent and hexavalent chromium. While Cr^{III} occurs naturally and is required in small doses by the human body to metabolize glucose, it is not regarded as being particularly mobile in the environment. Conversely, Cr^{VI} is soluble, mobile, and toxic, being a known human carcinogen. Although Cr^{VI} can be naturally present in the environment as a consequence of oxidizing conditions in the water, the highest concentrations tend to be those directly linked to human industrial activities such as steel manufacturing, electroplating, leather tanning, and the treatment of wood.

Case study: Hexavalent Chromium and the Town of Hinkley

Knowledge of the toxicity of Cr^{VI} in groundwater became widely known following what happened in the town of Hinkley in Southern California, which was widely discussed in the news and further publicized in the movie *Erin Brockovich*. Here, Pacific Gas and Electric (PG&E) dumped 370 million gallons of Cr^{VI} tainted water into unlined tailings ponds, which resulted in the leaching of Cr^{VI} into the towns groundwater. After much debate related to the possible carcinogenic effects of ingested Cr^{VI}, which at that time was not well understood, it was finally concluded that the waste produced by PG&E was responsible for the observed increased incidences of cancer in the region. This resulted in a $333 million settlement in 1996 and subsequent lawsuits with the last settlements being paid out in 2008. To date, PG&E has spent hundreds of millions of dollars in remediation and lawsuits, but the Cr^{VI} plume generated from these initial activities continues to spread in groundwaters in the region.

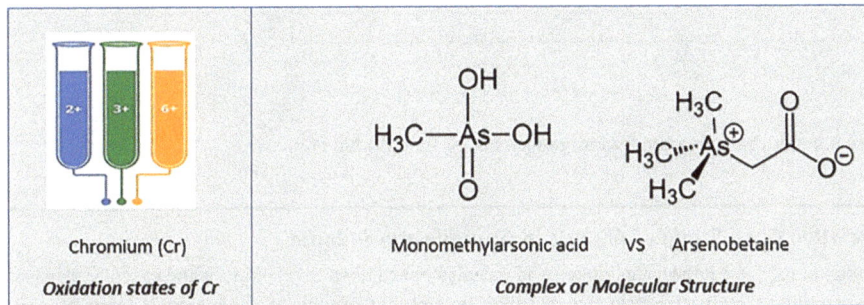

Chromium (Cr)

Oxidation states of Cr

Monomethylarsonic acid   VS   Arsenobetaine

Complex or Molecular Structure

Figure 8.3: Most common definitions when referring to "chemical speciation".

Factors affecting the predominant chemical species of a metal in solution, and consequently its toxicity, are complex and are largely governed by factors such as the redox parameters of the system, complexation with available ligands and the pH. As a result, this relationship is often described using Pourbaix diagrams (also known as Eh-pH diagrams), which have been discussed in Chapter 2. Pourbaix diagrams are often used to predict which chemical species is expected to be present under certain environmental conditions that are defined by activity of the electrons (Eh, oxidation/reduction) and the activity of the hydrogen ions (i.e., the pH and the acidity/alkalinity). The lines in a Pourbaix diagram define which species is prevalent at a specific Eh/pH (Figure 8.4).

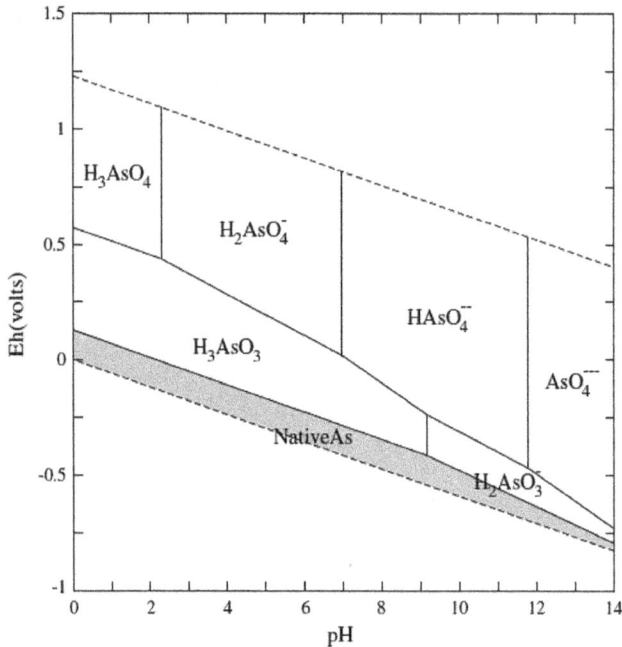

Figure 8.4: Pourbaix diagram for inorganic As at 25 °C and 1 bar [12].

Speciation Case Study of Arsenic in Groundwater in India

Arsenite, or As^{III}, is a known carcinogen and has also been linked to respiratory and vascular disease, diabetes, and an increased incidence of stillbirths. In terms of toxicity, the inorganic forms or arsenic, arsenite (As^{III}, AsO$_3^{3-}$) and arsenate (As^V, AsO$_4^{3-}$) are most prevalent in environmental waters and are regulated in drinking waters to <10 ppb total inorganic arsenic (iAs) in most countries around the world in accordance with WHO guidelines. Inorganic As can be released to the hydrosphere from both natural and anthropogenic sources. Arsenic concentrations in groundwater can be significantly elevated in areas which have sedimentary deposits resulting from volcanic rocks and sulfide mineral deposits. In India the allowable concentration of iAs in water is 50 ppb in the absence of an alternative source per the Bureau of Indian Standards [13], five times higher than the WHO guideline. Despite this relaxed regulatory guideline, the iAs concentrations in many states in India far exceed this limit, with some hand pumps delivering groundwaters which exceed the WHO guideline value 150-fold (>1,500 ppb iAs). Although there are a number of theories surrounding the source of iAs contamination in these waters, the prevailing theory is that the source is of non-point origins.

In the 1970s surface water was the primary source of drinking water in India. Due to increasing concerns around contaminated surface waters with waterborne pathogens that were causing gastrointestinal diseases and diarrhea, millions of tube wells were installed in the hopes that this would give the local population access to clean groundwater resources. Little did people know at the time that the groundwater was naturally contaminated with iAs which was released from arsenopyrite, an insoluble conjugate with iron, which was present in the underlying geology. The overuse of these groundwater reserves reduced the amount of water in the aquifer and exposed the sedimentary minerals to air, which in turn enabled the oxidation of arsenopyrite and was associated with the release of iron (II) and iron (III) sulfate, As^V and As^{III} into the groundwater. As a consequence, iAs concentrations in groundwater reserves soared, with almost 20% of India's total land area possessing toxic levels of iAs in its groundwater. The states of Punjab, Bihar, Gujarat, West Bengal, Jharkhand, Uttar Pradesh, Assam, Manipur, and Chhattisgarh are the most significantly impacted, exposing over 250 million people to this toxic element, which has caused over 1 million deaths within the past three decades [14]. Since over 28% of India's population live in poverty, many people are unable to purchase bottled water and continue to use these contaminated groundwater supplies.

High iAs concentrations in groundwater also pose a significant risk in areas in which groundwater is used to irrigate food crops. Consequently, in many areas iAs accumulates can accumulate in certain food crops, and for certain plant species may occur at levels that far exceed those in the soil or water alone, a process known as hyperaccumulation. This property of certain crops raises further concern and requires that food crops be increasingly scrutinized for their suitability for export. Unfortunately, those which fail export criteria often enter back onto street markets in India. To date there have been no effective solutions employed which are able sufficiently reduce the concentrations of iAs in India's groundwaters to acceptable concentrations.

8.5 Metal uptake mechanisms

In plants and animals, the major barrier to metal uptake are either the cell walls or the membranes, e.g., the cell wall of the plant leaf and in animals the membranes of the lungs or the gill surface, the cell membranes of the skin or the gastrointestinal tract, etc. The form of the metal in the bulk solution surrounding the biological barrier is determined by the physicochemical conditions discussed in the preceding section. Before the metal interacts with the outer surface of the cell wall or membrane it needs to pass through a thin diffusion layer (20–30 μm) and mucus layer (in the case of cell membranes of gill, lung, or gut tissue). In this process the form of the metal may be transformed, thus altering the state it was in the bulk solution (Figure 8.5). Therefore, the form of the metal that is taken up at the biological barrier is dependent on both the physicochemical conditions of the ambient environment as well as the micro-environment of the cell barrier [15].

In plants, dissolved metals are taken up through roots or leaves, while atmospherically derived aerosol metals may enter plants through the stomata on the leaf surfaces. The phospholipid bilayer structure of the animal cell membrane allows for metal uptake via processes such as facilitated diffusion, active transport, and pinocytosis (Figure 8.6) [16].

8.5.1 Simple passive diffusion

Although most metals in the environment are hydrophylic, there are some metal complexes e.g., methyl mercury (MeHg) that are lipid soluble that move across the

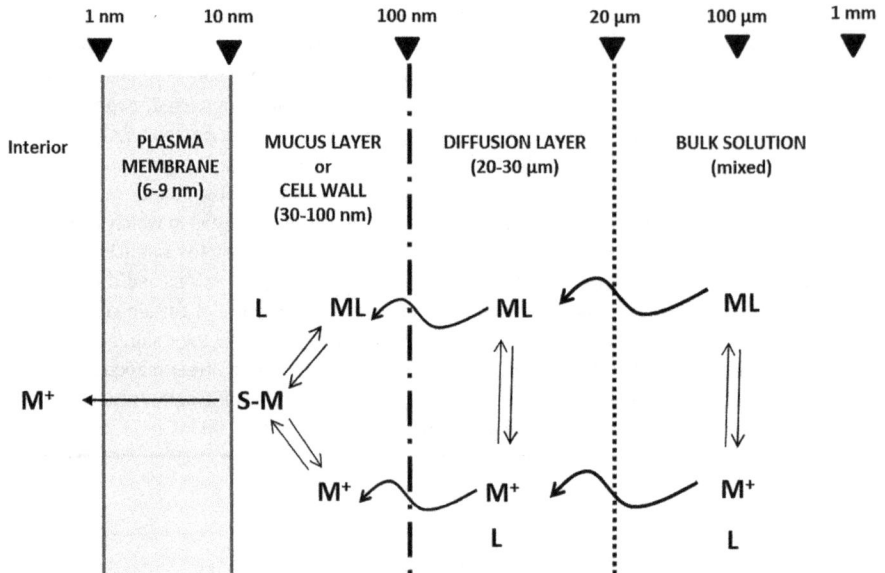

Figure 8.5: A conceptual model of the solution – cell barrier interface indicating potential transformation processes that metals may undergo when moving from the bulk solution to the cell barrier surface. The arrows at the top of the diagram represent distances (not to scale) between the internal organism environment to the surrounding medium (i.e., up to 1 mm from the interior of the cell). ML represents the metal and ligand and M^+ represents the metal ion and S represents the surface.

membrane along a concentration gradient until equilibrium is reached [16]. The uptake rates' limiting factors are the concentration gradient, thickness, and surface area of the membrane, the ionization state of the metal, counter ions and coordination ligand molecule(s) and the molecular size.

8.5.2 Facilitated diffusion

The presence of integral protein pores in the lipid bilayer structure of animal cells can mediate the passing of certain metals across the membrane in a concentration dependent manner. This process does not require energy, but it does require a molecule (integral membrane proteins) that facilitates diffusion. An example is the transport of Ca across the membrane by calmodulin, which is a Ca-specific binding protein.

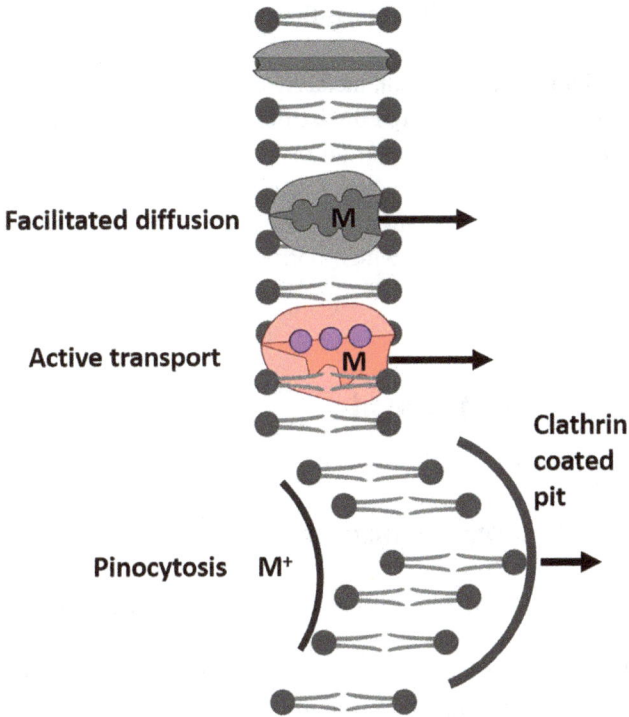

Figure 8.6: Uptake mechanisms across the phospholipid bilayer of an animal cell. Metal complexes are represented by M and the ionic forms of a metal by M^+.

8.5.3 Active transport

The transport is mediated by an "energy-requiring carrier system" and is often metal-specific. For example, a Na-K pump involves the binding of Na to the external surface of a voltage-gated protein channel. Through the action of the enzyme Na ATPase, the protein channel undergoes a conformational change to translocate the Na across the membrane into the cell. This typically takes place against a concentration gradient and may utilize the hydrolysis of a molecule ATP or the concentration gradient of another ion leading to symporter vs antiporter activity.

8.5.4 Pinocytosis

This process involves the involution of the membrane to envelope larger sized molecules that cannot pass through the membrane. This is the mechanism whereby airborne metal-bound particles are taken up across the surface of lung alveoli membranes. An example is the transport of Cd across the membrane through a clathrin-coated pit.

Regardless of the physical uptake mechanism, metals need to be present in a molecular form that allows for them to pass through the barrier from the external environment into the organism. Thus, the concentration of a particular metal in the environment is not the only determining factor when it comes to understanding how plants and animals take up metals from the environment. It is rather knowledge of the speciation of the metal and its concentration which both influence the uptake by biota.

8.6 Why do we care about metals in the environment?

8.6.1 Bioaccumulation and biomagnification

All living biota can accumulate metals irrespective as to whether they are essential or not. The resulting tissue concentrations can display a large variability for both metals and the different biota. Metal bioaccumulation is the net result of the bioconcentration (uptake) and excretion (elimination) of metals from the surrounding environment:

Bioaccumulation = [bioconcentration]-[excretion]

Bioaccumulation may also occur in the form of biomagnification, which is regarded as the uptake of metals via the food web where concentrations increase at each trophic level. Metal bioaccumulation is determined by both extrinsic, e.g., physico-chemical (i.e., metal speciation, environmental partitioning) and environmental factors (i.e., season, river runoff, etc.), as well as intrinsic factors such as species involved, sex, feeding behavior, and physiology. Once inside the organism, metals can be transported to target cells and bind to any molecule that has an affinity for the metal. Since most toxic metals have an affinity for nitrogen and sulfur-containing functional groups that are present in proteins and peptides, there are many potential available binding sites for metals in cells. The metals therefore remain metabolically available until the organism either inactivates them (e.g., HgSe is deposited in the kidneys of marine mammals) or excretes the metals via urine or bile.

Case study of single species impact: The mercury contamination saga of Thor Chemicals, South Africa

During the early 1990s the death of three workers and the subsequent public outcry resulted in an investigation of the Hg levels in a stream adjacent to a Hg incinerator complex. The plant had been processing Hg waste received from around the world since 1986 and the study found that spent Hg waste was being discharged into the Mngceweni River, South Africa. The Hg concentrations in sediments (up to 800 ng/g wet weight) and fish (mean of 670 ng/g wet weight) from the Mngceweni River were elevated and exceeded the USEPA guidelines for sediment (200 ng/g) and fish (500 ng/g) respectively [17]. To determine whether or not Hg was being taken up by the local population, non-invasive sampling, namely sampling of human hair, was employed. This sampling approach is often considered to be useful for identifying exposure to organic mercury species, such as methylmercury (MeHg) but is less useful for inorganic mercury (iHg) assessments. However, the Hg levels in human hair samples did not exceed international guideline values. Ongoing monitoring of Hg was recommended at that stage since concern was expressed that this river is a tributary of the uMgeni River, which supplies the reservoirs from which the eThekwini (the third largest metropolitan area in South Africa) receives its water. The processing activities were reportedly discontinued 13 years after the incident, but the untreated waste had remained stockpiled on the site. A follow-up study in 2007 (20 years after the incident) revealed that nearly 20% of the human population sampled had Hg concentrations (whole blood and hair) that exceeded the guideline levels of the WHO [18]. The elevated Hg levels were attributed to the long-term release of Hg from the contaminated sediments of the Mngceweni River that was transported to the Inanda Dam (water reservoir) during flooding events over the two-decade period. Thus, the human population in the vicinity of the reservoir was exposed to total Hg (iHg and MeHg) through the consumption of vegetables cultivated on the banks of the reservoir and contaminated fish from the reservoir. In 2020, an investigative journalism television team undertook more sampling and the results showed that the Hg concentrations remain high in sediment, animal and human hair samples from the region (unpublished). Despite the call for a long-term monitoring program, no further action has been taken.

Case study of whole ecosystem impact: Acid mine drainage and uranium contamination in the gold mining areas of South Africa

Extensive gold and coal mining activities in South Africa have resulted in the exposure of pyrite containing rocks to surface rainwater in the case of open cast mines and mine dumps or groundwater, when low-grade ore is left underground. The oxidation of the pyrite forms sulfuric acid and then either decants from the underground cavities or percolates through the mine tailings as acid mine drainage (AMD) water. During this process the low pH water remobilizes metals, including uranium (U), ultimately translocating the dissolved metals into surface water streams and rivers [19]. As gold mines near the end of their lifetime, the removal or pumping out of groundwater is discontinued, resulting in the filling of the mining cavities. Many regions where this highly acidic water (which is high in sulfate and metal concentrations) percolates to the surface, a process known as decanting, are characterized by river-banks covered in an rust-colored precipitate, which corresponds to iron oxides that precipitate when exposed to air.

During the gold mining process, vast amounts of ore are brought to the surface for processing since the gold concentrations are generally low (gold to waste ratio is 1: >100,000). This has resulted in more than 6 billion tons of mine tailings being deposited over an area of 400 km^2 during the past century. Since U is a by-product of gold mining, it has been estimated that more than 510,000 t of U has been deposited on mine dumps and tailing dams as gold tailings [20]. Seepage from these tailings dams results in approximately 240 t of dissolved U reaching surface and ground water per year. In regions where AMD is prevalent the remobilization of U has resulted in off-site U concentrations that can exceed the concentrations of uranium at the source. The implications are that the contaminated sediments become sources of U contamination rather than sinks. Although the U concentrations in surface and groundwater, and sediments are elevated in these regions, there are no long-term chemical or biological monitoring programs in place to determine the ecological consequences of the U exposure.

Case study of whole ecosystem impact: Metal contamination due to mine tailing dam failures

Globally the mining industry produces billions of tons of waste, which are typically contained in tailings storage facilities. It is estimated that there are approximately 534 billion m^3 of mine tailings in the estimated 29,000–35,000 global tailings storage facilities [21]. The current annual mining activities generate an additional 19 billion m^3 of tailings. The increasing global demand for metals has resulted in increased volumes of waste being generated due to the lower-grade ore that is mined. Coupled to the increased waste production and, in many cases, poorly maintained, outdated, and low-capacity tailing storage facilities, it is expected that the number of tailing dam failures will increase. Over the past 50 years there have been 63 major tailings failures worldwide [19]. In addition to the loss of life (2,375 deaths between 1961 and 2019) there has been extensive damage to the environment.

Two tailing dam failures in particular have become synonymous with mine-tailings disasters.

In April 1998, approximately 4 million m^3 of acidic water and 2 million m^3 of lead-zinc tailings were released into Río Agrio from a tailings dam in Aznalcollar (south western Spain). This toxic slurry entered the Río Guadiamar River. Given the rather flat area, the slurry covered over 4,000 hectares of farmland adjacent to the river at distances of up to 40 km downstream. Furthermore, the slurry covered more than 2,700 hectares of shoreline within the Doñana Natural and National Parks. This is a proclaimed United Nations World Heritage Area that serves as the largest natural reserve for bird species in Europe. The slurry, containing toxic metals such as As, Cd, Pb, Hg, and Tl at a pH of 3 caused mass mortalities of fish (73 tons) and posed an immediate risk to birds. In total 16,000 t of Zn and Pb, 10,000 t of As and 4,000 t of Cu were spread over the aforementioned areas [22].

Although most of the metals were in an insoluble sulfide form, oxidation to sulfate and the remobilization of metals occurred. To reduce the risk of metal mobilization, large scale mechanical removal of the slurry material was undertaken over a five-month period. Despite these cleanup efforts and other soil remediation exercises, the soils close to the mining area remain contaminated with high levels of As (4.7 µg/g), Cd (31 µg/g), Pb (15 µg/g) and Hg (101 ng/g), even increasing over time. A comparative assessment of metal biomagnification in lizards from the affected area revealed elevated As, Cd, and Hg concentrations compared to lizards from an uncontaminated site nearby. Thus, metal pollution was identified as the main cause for the decline in lizard populations in the region [23].

The second most iconic tailing dam failure was that in Minas Gerais State (Brazil), where within four years, two mine tailings dams failed. Collectively more than 52 million m^3 of tailings were released into nearby rivers. The largest of the two tailing dam failures occurred in 2015 when 40 million m^3 of mine tailings from the Fundão tailing dam (SAMARCO) affected more than 650 km of Gualaxo do Norte, Carmo, and Doce rivers and the related stretch of the Atlantic Ocean. This influenced the primary water and food resources of several communities and the slurry covered approximately 1,500 hectares of natural and indigenous reserve lands. Although the slurry contained mainly Fe and Si oxides, other metals such as As, Hg, and Mn were present at elevated concentrations as well. Based on elevated Cd, Pb, Cr, Zn, Cu, and As concentrations in sediments off the mouth of the Doce River, scientists have recommended that fisheries in the region should be closed until there is a better understanding of the ecological risks that these metals may pose. More recently in January 2019 a tailings dam in Brumadinho released approximately 12 million m^3 of iron ore sludge containing Ba, Pb, As, Sr, Fe, Mn, and Al, which directly affected the Paraopeba River [24]. The immediate effects were related to increased turbidity due to fine particulate matter and enrichment of metals in the sediments. While the medium- to long-term effects of both mining tragedies remain unknown, these events have renewed the global need to ensure safer designs of tailings storage facilities.

Case study of whole ecosystem impact: Human exposure to soils and dust contaminated by lead in Kabwe, Zambia

Ninety years of extensive mining for Pb and Zn in the Kabwe region of central Zambia have left a legacy of metal contaminated soils and a risk of contaminated waters. Between 1906 and 1994, mining and smelting operations generated over 5 million t of waste, which have ended up on largely unprotected mine tailings dumps. These old slag heaps, referred to as "Black Mountain," have not undergone any remediation and remain uncovered. The Kabwe region has the fifth largest population in Zambia and the communities surrounding the mining areas are exposed to metals via both direct contact as well as wind-borne dust [25]. The high metal levels in the fine dust fractions of <48 μm fractions which contains metal-rich Mn oxides has resulted in this region repeatedly being ranked in the world's top ten most polluted sites.

At the time of the study [26], the median levels of Pb and Zn were 282 mg/kg (maximum 51 g/kg) and 607 mg/kg (maximum 92 g/kg) respectively in Kabwe soils. In addition, levels of As, Cd and Cu were also elevated in this region. Studies have shown that more than 25% of the soils sampled in the region exceed 400 mg/kg of Zn. This has resulted in high levels of these metals in both domestic and wild animals from this region [26]. What is even more concerning is that studies showed that more than 200,000 people have elevated blood Pb levels above the level of 5 μg/dL. Over 95% of children were found to have Pb levels higher than 25 μg/dL and of them 50% were >45 μg/dL, the threshold above which medical intervention is required. A 2012 clinical study, based on questionnaires, showed that 10% of children revealed typical chronic Pb exposure symptoms such as anemia, headaches, memory problems, and seizures. It is therefore of the utmost importance that mitigation measures are put in place to suppress the formation of dust through, e.g., phytostabilization of the mining slag dumps. Further remediation of the metal contaminated dust through solidification and biocementation techniques are currently being tested [27].

Case study: The ecotoxicology of platinum group elements from a platinum mining region of South Africa

The need for platinum group elements (PGEs) has increased in recent years owing to their increased application in medical applications (e.g., cisplatin is an anticancer drug), and their use in chemical and electrical industries. Pt-mining has resulted in an increased presence of PGEs in the overall natural environment and surrounding mining sites, with increasing global environmental levels over the last 3 decades. The increased demand has resulted in a 20% increase in production over the past five years.

South Africa is the world's largest PGE producer responsible for over 73% of the global Pt production. The main PGE resources are situated within a narrow geological band of the Bushveld Igneous Complex in the northwest of South Africa. While there is a growing body of literature on the levels of PGEs in the environment due to automotive catalysts in road and airborne dust which enter or are deposited in soils, there is a distinct lack of data on PGE discharges from mining and production activities and the ecological effects of PGEs in the surrounding environment.

Recent studies were undertaken in the Hex River [28–30], which drains the largest and most productive PGE mining and smelting region in the world, to assess the ecotoxicology of Pt and other metals that are often associated with PGE deposits (e.g., Cr and Ni). The main sources of Pt and other associated metals that enter the Hex River are untreated mining water, seepage from tailing dams, runoff from waste dumps or air deposition from ore smelters and dust from waste dumps and tailing dams. The results showed that Cr, Ni, Cu, As, and Pt contamination of the aquatic ecosystem originated from the intensive mining activities, while urban runoff contributed to Zn, Cd, and Pb contamination.

Platinum and co-occurring Cr and Ni were found to be present in bioavailable forms as these metals had accumulated in tissues of sediment-dwelling aquatic invertebrates, such as snails and worms. Tube-building worms in the sediments from the Hex River system were able to bioaccumulate Pt (72 ng/g) orders of magnitude higher than freshwater clams from European systems that were exposed to PGEs from auto-catalysts in road dust [29]. The study found that water alkalinity was the main factor responsible for metal speciation and subsequent bioavailability in this system. Furthermore, bioaccumulation of Cr and Ni in fish from the Hex River system was also evident. The accumulated levels did not only pose a threat to fish health, but also posed a human health risk if the fish were consumed. The fish from the mining impacted sites (Cr – 2.6 µg/g and Ni – 1.2 µg/g) were associated with a 10–90% higher human health risk compared to fish from the reference site and could potentially cause several carcinogenic and non-carcinogenic effects [30].

An interesting finding in these studies was that the metal bioaccumulation in parasites found in the gut of the fish displayed higher metal concentrations than the levels in the fish host. Since these internal parasites do not contain a true digestive tract, the only metal uptake route is across their body surface. This indicates that the metals that are taken up by the fish through food or accidental sediment ingestion are in a bioavailable form in the gut-environment of the fish. Interestingly, the non-essential metals were bioaccumulated to a far greater degree by the parasites, compared to the fish host, while the fish had higher levels of essential metals. Similarly, to the results of metal bioaccumulation by the macroinvertebrates, the fish data also showed that Pt bioavailability from mining activities exceeds that from Pt in road dust from auto-catalysts.

These studies concluded that the water, sediment, and biota in the Hex River are exposed to multiple metal stressors but that the intensive mining activities were primarily responsible for the observed high concentrations of PGE metals. Other metal and nutrient contaminants were found to be largely due to urban and industrial effluents. The metals from PGE mining sources were bioavailable and bioaccumulated in several aquatic biota species. The bioaccumulation in fish particularly indicated a threat not only to the ecosystem health but also to humans relying on fish from this region as a source of sustenance.

8.7 Mitigation and remediation

8.7.1 Linear model

To date waste management has largely followed a linear model; that is, water is taken in from surface or groundwater, used by the anthropogenic activity; in some cases this is followed by the removal or partial removal of metals from waste discharges and the disposal into surface waters.

8.7.2 Strategies for treating wastewaters

At present, 80% of global wastewater is released into the environment without sufficient treatment [31]. For jurisdictions that do treat their wastewater, there are several strategies which can be used before it is discharged into the environment. Although we will not go into an exhaustive discussion on every type of wastewater treatment, we will cover a few of the key strategies which are currently employed by industry.

8.7.2.1 Activated carbon

Due to its porosity, surface functional groups and surface area, the addition of activated carbon (also referred to as charcoal) to water and wastewater, can be an effective method to remove contaminants, including metals from these matrices. Activated carbon is prepared using natural resources such as coal, wood, petroleum residues, nutshells etc. Unfortunately, these can have an adverse impact on the environment in other ways such as raising carbon emissions and requiring mining or processing in order to be effective. Moreover, once activated carbon has served its purpose, it is often disposed of in landfills, running the risk of releasing the absorbed chemicals back into the environment. Although more recent studies have started to investigate the treatment of spent adsorbents to avoid landfill disposal, this often requires the use of expensive acids and energy in the form of heat and can lead to other challenges related to treatment and discarding of acid wastes [32]. Therefore, although a seemingly attractive option, its prohibitive cost and waste management challenges make it only suitable for use for targeted applications and during specific stages of waste management.

8.7.2.2 Chemical precipitation

Chemical precipitation techniques are some of the most common techniques for the treatment of wastewaters containing metals and are particularly useful in removing soluble ionic species from solution. This approach aims to transform soluble compounds into an insoluble form via the addition of chemicals and differs from solidification/stabilization techniques in that the latter uses a chemical binder such as cement, fly ash, silicates and pozzolanic (where "pozzolanic" refers to fine-grained siliceous or siliceous-aluminous material which can chemically react with calcium hydroxide under room temperature to form cement-like compounds) materials to make the metal less susceptible to leaching [33]. Some of the metals typically treated using chemical precipitation methods include Pb, Co, Ni, and Zn. Some of the advantages and disadvantages of this approach are as follows:

Advantages:
- Relatively low cost per unit volume of wastewater treated
- Well-established, reliable, and only requiring the replenishment of the chemicals used.

Disadvantages:
- Stoichiometric control of the addition of chemicals is required
- Requires the disposal of sludge which has a high-water content
- May require two stages of precipitation
- Requires large continual flow

– Efficiency of co-precipitation agents is determined by the initial concentration of the contaminant and the surface area of the primary flocculant.

In some cases, co-precipitation may occur and results from soluble compounds adsorbing to the precipitate. Chemicals commonly employed in chemical precipitation techniques are hydrogen peroxide, sulfide, carbonate, xanthate (for precipitation reactions), or a combination thereof. Some of the main advantages and disadvantages of each technique are outlined in Table 8.1. In each case, once the metal precipitate has been formed, it can be removed by flocculation and sedimentation/filtration operations.

Table 8.1: Advantages and disadvantages of various precipitation techniques where • = poor/disadvantageous, •• = moderate and ••• = good/advantageous.

	Hydroxide	Sulfide	Carbonate	Xanthate
Cost-efficiency	•••	•	••	••
Proven and accepted technique	•••	••	••	•
Simple application	•••	•	•••	••
Precipitation efficiency	••	•••	••	•••
Time-consumption	•	•••	••	••
pH sensitivity of precipitate	•	•••	••	•••
Ease of water removal from sludge	•	•••	••	•••
Suitability for mixed wastes	•	••	••	•••
Completeness of metal removal	••	••	•	•••
Mass of precipitant-to-volume of waste ratio	•	•••	••	•••
Formation of toxic by-products	•••	•	•••	•
Density of precipitates	•	••	••	•••
Suitability for continual flow operation	•	••	•••	•••

8.7.2.3 Coagulation or flocculation

Coagulation or flocculation are commonly employed techniques to remove heavy metals from solution, where coagulation neutralizes the electronic charge on suspended particles, flocculation involves adding a chemical to the water and slowly mixing the waste solution, thereby encouraging agglomeration. Both have been found to greatly improve the removal of metals and are especially useful when used as a tandem process for more challenging metals such as Fe, Cr, Cu, Zn, and Ni, which are not effectively captured by traditional chemical treatment techniques.

8.8 Remediation of previously impacted environments

8.8.1 Phytoremediation

Many plants (including trees) which naturally occur in heavily contaminated areas employ strategies to either limit their uptake of metals or accumulate metals at higher concentrations than the surrounding soil, a process known as hyperaccumulation. Although this property can be of concern for crop plants, it can be taken advantage of during remediation efforts where hyperaccumulating plants can be grown along the sides of water-retention and tailings dams (an application referred to as phytoremediation – Figure 8.7). These not only stabilize the dam wall (phytostabilization) but also help to take in metals from the affected soils and transport the metals to the shoots, stem, leaves, seeds, and flowers (known as phytoextraction). This approach to pollutant removal is an attractive alternative to excavation and removal or chemical in situ stabilization and conversion due to its cost effectiveness and aesthetically pleasing outcome.

Phytoextraction can either be continuous or induced, where continuous phytoextraction relies on having hyperaccumulating plants while induced phytoextraction works to enhance a single extraction event by adding chelators such as EDTA to the soil to enhance the mobilization and subsequent accumulation of metals by plants. The latter approach can be challenging in that certain chelators may also increase water solubility, moving toxins deeper into the soil layer and potentially into the groundwater where the pollution may become more widespread. Phytoextraction techniques may serve to trap and sequester metals in the plant roots which in turn reduces the chance of these metals leaching into groundwater or run-off and entering into the food chain [34]. This is most applicable for slow-growing plants with high hyperaccumulation potentials such that the disposal of hazardous biomass may not be required in the short term. In these applications, plants with high transpiration rates are often preferred as they reduce the migration of water away from the site [35, 36].

For faster growing plants, regular harvesting and removal from the site is often required. From there, two potential avenues of dealing with the plant matter are available. The first would be to decompose or burn and bury the plant matter, but this is not ideal since it simply moves the pollution issue elsewhere. A slightly higher cost alternative is to employ phytomining strategies, i.e., burn the material in a furnace followed by metal extraction and recycling from the ash. The associated cost of this remediation approach can be somewhat offset by the income generated from the extracted metals.

The application of phytoextraction techniques is useful only in areas where the removal of metals from the soil is not urgent and where the contamination depth is

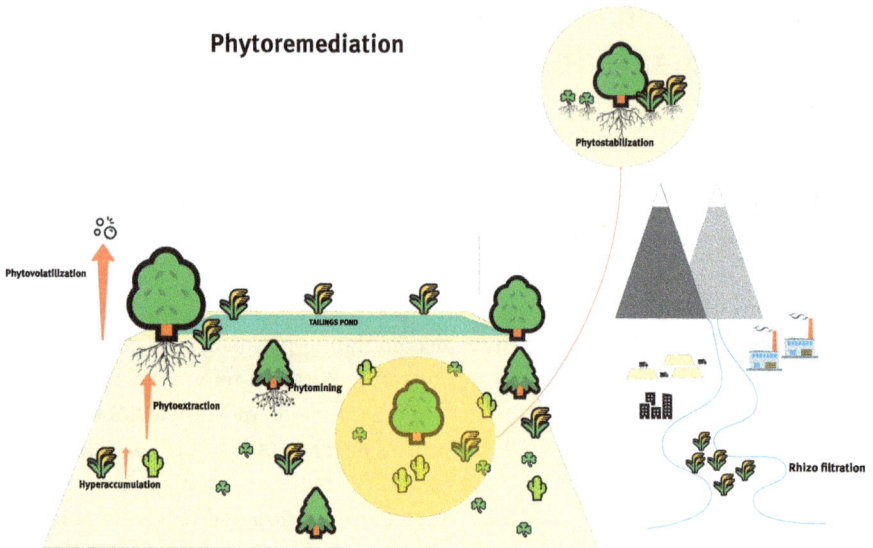

Figure 8.7: Phytoremediation approaches for water and soil remediation of metal contaminated sites.

less than 5 m [36], which is an important limitation to this technique. This depth is dependent on geology and soil type restrictions of vertical water movement and plants root depth. Consequently, phytoextraction is often regarded as complementary, and not a stand-alone approach when compared to other waste management methods.

Phytoremediation can also occur in the form of natural or artificial wetlands, which can act as natural filters to trap, sediment and hyperaccumulate the low concentrations of industrial pollutants from the water flowing through these systems. Unfortunately, many wetlands today are impacted or removed to make way for urban and agricultural development.

Artificial wetlands have been introduced in some areas as a direct form of phytoremediation of contaminated waters. Typically, the plants used are selected based upon the ability of their roots to remove pollutants from water (rhizofiltration). These plants are grown under controlled hydroponic conditions first, then exposed to polluted water to allow acclimation to the wastewater conditions, followed by transplantation [34]. Plants typically used for this application are hyacinth, duckweed, cattail, azolla, and poplar due to their large root surface area. The sources of pollutants typically targeted by rhizofiltration applications include industrial discharges, agricultural runoff, and nuclear material processing wastes.

A good example demonstrating the potential efficacy of rhizofiltration applications is that done by Norman Terry from the University of California, Berkeley, who found rhizofiltration to be effective at removing 89% of Se from selenite-contaminated wastewater released from oil refineries in the San Francisco Bay Area. Here water flowing into the wetland had 20–30 µg/L of selenite and the outflow had <5 µg/L of selenite.

In order for a plant to be used for phytoremediation, there are a few requirements which need to be met, namely that the plant should be:

– Able to hyperaccumulate metals
– Able to tolerate the toxicity of the metal
– Able to tolerate other pollutants, including organic pollutants at the remediation site
– Able to grow in the fine sediments if the application is the remediation of tailings dams
– Easily cultivated in the existing soil and metrological conditions
– Have a fast growth pattern, be simple to harvest, and should allow harvest several times a year
– Should not be a plant that will be eaten by the animals in the area
– Indigenous to the target area in order to ensure that alien plant species aren't introduced into the area
    – Local plants may be specifically and selectively bred/engineered to enhance hyperaccumulating traits and increase biomass
    – Indigenous plants tend to be well adapted to the local climate and are more resistant to local pests and diseases
    – The plant should not be cultivated as an edible crop plant and should preferably not have natural predators which could be impacted as a consequence of enhanced metal uptake
    – Any genetically engineered plants should also be bred such that they cannot naturally spread

There are, of course, a few limitations of phytoremediation which can be listed as follows:

– typically takes several years compared to excavation and removal strategies
– depends upon the bioavailability of the metal to the plants
– only suitable for applications where the contamination depth <5 m [37]
– dependent upon the **bioconcentration factor** (The "bioconcentration factor" is the ratio of the concentration of metals in the shoot of the plants to the surrounding soil concentration [35, 38].
– may volatilize hazardous metalloids such that they enter the air and thus can become more widely distributed (e.g., Se).

8.8.2 *Direct "liming" of rivers and lakes*

Though the direct addition of lime and other alkaline precipitants to surface water bodies is generally not recommended, in some drastic cases, such as those related

to acid rain and AMD, this approach may be necessary as a "lesser of two evils" approach. In these applications, lime is used to raise the pH and alkalinity of the river or lake and remove dissolved metals from the water column via precipitation processes. In recent years watershed liming, i.e., adding $CaCO_3$ or $MgCO_3$ to soils within the watershed has been used and is often favored as this helps to neutralize the pH of the water draining from that watershed and has been found in some cases to be more effective than the direct addition of lime to rivers. Limestone has also been directly added to many streams in Pennsylvania and other places around the globe. The pros linked with this approach are:
- after the addition of limestone no further maintenance is needed
- simple
- relatively inexpensive
- proven efficacy

The cons linked to this approach are:
- not effective at high-flows
- dosage-volume recommendations are not clear
- requires regular replenishment of limestone to prevent the remobilization of metals
- observed loss of biodiversity of benthic organisms due to stone covered substrates
- causes Al and other metals to precipitate and impact the benthic ecosystem reducing insect biodiversity
- increased pH and dissolved organic carbon (DOC), increases mobilization of metals such as mercury (Hg) into surface waters.

8.9 Circular model – circular economy

Simply put, circular economy principles are relatively new though the general concept has been known for some time. This economic model is vastly dissimilar to the current linear model of taking natural resources, making a product, and producing waste which can have ecological impacts and instead aims to develop sustainable consumption and production, focusing on low or no waste.

Some of the preliminary circular economy approaches focus on pumping treated wastewater back into the very processes which generated them, such as cooling water for power plants and process water for mines, industrial processes, irrigation, recreational use, and replenishing groundwater aquifers.

8.9.1 Agricultural fertilizers

Although many of the toxic metals are not necessarily targeted in existing circular models, there has been some focus on elements such as phosphorus and nitrogen which can be harvested from wastewater to create agricultural fertilizers. This approach prevents high phosphorus and nitrogen containing materials from entering natural waters, which would otherwise result in their eutrophication and potentially lead to toxic algal blooms. In terms of benefits, such approaches to wastewater management can:
- provide an additional source of revenue for companies
- reduce the demand for synthetic fertilizers
- reduce need for phosphorus mines which would otherwise increase land disturbance and the entry of metals into waterways.

8.9.2 Inorganic coagulants in circular economy

Inorganic coagulants, such as ferric chloride, can be used to treat wastewater. Some companies use by-products such as scrap iron from automotive and appliance manufacturers and spent pickling liquor (where "pickling liquor" is an acidic solution used in steel mills to remove impurities and contaminants from metals used in the manufacture of steel) to produce inorganic coagulants which can be used in wastewater treatment. In contrast to the disposal of these by-products, which can be costly, such recycling activities allow some economic gain while also mitigating the entry of certain metals into the environment.

8.9.3 Re-mining

In recent years, a significant number of companies are starting to mine computer dumps, landfills and old mining heaps for other metals which were not originally of interest. As an example, in South Africa gold was mined from the gold reef in the Witwatersrand and the mined soil placed in large dumps surrounding this area. For many years, uranium in these dumps had seeped into groundwater and had been dispersed as wind-blown dusts. More recently, the dumps have been re-mined for uranium and other metals. There are several benefits to re-mining activities, namely:
- reducing the need for the mining of virgin materials,
- reducing local contamination from the original pollution source,
- allowing a "second life" for mining dumps.

8.9.3.1 Conclusions and future outlook

In conclusion, metals are an integral part of the world that we live in, being both important for our physiology and necessary for driving technological advancement. Anthropogenic activities and processes release metals into the environment which would not necessarily have been available through natural processes alone. These can have cascade effects in ecosystems and in some cases result in their biomagnification in the food chain. Consequently, it is important to mitigate the entry of these metals into the hydrosphere in the first place, ensure that any wastes produced are adequately treated, and remediate previously impacted environments to prevent and reduce environmental degradation.

An important part of science is learning from the mistakes of those before us and to be more thoughtful about the impact our decisions make on the environment, ecosystems, and the world we live in. Knowing what you know now about metal contamination in the environment, think for a moment what changes you could make today to reduce your use and disposal of metals and metal-contaminated water. Would it be:

- reducing the amount of household wastewater that you produce?
- reusing grey water?
- recycling spent materials?
- reducing single use items?
- repairing items to increase their working lifetimes?
- changing purchasing habits to reduce your turnover of electronics and clothing?

What would you do if you were the lead environmental chemist or ecologist for a new industry or mine? For instance, let's take a look at the current development of the area known as the "Ring of Fire" in Northern Ontario, Canada which is said to possess the largest deposits of high-quality ferrochrome ore on earth [39]. What mitigative actions would you employ to reduce the chances of hexavalent chromium from entering local waters and the Great Lakes and impacting the local ecosystem? What procedures would you put in place to protect the local indigenous first nations communities which live in this area? And finally, how will you contribute toward a sustainable future for the generations to come?

Suggested readings

Carson R Silent Spring. Boston: Houghton Mifflin Harcourt. 1962.

Connell DW, Lam P, Richardson B, Wu R Introduction to Ecotoxicology. London: John Wiley & Sons, 1999.

World Economic Forum. From linear to circular-Accelerating a proven concept. Available at https://reports.weforum.org/toward-the-circular-economy-accelerating-the-scale-up-across-global-supply-chains/from-linear-to-circular-accelerating-a-proven-concept/

The circular economy imperative. Available at https://youtu.be/yPZFNvrnO4E

Circular Economy and Material value Chains. Available at https://www.weforum.org/projects/circu lar-economy

Towards the circular economy: Accelerating the scale-up across global supply chains. Available at https://www3.weforum.org/docs/WEF_ENV_TowardsCircularEconomy_Report_2014.pdf

References

[1]   Printable periodic table [Internet; Cited 2021 August 10]. Available from: https://science notes.org/printable-periodic-table

[2]   Blanchette ML, Lund MA. Pit lakes are a global legacy of mining: An integrated approach to achieving sustainable ecosystems and value for communities. Curr Opin Environ Sustainability. 2016, 23, 28–34.

[3]   Groundwater Awareness Week [Internet; Cited 2021 April 2]. CDC. 2019. Available from: https://www.cdc.gov/healthywater/drinking/groundwater-awareness-week.html

[4]   Duaime TE. Butte Mine Flooding Monthly Report, September 2014. Montana Bureau of Mines and Geology. 2014, 59.

[5]   Lenssen N. Nuclear Waste: The Problem that Won't Go Away", Worldwatch Institute, Washington, D.C., 1991, 15.

[6]   Chelyabinsk-65/Ozersk. Global Security Organization [Internet; Cited 2021 June 5]. Available from: https://www.globalsecurity.org/wmd/world/russia/chelyabinsk-65_nuc.htm

[7]   IAEA. Radioactive contamination of the Tech River, South Urals, Russia [Internet; Cited 2021 May 20] Available at: https://www-ns.iaea.org/downloads/rw/projects/emras/emras-aquatic-techa.pdf

[8]   Hacon Sds, Oliveira-da-costa m, Gama CdS, Ferreira M, Basta PC, Schramm A, Yokota D. Mercury Exposure through fish consumption in traditional communities in the Brazilian northern Amazon. Int J Environ Res Public Health. 2020, 17, 5269.

[9]   WWF report. Healthy Rivers, Health People [Internet; Cited 2021 July 4]. 2018. https://wwfint. awsassets.panda.org/downloads/healthy_rivers_healthy_people.pdf

[10] Bradl HB. Heavy Metals in the Environment. London: Elsevier, 2005.

[11] Templeton DM, Fujishiro H. Terminology of elemental speciation – An IUPAC perspective. Coord Chem Rev. 2017, 352, 424–431.

[12] Lu P, Zhu C. Arsenic Eh-pH diagrams at 25 °C and 1 bar. Environl Earth Sci. 2010, 62, 1673–1683.

[13] Bureau of Indian standards. Indian standard drinking water – Specification Rev 2 [Internet; Cited 2021 November 10]. 2012. https://www.indiawaterportal.org/sites/default/files/2020-11/bis_10500-2012_wq_standards_0_0.pdf

[14] Paul D, Kazy SK, Banerjee TD, Gupta AK, Pal T, Sar P. Arsenic biotransformation and release by bacteria indigenous to arsenic contaminated groundwater. Bioresour Technol. 2015, 188, 14–23.

[15] Simkiss K, Taylor MG. Trace element speciation at cell membranes: Aqueous, solid and lipid phase effects. J Environ Monit. 2001, 3, 15–21.

[16] Connell DW, Lam P, Richardson B, Wu R. Introduction to Ecotoxicology. London: John Wiley & Sons, 1999.

[17] Oosthuizen J, Ehrlich R. The impact of pollution from a mercury processing plant in KwaZulu-Natal, South Africa, on the health of fish-eating communities in the area: An environmental health risk assessment. Int J Environ Health Res. 2001, 11, 41–50.

[18]   Papu-Zamxaka V, Mathee A, Harpham T, Barnes B, Röllin H, Lyons M, Jordaan W, Cloete M. Elevated mercury exposure in communities living alongside the Inanda Dam, South Africa. J Environ Monit. 2010, 12, 472–477.

[19]   McCarthy TS. The impact of acid mine drainage in South Africa. S Afr J Sci. 2010, 107(5/6), 712.

[20]   Winde F, Sandham LA. Uranium pollution of South African streams – An overview of the situation in gold mining areas of the Witwatersrand. GeoJournal. 2004, 61, 131–149

[21]   Owen JR, Kemp D, Lébren É, Svobodovaa K, Murillo G-P. Catastrophic tailings dam failures and disaster risk disclosure. Int J Disaster Risk Reduct. 2020, 42, 101361.

[22]   Grimalt JO, Macpherson E. The environmental impact of the mine tailing accident in Aznalcollar (South-West Spain). Sci Total Environ. 1999, 242, 1–332.

[23]   Fuentes I, Márquez-Ferrando R, Pleguezuelos JM, Sanpera C, Santos X. Long-term trace element assessment after a mine spill: Pollution persistence and bioaccumulation in the trophic web. Environ Pollut. 2020, 267, 115406.

[24]   Dos Santos Vergilio C, Lacerda D, Vaz de Oliveira BC, Sartori E, Campos GM, de Souza Pereira AL, Borges de Aguiar D, Da Silva Souza T, Gomes de Almeida M, Thompson F, de Rezende CE. Metal concentrations and biological effects from one of the largest mining disasters in the world (Brumadinho, Minas Gerais, Brazil). Sci Rep. 2020, 10, 5936.

[25]   Ettler V, Štepánek D, Mihaljevič M, Drahota P, Jedlicka R, Kríbek B, Vanec A, Penížek V, Sracek O, Nyambe I. Slag dusts from Kabwe (Zambia): Contaminant mineralogy and oral bioaccessibility. Chemosphere. 2020, 260, 127642.

[26]   Toyomaki H, Yabe J, Nakayama SMM, Yohannes YB, Muzandu K, Liazambi A, Ikenaka Y, Kuritani T, Nakagawa M, Ishizuka M. Factors associated with lead (Pb) exposure on dogs around a Pb mining area, Kabwe, Zambia. Chemosphere. 2020, 247, 125884.

[27]   Bose-O'Reilly S, Yabe J, Makumba J, Schutzmeier P, Ericson B, Caravanos J. Lead intoxicated children in Kabwe, Zambia. Environ Res. 2018, 165, 420–424.

[28]   Ruchter N, Zimmermann S, Sures B. Field studies on PGE in aquatic ecosystems. In: Platinum Metals in the Environment (Zereini F, Wiseman CLS Eds.). Berlin: Springer, 2015.

[29]   Erasmus JH, Malherbe W, Zimmermann S, Lorenz AW, Nachev M, Wepener V, Sures B, Smit NJ. Metal accumulation in riverine macroinvertebrates from a platinum mining region. Sci Total Environ. 2020, 703, 134738.

[30]   Erasmus JH, Wepener V, Nachev M, Zimmermann S, Malherbe W, Sures B, Smit NJ. The role of fish helminth parasites in monitoring metal pollution in aquatic ecosystems: A case study in the world's most productive platinum mining region. Parasitol Res, 2020, 119, 2783–2798.

[31]   UN Water.org. Water quality and wastewater [Internet, Cited 2021 June 17]. https://www. unwater.org/water-facts/quality-and-wastewater/

[32]   Hwang SY, Lee GB, Kim JH, Hong BU, Park JE. Pre-treatment methods for regeneration of spent activated carbon. Molecules. 2020, 25, 4561, doi:10.3390/molecules25194561

[33]   Peters RW, Shem L. Separation of heavy metals: Removal from industrial wastewaters and contaminated soil [Internet; Cited 2021 July 2]. United States. 1993. https://www.osti.gov/servlets/purl/6504209.

[34]   Yan A, Wang Y, Tan SN, Yusof MLM, Gosh S, Chen Z. Phytoremediation: Promising approach for revegetation of heavy metal-polluted land. Front Plant Sci. 2020, 11, 1–15.

[35]   Peer WA, Baxter IR, Richards EL, Freeman JL, Murphy AS. Phytoremediation and hyperaccumulator plants. In: Tamas MJ, Martinoia E (eds) Molecular Biology of Metal Homeostasis and Detoxification. Topics in Current Genetics, 14. Springer, Berlin, Heidelberg. 2005.

[36]   Suresh B, Ravishankar GA. Phytoremediation- a novel and promising approach for environmental clean-up. Crit Rev Biotechnol. 2004, 24, 97–124.

[37] Schnoor JL, Licht LA, McCutcheon SC, Wolfe NL, Carreira LH. Phytoremediation of Organic and Nutrient Contaminants: Pilot and full-scale studies are demonstrating the promise and limitations of using vegetation for remediating hazardous wastes in soils and sediments. Environ Sci Technol. 1995, 29, 318–323.

[38] McGrath SP, Zhao F-J. Phytoextraction of metals and metalloids from contaminated soils. Curr Opin Biotechnol. 2003, 14, 277–282.

[39] MiningWatch Canada. Potential Toxic Effects of Chromium, Chromite Mining and Ferrochrome Production: A literature Review [Internet; Cited 2021 June 3]. Available at: https://mining watch.ca/sites/default/files/chromite_review.pdf

Raymond J. Turner
Chapter 9
Toxicity of nanomaterials

9.1 Introduction

Nanoscience and nanotechnology represent the world of "very small materials", impacting many different organisms and ecosystems because of their use in biomedicine, energy, environmental engineering, chemistry, material science, optoelectronics, and life science, just to name a few. Here we provide an overview of what nanomaterials are, what kinds of nanomaterials exist and what they are used for. This overview will distinguish natural/biological nanomaterials from engineered nanomaterials and nanomaterial-enhanced products. The overall goal is to provide the reader with information as to where nanomaterial pollutants come from and to understand their subsequent lifecycles in ecosystems. The toxicity of selected nanomaterials will be briefly discussed based on their structure and chemical composition. Analytical methods to study nanomaterials will be briefly discussed as well. The fields of nanoscience, nanoengineering, and nanotechnology are now maturing, with many new materials and products being produced and used yearly. This chapter will provide a general overview and should not be considered to be a comprehensive introduction to all nanomaterial types that are available today.

9.2 Nanomaterial background

9.2.1 What are nanomaterials?

The prefix *nano* refers to structures defined to have at least one dimension on the nanometer scale (1 nm = 10^{-9} meter). This gives the nanomaterials (NM) sizes that are intermediate between molecules and bulk materials [1]. Nanotechnology is defined as the field that studies the manipulation of matter at the nanoscale (1–100 nm) toward technological materials for industrial applications. Although, now many useful nanomaterials extend into the 100s of nm size range. The infinitesimal size of nanostructures

Acknowledgments: The author would like to acknowledge funding from the Natural Sciences and Engineering Research Council of Canada

Raymond J. Turner, Department of Biological Sciences, University of Calgary, Calgary, Alberta, Canada, e-mail: turnerr@ucalgary.ca

https://doi.org/10.1515/9783110626285-009

(NSs) confers on them unique physical-chemical features in part due to their large surface-to-volume ratio, spatial confinement, high surface area, and energy. Essentially, "macro" chemistry and physics does not apply to the nanoscale. Many nanoscale atom interactions generate fascinating quantum effects now referred to as plasmons. This is a quantum phenomenon that represents waves of collective electron oscillations on the surface of a conducting material that provide properties for many applications. Overall the size-dependent properties change and enhance their mechanical, chemical, electrical, optical, and magnetic features [2].

Nanotechnology is classified by the FDA as a field that includes: (i) research and technological development or products that are at the atomic, molecular or macromolecule levels where at least one dimension that affects the behavior is on the length scale range of 1–100 nm, (ii) creating and using structures, devices and systems that have novel properties and functions due to their nm size/dimension; (iii) ability to control or manipulate matter at the atomic size.

The idea of nanoscience has been around for about 60 years. In fact, in a presentation in 1959, Richard Feynman mentioned that "there is plenty of room at the bottom", and he suggested the synthesis of molecules via the direct manipulation of atoms. N. Taniguchi was the first to use the term "nanotechnology" in 1974. However, the concept of nanotechnology was not really appreciated until 1986 when D.E. Dexler defined that the era of nanotechnology was upon us and the first successful nanoscale assembly of fullerenes were reported in 1985 ("buckyballs" ~ 1 – 2 nm diameter). The field began to grow throughout the 1990s and since 2001 the field has virtually undergone explosive growth in terms of dedicated research aimed to develop suitable NM synthesis processes. Based on both the use of innovative physical-chemical methods and the availability of new and advanced tools for their characterization (e.g., spectroscopic and electron microscopy techniques) novel nanomaterials continue to be produced [3, 4].

Among the diverse morphologies of NMs, nanoparticles (NPs) that have a spherical shape are considered to be the most common and they are referred to as 0-dimensional (0-D) nanocrystals and quantum dots (QDs). Next, we have the 1-dimensional (1-D) nanorods (NRs), nanowires (NWs), nanotubes (NTs), and nanobelts (NBs). One also needs to consider 2-dimensional (2-D) arrays of nanoparticles (NPs) which can be used to generate thin films, such as single atom layered sheets or plates (e.g., graphene). Finally, there can also be 3-dimensional (3-D) structures (superlattices and dendritic patterns) [3] (Figure 9.1).

Nanomaterials may be of uniform atom type or a mixture of atoms and formulations. For many types of nanomaterials their stability (and often their features) is dependent on the molecular framework on the surface that surrounds their atomic cores. This is called the cap or coating and may be composed of a different elemental atoms or defined molecules that may or may not originate from the synthetic process. When one considers the toxicity of a NM, one therefore has to consider the effect of the NM as a nanostructure to cells, but also the toxicity of capping molecules and, if the NM decomposes, the toxicity of the core atoms that are released.

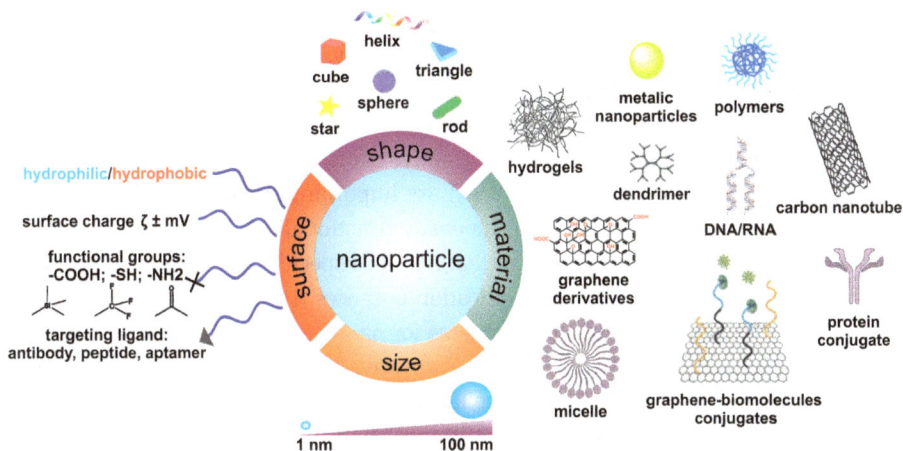

Figure 9.1: Examples of types, shapes, and properties of some nanomaterial (adapted from Centrum Nanobiomedyczne, University Adam Mickiewicz, Poland (http://cnbm.amu.edu.pl/en/nanomaterials)).

9.2.2 Types of nanomaterial

Here we briefly describe the different kinds of nanomaterials with an emphasis of considering their most common building blocks.

Inorganic nanoparticles. These are typically of single or binary elements and may be metal or metalloid based. Perovskites are inorganic crystals that follow the general structure of $CaTiO_3$, but many different cations and halides can be incorporated, and Ti may be replaced with other elements, such as Si. These are gaining attention for catalytic nature and electrode modifiers in fuel cells, sensing, and solar cells [5].

Polymer based. Silica can generate a hydrogel for encapsulating other NMs or molecules. A promising application are antimicrobial coatings [6]. A variety of organic polymers can also be tuned to form NPs of various sizes. The polymer chemistry can be functionalized to be photo or pH responsive for drug delivery [7]. Overall conjugated polymer NPs are very attractive due to their comparative ease of synthesizing and functionalizing for a wide variety of applications [8]. A group of macromolecular complexes referred to as dendrimers, are highly ordered and branched polymers [9] that provide features for diverse applications, particularly in the medical field.

Amphiphiles and lipids. These molecules contain a hydrophobic and a hydrophobic end which results in a very large separation potential when they are introduced into water. A powerful physical phenomenon of amphipathic molecules is their tendency to spontaneously form spherical structures (micelles, vesicles, liposomes) as this minimizes their energy when the hydrophobic moieties are repelled by the water. Their overall size is dependent on how polar and hydrophobic the components are and by the length of the hydrophobic moiety.

Biological molecules. Apart from the engineered NMs, Nanostructures are present in a wide range of living organisms [10]. In fact, the dimensions of bacterial phages and viruses are on the nanoscale and certainly, many biomolecules are also "nanosized". Nucleic acid polymers (DNA, RNA) are important functional nanomaterials, which are tunable to length and can be easily functionalized. DNA is essentially a nanowire with a 2.5 nm diameter. Given that DNA has specific hybridization properties, it can be used for very specific nanofabrication purposes [11]. Enzymes are bionanocatalysts. Large biomolecules are also templates for nanomachines, such as the F_oF_1 ATPase and flagella are rotors. One can also take advantage of natural self-assembling protein-based polymers (silk, keratin, collagen, elastin, actin, pilin, etc.), which can then be functionalized for various applications [12]. Another group that fits into this category is nanocellulose, which may be one of the most prominent green materials. In fact, cellulose nanocrystals are increasingly used in applications such as emulsifiers, adhesives, and applications in wastewater treatment and biomedical applications [13]. Chitosan is a glucosamine carbohydrate polymer that is explored for similar applications. Finally, naturally produced polyhydroxyalkanoate-based nanoparticles are of interest for biomedical applications as they are very biocompatible and biodegradable while offering good mechanical flexibility [14].

Natural nanostructures are employed in biological systems for a variety of functions, including the color and pattern of butterfly wings, peacock feathers, and other insect colorations. These colors, in fact, do not come from pigments, but rather from nanomaterials that uniquely reflect light [15], which can be used for coloration purposes while avoiding more toxic pigments/paints/dyes. Another example of biological NMs is nano-propolis, which is bee 'glue' which has a wide range of applications for animal health in agriculture [16]. For the most part such natural nanomaterials are not considered to be toxic.

Complex nanomaterial. An interesting class of nanomaterial worth mentioning are nano-conjugates (chemically crosslinked mixed systems) and sometimes referred to as nanobots (nano-robots), which are 'machines' with components at or near the nm size range. A simple view of a nanobot is a nanomaterial composite that is "programmed" to carry out a specific targeted function. An example would be coupling an enzyme into to a NP. This can be done by adding the enzyme into a micelle which is decorated with antibodies for specificity. In this way the antibody targets a cell and the vesicle fuses and releases the enzyme that can then change or repair a process within the cell. In this way the nanobot is "programmed" to transport a specific molecular payload to a target site within a cell.

9.2.3 Synthesis chemistry and physics of nanomaterials

Very small NPs (clusters of a few atoms) are the building blocks for the generation of various and diverse NSs [4]. Overall nanomaterials can be synthesized based on three principal approaches (Figure 9.2):

(i)   bottom-up approaches, where the conversion of the metal ion-precursors into NSs is achieved by specific chemical reactions (e.g., reduction, hydrolysis, oxidation) or hydrothermal and solvothermal processes.

(ii) top-down processes, by which NMs can be obtained following the addition of a bulk precursor to a solvent that, under receiving energy (light irradiation (pulsed laser ablation, laser-induced melting/vaporization, photocatalysis) microwaves, pressure), mediates the bulk material's melting or transformation and dissociation into smaller entities.

(iii) template-based procedures that involve the use of different physical or chemical templates to convert the precursors into NSs [2, 17].

Considering the fundamentals that size and shape gives material at the nanoscale range, various synthesis procedures have evolved to obtain monodisperse and stable metalloid NPs. Conversely, 1D metalloid NSs are generally produced by (i) varying crucial parameters of existing synthesis methods or (ii) exploiting the strong anisotropic nature of specific elements which spontaneously grow by adding atoms along one only axis [18]. The choice between the different processes to generate NMs mainly relies on their ease and versatility, as some of them are typically used to tune the production toward specific nano morphologies. However, for those

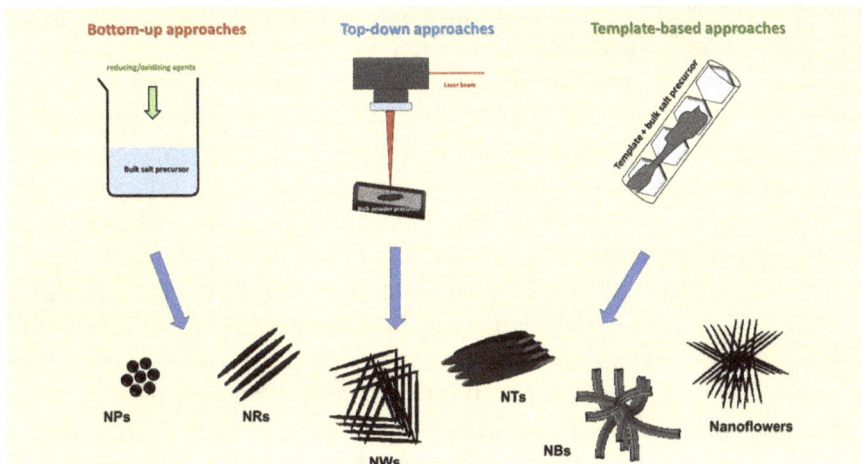

Figure 9.2: Overview of bottom-up, top-down, and template-based chemical and physical methods commonly used to produce various nanomaterials (adapted from [20]).

elements that have a strong colloidal nature, the production of NMs is associated with several challenges, such as obtaining homogeneous populations (monodisperse) with good thermodynamic stability [19].

Before we consider the toxicity of nanomaterials themselves, it is instructive to reflect on the manufacture processes. What kind of pollutant wastes are produced as a result of nanomaterial manufacturing? High-quality metal NPs are mainly produced by embracing chemical bottom-up approaches due to their competitive cost and their short synthesis times [17]. Bottom-up processes to make metal based NMs are generally based on reduction reactions that occur between a suitable metal salt or chelate precursor and a reducing agent. The choice of reducing agents is broad and includes harsh chemicals, such as hydrazine (N_2H_4), acetic acid, oxalic acid, sodium ascorbate/thiosulfate, or a mixture of phosphoric acid (H_3PO_4), hydrochloric acid (HCl) and iron(II), and potassium borohydride (KBH_4).

Chemical reductions have also been applied for the production of 1D metalloid NSs either by increasing the amount of reducing agent generally used for Se or Te NPs synthesis, which improves the reduction kinetics and, therefore, the number of metalloid atoms that are available for the growth of NM in one direction to produce a rod or wire [21]. Another approach is to use a variety of solvents and adding different alcohols (e.g., n-butyl alcohol) to an aqueous system, which can result in the accumulation of NPs chains at the water/alcohol interface [22].

The chemogenic approaches to synthesize NMs that involve the use of hazardous and harsh chemicals are then converted in toxic waste, which has to be disposed [23, 24]. Moreover, methodologies such as chemical vapor deposition rely on the addition of metal catalysts to the reaction system, which can eventually lead to either the generation of an explosion if a spill event occurs, or the incorporation of their residues into the final NM products, making them potentially toxic for humans. To avoid the use of harsh chemicals, there is now a trend toward top down and template methods, however, given the energy required to run such experiments they are in no part eco-friendly. Thus, considerable efforts in the past decade have been directed to use various organisms and biomolecules to synthesize metal NMs. Examples of such biochemicals include glutathione, cysteine, dextrose, ascorbic acid, L-asparagine, variety of reducing sugars, and oxidoreductase enzymes particularly for metal nanomaterials as well as template-agents biologicals such as chitosan, oligosaccharides, proteins, and even bacterial phage. Beyond using an organisms' biomolecules, the organism itself can be used to achieve NM biosynthesis. To this end, there are plant-based approaches, where plant by-products (e.g., banana peels, tea leaves) are used as a source of reducing sugars [25]. Fungi and bacteria are becoming increasingly popular for the eco-friendly synthesis of nanomaterials, particularly Ag NPs for use as antimicrobial applications [see reviews [26–28]]. These biogenic approaches are considered *greener* and *eco-friendlier*.

9.2.4 Stability of nanomaterial

The so-called Derjaguin-Landau and Verwey-Overbeek (DLVO) theory, formulated in the 1940s, describes the stabilization of colloids, which is defined as a two-phase system consisting of particles that range from 1 nm to 1 μm in size and can be dispersed in a continuous phase (dispersion medium) [29]. The DLVO theory predicts the probability of two identical particles to aggregate, which is determined by the attractive/repulsive total potential energy of interaction. NMs can undergo only two types of aggregation: homo- or hetero-aggregation, which refers to the aggregation of similar or dissimilar (e.g., different size, shape, or composition) particles, respectively [30]. Since the classical DLVO theory explains only the stability of homogenous NMs, it was subsequently extended (XDLVO) to describe the hetero aggregation phenomenon through the addition of forces affecting the interaction between NMs, namely: magnetic and bridging attractions, hydrophobic and Lewis acid-base forces, osmotic and elastic-steric repulsions [30]. Based on the XDLVO theory, chemogenic metal NMs are stabilized through 3 major mechanisms: (i) electrostatic stabilization, which depends on the formation of a charged layer around the NS surface; (ii) steric stabilization, in which a protecting layer surrounding the NM is formed by the addition of a polymer adsorbed to the NS surface; and (iii) electrosteric stabilization, which is a combination of the two aforementioned approaches [31] (Figure 9.3). It is critical to appreciate the theory and processes of NM stability, as it contributes to the residence time and their state in biological fluids, and as such their toxicity.

For chemogenically produced NMs, the chemical agents used are critical, as often this molecule becomes part of the final NM as the cap or coat and thus influences the stability of the NS. In the case of biogenically produced NM, biomolecules from the organism inevitably become part of the coating. Various biochemicals (amino acids, fatty acids, nucleotides, metabolites) and macromolecules (proteins, peptides, lipids, nucleic acids (DNA or RNA) and carbohydrates) have been found as components of the organic material that surrounds NSs. These seem to also participate very well as contributors to their stabilization, yet the chemical-physical mechanisms behind the ability of these molecules to provide stability to biogenic NMs have not been elucidated yet, but likely involve combinations of various stabilization parameters (Figure 9.3).

The ability of several microorganisms to synthesize nanomaterials (Biogenic-NMs) as a result of their diverse metabolism is now recognized as a valid eco-friendly and cost-effective alternative to chemical synthetic methods. Although the mechanisms of biogenic-NM synthesis are not completely elucidated, the natural stability of these NMs represents one of the most important advantages in using biological systems over the well-known and established chemical synthesis procedures.

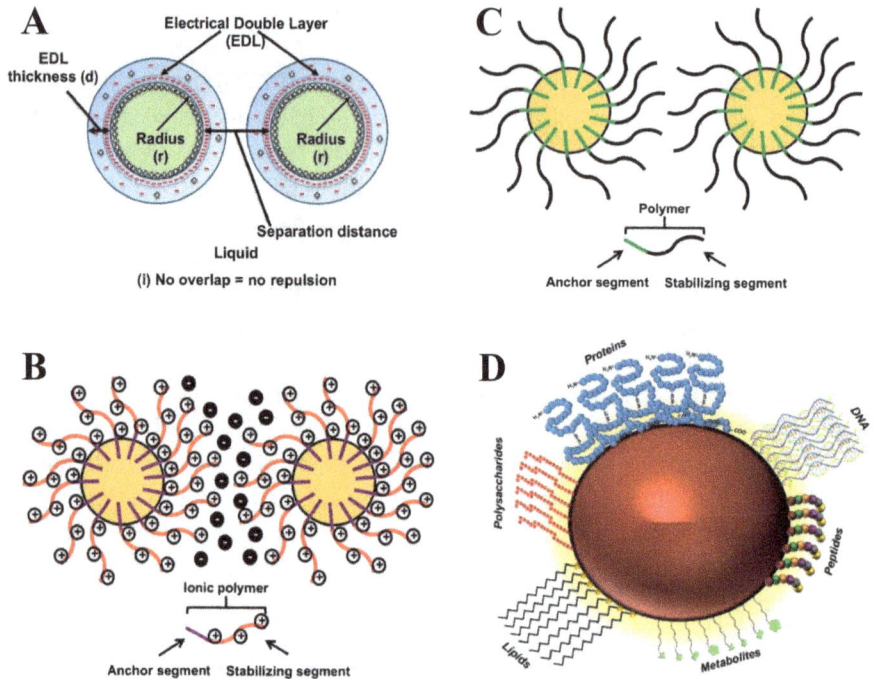

Figure 9.3: Schematic illustration of the various nanomaterial stabilization processes.
A. electrostatic repulsion forces between two NSs approaching each other as function of their separation distance. B. steric repulsion force generated by the adsorption of polymer molecules onto the NM surfaces. C. electrosteric repulsion forces generated by the adsorption of ionic polymer molecules onto NM surfaces. D. Biomolecule classes involved in the thermodynamic stabilization of a biogenic NMs, which can be mediated by the occurrence of both electrostatic and steric interactions as function of the type of biomolecules considered. Overall, electrostatic repulsion can be mediated by DNA, proteins, peptides, charged lipids, or metabolites, while neutral lipids, polysaccharides, and metabolites can be responsible for the steric contributions to the stability of biogenic NMs. Adapted from [32].

9.3 Uses of nanomaterials

Nanotechnology offers benefits because of the ability to tailor NSs at small scales to achieve specific and desirable physicochemical properties. Materials can be made stronger, lighter, more durable, reactive, conductive, and sieve like compared with their macro-sized material counter parts. There are many commercial products on the market now that are in daily use that rely on NMs. The addition of nanomaterials provides enhanced general properties in terms of weight, durability, thermal/biological/UV stability, electrical conductivity, catalysts, and solvents as well as antimicrobial activity. Below, a brief overview of uses and applications of nanomaterials (Figure 9.4),

APPLICATIONS OF NANOPARTICLES

Heat retaining textiles

Natural/synthetic polymer hybrid fibres

Wound dressing
Dental ceramics

Paint-on solar cells
Anti-stain textiles
Technical textiles
Medical textiles
Bone growth
molecular tagging

Dye sensitized solar cells
Lithium ion battery electrodes

Drug controlled release
Biomakers

Fuel additive catalysts
Hydrogen production photocatalysts
TEXTILES
Cancer therapy
Hyperthermic treatment

Hydrogen storage materials
Fuel cell catalysts
RENEWABLE ENERGY
BIOMEDICAL
Drug delivery
MRI contrast agents

UV blocking textiles
Environmental catalysts
IR contrast agents

Self-cleaning textiles
Waste water treatment
ENVIRONMENT
Nano particles
HEALTH CARE
Imaging
Sunscreens

Automotive catalysts
Quantum computers
Antibacterial
Antioxidants

Pollutant scavengers
High density data storage
UV protection
Interactive food

Pollution monitoring sensors
ELECTRONICS
FOOD AGRICULTURE
Nutraceutical
Food processing catalysts

Ferro fluids
Fungicides
Food quality/safety analysis sensors

Quantum lasers
Nanoscale patterning of electronic circuits
Food packaging
Gas-barrier coatings

High power magnets
Refractive index engineering
INDUSTRIAL
Functional nanocomposites
UV blocking coatings

Single election transistors
Superplastic ceramics
Industrial catalysts
Wear resistant coatings

High sensitivity sensors
Chemical sensors
Paper additives And coatings
Reinforced plastics
Antimicrobial coatings

Gas sensors
Super thermal-conductive liquid
Nano pigments
Self-cleaning building surface

Chemical mechanical planarization
Nano-phosphors for display

Nano-inks
Transparent conductive polymer films
Antifouling coatings

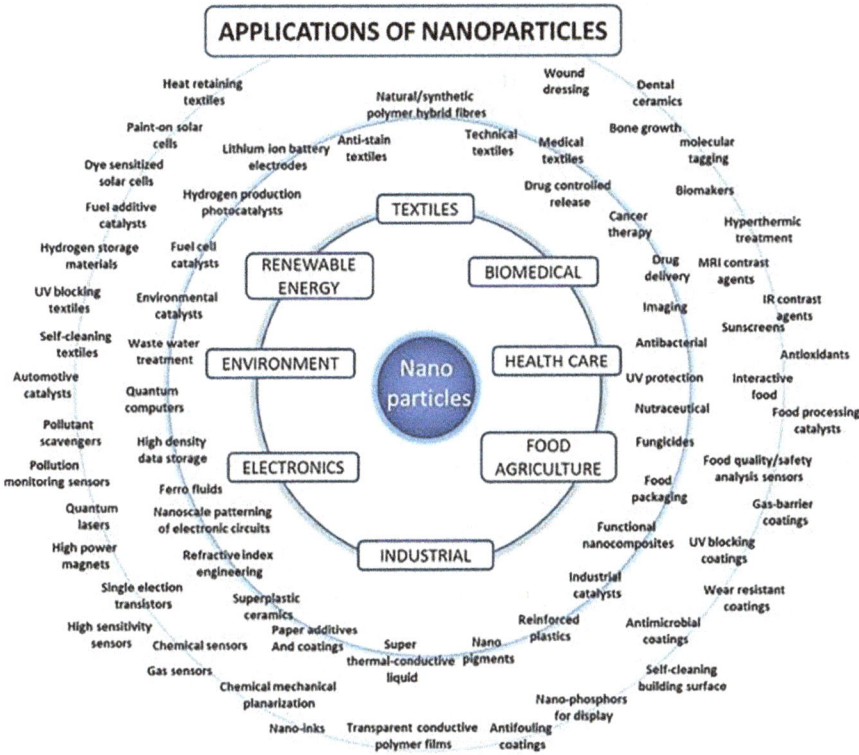

Figure 9.4: Nanomaterials are everywhere and used now in almost all fields and applications. Adapted from [33].

but it is by no means exhaustive and is intended to highlight the diversity of applications of NMs for students to contemplate potential sources of environmental release. For a more extensive list and information see the National Nanotechnology Initiative site (www.nano.org). The discussions below are to solidify in the readers' mind that nanomaterials are now ubiquitous in our lives and daily contact with them is now essentially inevitable. Given that nanomaterials are now everywhere, this equates to increased manufacturing waste chemicals as well as the release of NMs into the environment either during the use of the product or at the end of a products life and its disposal.

9.3.1 Electronics

This is the field we are seeing the most applications of NMs in our daily lives, such as ultra-high-resolution displays in televisions, computer monitors, pads, and cell phones. Almost small electronics of the electronic components now include many different types of NMs in forms of transistors capacitors and magnetic random-access memory. Consider your smartphone and watches for the amount of "miniaturized" electronics. Conductive inks for tattoo sensors. Many sensing technologies, either as light (cameras), ultra-sensitive/responsive hearing aids or for chemicals and gases will also contain interfaces of coatings of NMs. Sensors for temperature, pressure, and/or stress use NMs that are now used within the city and transportation infrastructure for monitoring purposes.

9.3.2 Unique chemistries

Metals are used as catalysts in a plethora of synthetic reactions. As NMs have high surface area per volume, they provide a large catalyst potential. Additionally, various crystalline NM have shapes that provide enhanced reactivity, which allows for far less catalyst to be used. We see a variety of cleaning materials (anti-spotting in dishwashers) which contain NM catalysts. An important application relevant to this text are NP catalysts to enhance the degradation of pollutants, such as halogenated or textile/paint dyes.

NMs are also being applied in energy to help address climate change. Various NMs and nanotechnologies have been applied to improve oil and gas extraction. Solar electric panels also use NPs, with recent advancements of types of nanomaterial on nanostructured supports continue to improve efficiencies and applications. We see increased efficiency of electricity transmission lines, lighting systems, and daily electronics to reduce power loss and usage. In addition, applications for NMs to be used in carbon capture are starting to emerge.

In the cosmetics industry various NMs are used as preservatives, colorants, and color enhancers. Vesicles and micelles are used to deliver beauty and health enhancers in skin products. Sunscreens now include a variety of NPs. In personal hygiene we see NMs in toothpaste as whiteners and as antimicrobials.

9.3.3 Materials

Adding nanomaterials of various compositions can help to change the properties of common materials. This has led to lightweight building materials incorporated into transportation vehicles (e.g., carbon fiber). They are found as product enhancements in athletic materials such as sports rackets and helmets. A variety of nanomaterials like graphene and cellulosic materials are templates for flexible, bendable, rollable, and stretch electronics toward wearable applications. NMs are added to existing traditional materials to improve their properties. Example of coatings on visual surfaces (eyeglasses, monitors, camera lenses, windows) can enhance water and residue repellency, antifog, anti-reflective, light filtration, and abrasion resistance. Further enhanced electrical properties and greater durability and effectiveness are provided by NMs incorporated into power tools, building materials, plastics, and rubber products.

In the textile industry NMs help to reduce wrinkling, and provide anti-static, anti-staining, insect and water repellency, UV blocking, and bacterial inhibition and thus odor control [34]. Production of washable and durable "smart fabrics" are now possible with sensors for health with electronics that have their power collection through movement of the wearer.

9.3.4 Agriculture

Nanotechnology in plant agriculture is gaining a lot of interest [35, 36]. Pesticides and fertilizers manufactured to the nanoscale appear to have higher efficacy and may help prevent runoff. Nanotechnology can be used for genetic engineering with NPs carrying genetic material to modify the plants biology (genetically modified organisms), which is more efficient than traditional methods. NMs can also be used to augment the soil microbiome (targeted antimicrobial activities; fungicide, virucide) and for soil conditioning. NM seed coatings can facilitate their stability, prevent rot, and enhance the probability for seed germination. Plant related nanotechnologies are considered as a key solution with regard to the United Nations Sustainability Goals to decrease environmental issues around agriculture and at the same time increase yields to deal with the primary threat to the planet, overpopulation! An exciting opportunity is to use plants as sensors to nanomaterial pollutant levels in the environment. Beyond plants, agricultural animal health has also been improved through biomedical applications.

9.3.5 Biomedical applications

Beyond issues around toxicity, the cellular accessibility of ENMs, make them excellent nanocarriers for drug delivery. The field of nanomedicine applies nanomaterials to deliver cellular components (DNA, proteins, vesicles, etc.) to produce precision medicine for disease treatment, prevention, and diagnosis [37]. We see both lipid-based NPs and polymer based NMs being used for nucleic acid and drug delivery purposes. In cancer treatment NM photosensitizers are being used for photodynamic therapies against cancer cells. In disease diagnostics, magnetic NPs and various other NMs are useful agents for higher contrast in resolution resonance imaging (MRI). Certainly, we have seen micelle NSs being used for drug delivery and during the COVID-19 pandemic, carriers of nucleic acids as a vaccine. Injury treatment is seeing a lot of advances that involve nano materials as tissue and bone regeneration scaffolds and to foster nerve regeneration. Various metal NPs have strong potential as antimicrobials in infection control. An amazing genomic technology is nanopore sequencing for reading the DNA code.

9.4 Sources of nanomaterial pollutants

An important distinction that has to be made is to be aware of the difference between naturally formed nanomaterials and anthropogenically engineered nanomaterials (ENM). Such distinctions are similar to naturally occurring organic molecules and xenobiotics. Product formulations containing ENMs are now often referred to as nano-enabled products (NEPs) or nano enhanced materials (NEMs). The rapid advancement of nanotechnology has seen an exponential increase of engineered nanomaterials that are incorporated into a wide variety of products in pursuit of superior properties. Back in 2005 there were only about 50 NEPs on the market, whereas in 2020 there were > 5,000 [38]. Markets for NEPs are currently dominated by health and fitness products such as active wear, sunscreens, cosmetics, and sporting goods, but electronic and food processing applications are catching up.

9.4.1 Issues around policies and regulation

Despite the indisputable usefulness of NMs in a plethora of applications, there is a constant and justified concern that NEPs and ENMs may have unwanted ecological and biological effects on cells and organisms that have not yet been discovered or may be resulting in chronic health problems. Thus, there is an urgency for a more systematic understanding of the impact of NMs on health and ecosystems [39]. However, there seems to be no specific international regulations or internationally agreed upon protocols or even legal definitions for the production, handling, labeling, toxicity testing, and the environmental impact assessment of NMs [10]. Regardless, there are some key considerations of NMs as pollutants that are starting to be described in several policy documents from various regional organizations (FDA [40], OECD [41], ECHA [42]) that tend to identify similar issues and required measurements. These include:
- Identification of the most relevant material form and exposure routes for human health and the environment.
- Estimation of environmental compartmental distribution and fate.
- For soil and air, estimations of residence times and decomposition rates.
- For aquatic systems, estimation of removal during wastewater treatment.
- Estimation of transit from exposure site to systemic circulation in humans and organisms.

The Environmental protection agency (EPA) has been approving various ENMs, particularly nanosilver, for well over 15 years. The initial approval of the use of ENM is based on their enhanced performances. The approval is "conditional", meaning that the manufacturer must provide test results (within four years) showing how the NMs interact with the environment. However, given the backlog in this agency, follow-up is typically slow and insufficient.

It will likely take years for any toxicity issues of ENMs in humans to become apparent, and it is reasonable to predict that certain individuals will be more susceptible than others. Thus, as with other environmental toxins (e.g., DTT, dioxins, Hg, and Pb) it will take time until the effects of the use of PR-ENMs on aquatic, terrestrial, and human health are comprehensively understood.

9.4.2 Manufacturing exposure

Exposure to toxins in NM manufacturing processes is multifactorial. One has the reagents and the wastes from the synthesis protocols of producing the NMs. Then one has the exposure to the NMs produced. This is then met with the toxicity around the manufacturing of PNMs. There are also additional risks in the laboratory research exploring new NMs that would have unknown toxicities and yet explored properties. This means all types of exposure are possible not only to the NMs but the reagents and solvents used. For each NM and product protocols of risk and use are put in place following best research and manufacture processes. Yet this is in part a trial and error process and caution must be taken with such novel materials that have very different properties than their macro counterparts. Such agencies as the FDA (www.fda.gov), OSHA (www.osha.gov), and OECD (www.oecd.org; see also NANOMET (www.oecd.org/chemicalsafety/nanomet) are constantly adding and editing guidelines.

9.4.3 Product-released engineered nanomaterials

The ENM within NEPs are typically surface-bound or suspended in liquids, creams, or emulsions. Thus, an issue with such NEPs is that the ENMs can be easily released into the environment during product use and/or disposal. This type of pollution has received the name product-released (PR) ENMs or PR-ENMs and we are seeing their accumulation in wastewater treatment plants (WWTP) [43]. This means that these entities enter aquatic/marine systems, which can result in the environmental exposure of organism either via WWTP effluent or directly through industrial discharge (Figure 9.5).

Data on the environmental risks that are associated with PR-ENMs are very limited as exposure levels and accumulation levels in relevant organisms have not been thoroughly characterized. An issue is that there may be very low levels of PR-ENMs of nano-Ag and TiO_2 that are released from the textiles, paints, and cosmetics perhaps as low as 0.5% [44–45], but this value may vary depending on product and NM type. In fact, although they get a lot of attention, the EU commission found that only 1.5% of all cosmetic products contain ENMs [46], yet this number is likely to increase. This trend demands increased efforts to understand acute effects, but more so for environmental chronic effects through bioaccumulation in aquatic organisms. The primary NEMs in NEPs are nano forms of TiO_2, ZnO, Ag, SiO_2, and range in global production from 420 to 1.4 million tons/year [47]. These also present a reasonable risk for environmental exposure as they are emitted into environment [48].

TiO_2 and ZnO are widely used as active ingredients in sunscreens and cosmetic products. Both are used widely in textiles, paints, and both ZnO and Ag NPs are found in a wide variety of antimicrobials. Similarly, SiO_2 NPs are use in a variety of industrial activities including anti-caking, absorption, self-cleaning, paints, paper, and textiles as well as cosmetics (creams and hair care products). Overall, the majority of NPs in aquatic systems arise from PR-ENMs from sunscreens, personal care products, paints, and clothing and textiles [48]. Thus, simple activities of personal body hygiene through cleaning and washing clothes readily release these ENMs that subsequently enter WWTPs.

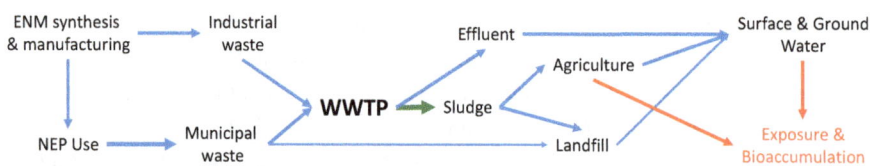

Figure 9.5: Product release of engineered nanomaterials into surface water. An illustration of the wastewater treatment plant (WWTP).

Overall, the levels of PR-ENMs in the environment are still very low. Release rates from various materials range considerably. Release from sunscreen is on the order of 20–40% during use and subsequent washing releases NMs to WWTP. Yet the remaining would be within the carrying agent (cream/lotion) that will also visit the WWTP. The amount of AgNP loss from various products range considerably and may be as high as 35% of the Ag in the product released toward WWTP or surface water [48].

From Figure 9.5 we see that the WWTP acts as a major gateway between the release of NMs from NEPs and the environment. Observations show that 70–90% of the NMs attach and settle with the biomass sludge of WWTP [49, 50], with the remainder staying in the effluent stream. The process relies on the NMs and metals binding to the organic waste and being sequestered with the sludge. For binding of NMs this depends on the NM coating and stability that in part defines how long the particle will stay with the sludge vs discharge to surface water [51]. If we consider that 50–60% biosolids are recovered and applied to agriculture fields as fertilizer and soil amendments, we can envision their bioaccumulation in agriculture food sources. Furthermore, the remaining biosolids are either incinerated or put into a landfill. In the case of incineration, the metals are not destroyed and rarely captured, but are often simply volatilized into the atmosphere potentially leading to inhalation exposure or again collected via rainwater and to surface water ecosystems. In land-fills, this can result in NM leaching to surface or ground water. So, the released concentrations of NMs are low, but they are within the range that may affect some sensitive biota. Microbial communities of aquatic and marine periphyton can certainly be affected by the antimicrobial Ag and ZnO NMs that were manufactured as antimicrobials in the first place. Thus, it is clear that for our surface water, there is the potential for ecosystem bioaccumulation and biodiversity damage.

It is clear that far more research is required to understand the release, transport, fate, and effects of PR-ENMs. Monitoring for ENMs in WWTPs and the general environment is not routine at this time. Certainly, the low concentrations do not seem to induce any significant acute effects on test organisms at present [52, 53]; however, studies that address their bioaccumulation and subsequent chronic exposure are lacking.

Moloi et al., [48] performed predicted-no-effects-concentration (PNEC) analysis (see Chapter 1) using data from the literature. They found that textile originating PR-Ag NPs presented the highest risk quotient for fish. Sunscreen-released PR-ENMs showed potential adverse risk quotients for crustacea, echinoderms and algae. Furthermore, the increase of ENMs in cosmetics is expected to pose a significant risk to the aquatic environment and the organisms therein. There are still considerable gaps on popular ENMs, and there are constantly new ENM produced and released (such as carbon nanomaterials). This leads to considerable challenges to any modeling studies [54].

9.4.4 Agriculture exposure

Given that most of the PR-ENMs are binding to the solid component of WWTP and in many jurisdictions these biosolids are spread on agricultural fields, thus transferring the ENP for availability of plant exposures. Exposure in plant agriculture is dependent on the type and morphology, size, agglomeration, crystal structure, and coating physicochemical properties and of course concentration. Plants can absorb NMs via their roots, shoots, the phloem/sap, and their leaves. Location has implications on how herbivores, insects, birds interact with the flora. NMs can affect root, stem, leaf morphology and size, product quality and yield, nutrient levels, and types (lipids, carbohydrates, and protein content) and secondary metabolites [36].

Beyond the ENM pollutant exposure to the plants, agriculture runoff translocates PR-ENMs into surface water ecosystem. Additionally, the increased use of beneficial ENMs in agriculture also leads to increased levels through runoff. The complexity in understanding the dynamic behavior of MNs in various ecosystems and the possibility of different environments leading to a change in the NM physicochemical properties makes exposure evaluation an extremely challenging task. Thus, to assess potential hazards by NMs in agriculture one needs to go beyond standard strategies for assessing conventional chemicals and consider the unique properties that NMs exhibit under different settings [55].

9.5 Toxicity

When it comes to considering the toxicity of NMs, various factors around the NM and the organism/cell must be considered. Overall, the cytotoxicity of NMs may be induced by the nature of the particle or the contents of the particle. Overall, it is possible to infer important toxicity variables if we know the following about the NMs [56, 57].

i.   Intended use. Will the application characteristics lead to cellular damage or lack thereof? What this means is one needs to consider if the presence of the NM in the product changes the product in a way that makes it inherently more dangerous.
ii. Nanoparticle properties. Size, size distribution, structure, and shape. How are the atoms distributed in the NMs?
iii. Surface properties. Surface area and surface chemistry. Surface ionization capacity, functional groups, and the presence of residual reagents (acid, base, reductants, oxidizers).
iv. Physical properties. Solubility (aqueous and other solvents), stability (melting temperature, decomposition half-life).
v.   Overall chemical composition/constituents. There should be a clear understanding of the mass/mole % of each of the elemental components. Both the NP core elements and capping molecules need to be considered.

Beyond what is listed here, of course regular toxicology factors play a role as well, including the dose and the overall exposure time. At this point, few studies have assessed the localization of NMs in organs or cells as well as their bioaccumulation and eventual sinks in the environment. Slowly resources are building, one of which is the Nanomaterial-Biological Interactions Knowledgebase (http://nbi.oregonstate.edu).

9.5.1 Nanomaterial properties in relation to toxicity

The size and surface area of NMs are intertwined. As a biochemist we can look at the molecular level to investigate if the initial cell contact is between the NM surface and certain sub-cellular components. As the NM diameter decreases the surface area of the particle increases on an exponential scale. So, if we consider area of interaction alone, if the cytotoxicity is induced by the NP interaction, smaller particles will be more toxic. But this is only one factor as the surface chemical reactivity has to be considered as well. NM characteristics of high reactivity and decreased melting temperature, which are products of the large surface area, can actually give higher cytotoxicity. This feature of higher reactivity will thus increase its biological interactions which can subsequently lead to biomolecule damages. But larger surface area is thermodynamically unstable and thus particles will aggregate if not well capped for stability. This aggregation may decrease reactivity, but could result is issues around their clearance from the organism.

To generate an inflammatory response the NPs need to migrate across the epithelial barrier. Thus, the inhalation of small NPS may penetrate deep into the lung parenchyma. NPs will be well distributed in the gas phase, but once they are in an aqueous phase they may be incorporated into cellular and bodily fluids, where their hydrodynamic radius and diffusion characteristics will come into play. Thus, one could see all sizes of NPs in the lungs, but smaller particles will have better diffusion properties into the alveoli to facilitate their infiltration into the bloodstream.

Regardless of the organ, once the NP comes into contact with cell surface, they can enter the cell through pathways, such as phagocytosis and pinocytosis [58]. These pathways allow a large size to range from 10 to 500 nm. Internalized particles typically enter lysosomal vesicles for potential degradation and elimination. Yet, this can lead to the decomposition of the NP due to the difference in the organelle pH, ionic strength, and redox potential. This environment will lead to the release of the NM atoms and potentially result in cell damage through ROS production, which may adversely affect key cell process particularly mitochondrial damage.

The NM surface chemistry of polarity and charge influences the biocompatibility and subsequently effective cellular internalization. Certainly, different surface charges lead to a variability in their biodistribution, which can overturn size effects. Thus, this factor is on the mind of drug discovery and delivery researchers. Regardless of the synthetic approach (chemogenic or biogenic), there is typically some form of molecular coating (i.e., the cap), which fundamentally determines the surface chemistry of the NM. The cap molecules play a major role in stabilizing the NM. One should also consider if the NM once it decomposes will release these cap molecules. Thus, one has to consider what their cytotoxicity characteristics are. Many NMs for biomedical applications are coated or decorated with polyethylene glycol (PEG), which is reasonably inert and can facilitate their clearing from the organism as well.

If the particles decompose, their core elements will be released. For metal-based NPs the consequence is a bit more obvious, as breakdown of the NP leads to metal ions being produced. This will lead to similar toxicological problems that have been discussed in other chapters in this textbook. However, as opposed to metal ion exposure from the external environment, the collapse of NPs can release a localized large dose of metal ions, which may result in metal toxicity. Regardless of the composition of the NMs, whether they are comprised of heavy metals, inorganic component, or organics. The breakdown of NMs will release these molecules and ions locally at high local concentration giving the opportunity to bind to proteins, nucleic acids and lipids and affect their functions.

Size and surface chemistry can also influence the pharmacokinetics. Studies have found that NPs with a diameter >6 nm cannot be excreted by the kidneys and accumulate in other organs. NP accumulation in key organs such as liver and spleen can lead to serious damage depending on the elemental cores. NPs >50 nm are found in the blood. Many NPs will eventually make it to the bowel or simply never be effectively absorbed and are therefore cleared through binding to stool components. Overall considering size with exposure, the NP distribution in organs follows and approximates order of higher load of: blood > liver > spleen > lung, liver > testis, thymus, heart > brain. Certainly, in most *in vivo* rodent studies most NMs eventually accumulate in the liver where we see metabolism and detoxification of other xenobiotics. Whether the NM is ingested, inhaled, absorbed by the skin, intravenously injected, or released from medical indwelling device coatings, they will eventually find their way to the liver. However, it is surprising that rather few studies on their hepatotoxicity have been reported in the literature.

To date, the genotoxicity of NMs has not been thoroughly investigated, which may have to do with the fact that little attention has been directed at their chronic exposure. The latter is also attributed to the limited time that ENM and NMP have been on the market. To date, genotoxicity studies tend to be inconsistent where some studies for TiO_2 or Ag NPs suggest genotoxic properties, whereas others show no effect. Regardless, one can generalize that genotoxicity follows similar trends as cell toxicity with the smaller the particle sizes the more genotoxic they are, but this may simply be due to cell absorption efficiency. See Table 9.1 for an overview of Shape on biological response.

9.5.2 Toxicity of specific types of nanomaterials

9.5.2.1 Metal nanoparticles

There are essentially five general mechanisms/paths of metal NM toxicity to a cell [60].
1. The particle itself partitioning into the membrane causing membrane dysfunction leading to ROS production.
2. The capping material of the NP interacting with cell biomolecules and adsorption onto the NP surface.
3. Particle dissolution/decomposition and thus subsequently releasing a large local concentration of metal ions leading to element specific toxicity.
4. Molecular structure–related effects that facility photochemical and redox properties
5. Carrier or Trojan horse effects of facilitating competing metal ion entrance to the cell.

Table 9.1: Effect of nanoparticle shape has on biological response.

NM shape	Toxicity mechanism	Physiological response
Spherical (homogeneous)	Internalization/membrane disruption	Dysfunction of cell division, disturbed cellular trafficking,
Fibrous (homogeneous)	Membrane disruption, inhibited phagocytosis, transport, distortion of cell shape	Chronic inflammation due to frustrated phagocytosis, mesothelioma formations.
Non-spherical (homogeneous)	Disruption of membrane integrity	Inflammation and impaired phagocytosis
Agglomerate (homogeneous)	Interaction with macrophages	Increased retention time
Agglomerate (heterogeneous)	Disruption of cell membrane, cell aggregation	Combinational effects similar to fibrous and agglomerates
Heterogeneous concentric	Membrane disruption	Cell shape and cell division disturbance.
		Adapted from [59]

An example of the dissolution problem can be seen with the increasingly used ZnO NPs. While several studies provide a wide recognition of ZnO NPs dissolution [61], it is not yet well understood to which extent the dissolved Zn^{2+} contribute to ZnO NP toxicity and what the underlying mechanisms entail. It continues to be difficult to separate the particle dependent effects vs dissolved metal ion effects. The mechanisms are likely separate as we have a good understanding of metal ion exposure (see other chapters in this textbook). But does the presence of the NP enhance or change the overall physiological effects and primary biochemical targets vs tertiary effects. This is a similar problem for most metal NPs such as Ag and Cu. The issue is likely different with TiO_2 where a lot of toxicity from the NP form is via light activated ROS production, the decomposed and released Ti^{3+} ions likely still provide a level of toxicity that is somewhat related to the mechanism of Al^{3+} [62].

Overall smaller metal NM will exhibit toxic effects due to their membrane interactions. Those materials that are less stable and thus dissolve into free ions will result in metal ion specific toxicity. Certainly, the mechanisms of action of metal nanoparticles will be different for different living species under different external physicochemical conditions of exposure.

9.5.2.2 Toxicity of nanofibers

The structure, shape, and morphology play a role in a ENM toxicity, particular for inhalation exposure. One class of concern is the nanofibers (NT, NW, NR) which cause lung inflammation and can cause cell lesions. Carbon nanotubes have been found to be more toxic than other ultra-fine carbon or silica dusts. The spherical fullerenes (buckyballs) also cause lung damage and in bioassays fish brain damage and insect death have been observed. The process of cellular toxicity is most likely a function of their hydrophobicity and subsequent partitioning into cell membranes thus destroying the cell integrity. In many cases prolonged exposure has been shown to cause several cancers, likely due to the high ROS exposure (reviewed in [56]).

9.5.2.3 Toxicity of lipid-based nanomaterials

Lipid based nanoparticles (LNP) have been explored extensively for drug delivery, as MRI contrast agents, and as mRNA vaccine delivery vehicles. Depending on the lipid that is used they can have a variety of sizes and shapes (cubosomes, liposomes, lipid tubes). Their interaction and uptake with a given cell will be dependent on the LNP and the lipid composition of the given cell type. They may interact through membrane fusion or by phagocytosis. It was concluded that cellular toxicity is dictated more by the lipid composition of the LNP rather than by the particle nanostructure or the uptake mechanism [63]. This important information allows for cell targeting for use of LNPs as pharmaceutical agents.

9.6 Analytical approaches and monitoring

There are of course new types of NMs synthesized and engineered all the time leading to new and novel applications. This leads to challenges to monitor and quantitate as well as evaluate acute but especially potential chronic issues from chronic exposure.

The limitations of current analytical tools to probe the release of metals from NM characterization in complex media represent a serious challenge toward understanding the fate of NMs in the environment. Various organizations have begun to put guidelines in place. US Food and Drug Administration [40], the European Chemicals Agency [42], and the Organisation for Economic Co-operation and Development [41].

9.6.1 Tools to characterize NMs

There are now a number of tools used in nanometrology [64]. The primary approaches to study and characterize nanomaterial include Electron Microscopy (scanning and transmission), Dynamic Light Scattering (DLS), Zeta potential, Voltammetry, Particle counters (CPS), Nanoparticle Tracking Analysis (NTA), and Correlation spectroscopy (PCS). These give information on size, shape, morphology, polydispersity, surface, and redox potential and diffusion properties. Instrumental bias and data processing can lead to very different size distributions from different methods. Unfortunately, some of these methods have issue with bias toward particles of larger size making interpretation difficult in complex samples.

Various spectroscopic approaches are also used to understand the chemical nature of the NMs, including but not limited to: Infrared, Ramon, Fluorescence, Nuclear Magnetic Resonance (NMR) Spectroscopy to follow various nuclei, as well as energy scattering and absorption methods, such as selected-area electron diffraction, X-ray absorption near edge structure, electron diffraction (ED), energy-dispersive X-ray spectroscopy, and X-ray diffraction (XRD).

Analytical methods are key to obtain concentration and speciation information of ENMs. For any metal/metalloid-based NM, the go to approach is liquid chromatography linked to inductively coupled plasma and mass spectrometry detection (LC-ICP-MS). The chromatography approach is typically a size-exclusion chromatography, but normal or reverse phase chromatography could be employed as well. A powerful technique that is evolving is that of inductively coupled plasma time-of-flight mass spectrometer (ICP-TOF-MS) that is operated in single-particle mode [65]. This approach helps in the identification and quantitation of ENMs against the high background of natural nanosized particles in soils, WWTP, and turbid waters. This takes the LC-ICP-MS to a higher level of analysis and helps to analyze multiple NM types at the same time.

Other issues dealing with the measuring and evaluating ENMs in the environment or biological systems typically revolve around sample integrity. Various centrifugation methods coupled with ultrafiltration and solvent extractions can be used, each with varying degrees of success. The inherent complexity of samples along with the very low concentrations of the ENMs remains a major hurdle to the field.

9.6.2 Biological systems used in assays and monitoring

To evaluate the toxicology of NMs in biological living systems, there are constant changes to the biological systems used as we learn what works best for different NMs. The dedicated volume of Wagener in *Bioanalytical Reviews* [66] contains several articles on measuring the biological impacts of NMs.

As in toxicology studies of macro pollutants and toxins a variety of systems are used. Examples include but are not limited to: fish (zebra fish (*Danio rerio*), trout), crustaceans (*Daphnia magna*), earth worms or *Caenorhadbitis elegans*, rodents (mice, rats), aquatic microbes, a variety of organ tissue cell lines as well as epithelial, fibroblast, and macrophage cell lines. Beyond cell growth/division toxicity values, various biochemical processes are measured including but not limited to cell integrity/lysis, motility, cytokine levels, oxidative response, mitochondrial electron transport chain activity, various enzymatic activities in key metabolic pathways, DNA replication rates, protein translation rates. A common tool also is to measure the coagulation time of blood and to observe erythrocyte lysis. Genotoxicity of NMs is evaluated by evaluating telomer sizes, DNA strand breaks, oxidative damage, and mutations. Likely the most common genotoxic tests performed on organisms is the Ames test (assessment of the mutagenic potential of toxin) or the Comet test (evaluation of DNA strand breaks). As mentioned above, plants may become very good environmental sensors of ENM exposure as many visible features of the plant can be used to assess plant health.

Moving forward it will likely be impossible to assess the *in vitro* or *in vivo* toxicity of all ENMs on a case-by case basis due to the exponential growth of new materials and uses. Thus, we will see computational approaches of (Q)SAR(quantitative structure activity relationship) for NMs. A challenge is that computational studies mostly focus on cell/organism viability which typically comes from acute studies. A recent study that evaluated the research literature uncovered huge a variation for in vitro nanotoxicological assays that investigated a large number of biological effects, such as cytotoxicity, inflammatory response, oxidative stress, and genotoxicity [67]. Therefore, at present there is a disconnect between what data is being produced and thus challenges toward building computational models for Human health risk assessments. However, some interesting approaches are being explored to simultaneously evaluate against multiple species and animal cell types [60].

9.7 Summary

Obtaining a clear view of nanomaterial toxicity at this time is quite a challenge as there are still a lot of knowledge and monitoring gaps. Regardless, key risk assessment considerations are becoming clear:

i)   Particle characteristics affect toxicity. These parameters will affect the cytotoxicity at the cell, organ, and organism level, and further to the ecosystem damage.
ii) Fate and transport through the environment will be different for NM compared to macroscale counter parts.
iii) Routes of exposure and metrics by which exposure is measured need to be carefully defined.

iv) Mechanisms of movement to different parts of the body need to be defined and characterized.

v) Finally, more effort on the biochemical mechanisms of toxicity and disease for ENMs is required.

Moving forward with regulations will require more data collection, less hype, and more solid published studies to provide data of reference. One thing that must be remembered about nanomaterial toxicology is that they cannot be considered just another chemical. Nanomaterial behaves VERY differently than the bulk material of similar composition. Given the merits of NMs make them superior in many applications we are undoubtedly going to see more ENM and NMP.

Recommended reading

Boyes WK, van Thriel C. Neurotoxicology of Nanomaterials. Chem Res Toxicol. 2020, 33, 1121–1144.

Daima HK, Kothari SL, Kumar BS. Nanotoxicology: Toxicity Evaluation of Nanomedicine Applications. CRC Press, Taylor & Francis Group. 2021.

Dhawan A, Anderson D, Shanker R. Nanotoxicology: Experimental and Computational Perspectives. Issues in Toxicology No. 35. Royal Society of Chemistry, CPI group Ltd, UK. 2017.

Duran N, Guterres SS, Alves OL. Nanotoxicology: Materials, Methodologies, and Assessments. Springer, New York, 2014.

Ganguly P, Breen A, Pillai SC. Toxicity of Nanomaterials: Exposure, Pathways, Assessment, and Recent Advances. ACS Biomater Sci Eng. 2018, 4, 2237–2275. (*a good overview highlighting some basics*)

Jeevanandam J, Barhoum A, Chan YS, Dufresne A, Danquah MK. Review on nanoparticles and nanostructured materials: History, sources, toxicity and regulations. Beilstein J Nanotechnol. 2018;9:1050–1074.

Keller AA, McFerran S, Lazareva A, Suh S. Global life cycle releases of engineered nanomaterials. J Nanopart Res 2013, 15, 1692.

Van der Merwe D, Pickrell JA. Chapter 18: Toxicity of Nanomaterials. In: Veterinary Toxicology 3rd ed. Gupta RC editor. Academic Press. 319–326, 2018.

Wegener J (editor). Bioanalytical Reviews 5: Measuring Biological Impacts of Nanomaterials. Springer, Switzerland, 2016 (*a number of extensive reviews covering in more details topics discussed in this chapter.*)

Wu D, Ma Y, Cao Y, Zhang T. Mitochondrial toxicity of nanomaterials. Sci Total Environ. 2020, 702, 134994.

Bibliography

[1]   Yuwen L, Wang L. Nanoparticles and quantum dots. Chapter 11.5, In: Devillanova F, Du Mont WW, eds. Handbook of Chalcogen Chemistry: New Perspectives in Sulfur, Selenium and Tellurium, 2nd ed. Cambridge: The Royal Society of Chemistry. 2013, 232–260.

[2]   Cao G. Nanostructures & Nanomaterials, Synthesis, Properties and Applications. London: Imperial College Press. 2004.

[3]   Rao CN, Muller A, Cheetham AK eds. The Chemistry of Nanomaterials: Synthesis, Properties and Applications. Weinheim: WILEY-VCH Verlag GmbH & Co. 2004.

[4]   Horikoshi S, Serpone N, eds. Microwaves in Nanoparticle Synthesis: Fundamentals and Applications. Weinheim: Wiley-VCH Verlag GmbH & Co. 2013.

[5]   Zeng Z, Xu Y, Zhang Z, Gao Z, et al., Rare-earth-containing perovskite nanomaterials: Design, synthesis, properties and applications. Chem Soc Rev. 2020, 49, 1109–1143.

[6]   Bernardos A, Piacenza E, Sancenon F, Mehrdad H, Maleki A, Turner RJ, Martinez-Manez R. (2019) Mesoporous silica-based materials with bactericidal properties. Small, 1900669.

[7]   Deirram N, Zhang C, Kermaniyan SS, Johnston APR, Such GK. pH-Responsive Polymer Nanoparticles for Drug Delivery. Macromol Rapid Commun. 2019.,40,e1800917.

[8]   Tuncel D, Demir HV. Conjugated polymer nanoparticles. Nanoscale. 2010. 2, 484–494.

[9]   Abbasi E, Aval SF, Akbarzadeh A, et al. Dendrimers: Synthesis, applications, and properties. Nanoscale Res Lett 2014, 9, 247.

[10] Jeevanandam J, Barhoum A, Chan YS, Dufresne A, Danquah MK. Review on nanoparticles and nanostructured materials: History, sources, toxicity and regulations. Beilstein J Nanotechnol 2018, 9, 1050–1074.

[11] Udomprasert A, Bongiovanni M, Sha R, et al. Amyloid fibrils nucleated and organized by DNA origami constructions. Nature Nanotech 2014, 9, 537–541

[12] DeFrates K, Markiewicz T, Gallo P, Rack A, Weyhmiller A, Jarmusik B, Hu X. Protein Polymer-Based Nanoparticles: Fabrication and Medical Applications. Int J Mol Sci. 2018, 19, 1717.

[13] Trache D, Tarchoun AF, Derradji M, Hamidon TS, Masruchin N, Brosse N, Hussin MH. Nanocellulose: From Fundamentals to Advanced Applications. Front Chem. 2020, 8, 392.

[14] Lu XY, Wu DC, Li ZJ, Chen GQ. Polymer nanoparticles. Prog Mol Biol Transl Sci. 2011, 104, 299–323.

[15] Gebeshuber IC, Lee DW. Nanostructures for Coloration (Organisms Other Than Animals). In: Bhushan B (eds) Encyclopedia of Nanotechnology. Springer, Dordrecht. 2016.

[16] Seven PT, Seven I, Baykalir BG, et al, Nanotechnology and nano-propolis in animal production and health: An overview. Ital J Anim Sci. 2018, 17, 921–930,

[17] Wang Y, Xia Y. Bottom-up and top-down approaches to the synthesis of monodispersed spherical colloids of low melting-point metals. Nano Lett. 2004; 4, 2047–2050.

[18] Gautam UK, Rao NR. Controlled synthesis of crystalline tellurium nanorods, nanowires, nanobelts and related structures by a self seeding solution process. J Mater Chem. 2004, 14, 2530–2535.

[19] Chaudhari S, Umar A, Mehta SK. Selenium nanomaterials: An overview of recent developments in synthesis, properties and potential applications. Progr Mater Sci. 2016, 83, 270–329.

[20] Piacenza E, Presentato A, Zonaro E, Lampis S, Vallini G, Turner RJ. Selenium and Tellurium Nanomaterials. Phys Sci Rev 2018, 3, 20170100.

[21] Chen H, Shin DW, Nam GJ, Kwon KW, Yoo JB. Selenium nanowires and nanotubes synthesized via a facile template-free solution method. Mater Res Bull. 2010, 45, 699–704.

[22] Song JM, Zhu JH, Yu SH. Crystallization and shape evolution of single crystalline selenium nanorods at liquid-liquid interface: From monodisperse amorphous Se nanospheres toward Se nanorods. J Phys Chem B. 2006, 110, 23790–23795.

[23] Matus KJM, Hutchison JE, Peoples R, Rung S, Tanguay RL. Green nanotechnology challenges and opportunities. A white paper addressing the critical challenges to advancing greener nanotechnology issued by the ACS Green Chemistry Institute in partnership with the Oregon Nanoscience and Microtechnologies Institute. ACS Nono. 2011. Available at: https://greennano.org/sites/greennano2.uoregon.edu/files/GCI_WP_GN10.pdf.

[24] Department of Health and Human Services (DHHS), Centers for Disease Control and Prevention (CDC), National Institute for Occupational Safety and Health (NIOSH). Current strategies for engineering controls in nanomaterial production and downstream handling processes. publication. 2014–2102. 2013, Available at: https://www.cdc.gov/niosh/docs/2014-102/pdfs/2014-102.pdf.

[25] Patil S, Chandrasekaran R. Biogenic nanoparticles: A comprehensive perspective in synthesis, characterization, application and its challenges. J Genet Eng Biotechnol. 2020, 18, 1–23.

[26] Dhillon GS, Brar SK, Kaur S, Verma M. Green approach for nanoparticle biosynthesis by fungi: Current trends and applications. Crit Rev Biotechnol. 2012, 32, 49–73.

[27] Fawcett D, Verduin JJ, Shah M, Sharma SB, Poinern GEJ. (2017) A Review of Current Research into the Biogenic Synthesis of Metal and Metal Oxide Nanoparticles via Marine Algae and Seagrasses. J Nanosci. 2017, 1–15.

[28] Singh A, Gautam PK, Verma A, Singh V, Shivapriya PM, Shivalkar S, Sahoo AK, Samanta SK. Green synthesis of metallic nanoparticles as effective alternatives to treat antibiotics resistant bacterial infections: A review. Biotechnol Rep (Amst). 2020. 25, e00427.

[29] Hiemenz P, Rajagopalan R, eds. Principles of Colloidal and Surface Chemistry 3rd edition. New York, Marcel Dekker, 1997.

[30] Hotze EM, Phenrat T, Lowry GV. Nanoparticle aggregation: Challenges to understanding transport and reactivity in the environment. J Environ Qual. 2010, 39, 1909–1924.

[31] Segets D, Marczak R, Schafer S, Paula C, Gnichwitz JF, Hirsch A, Peukert W. Experimental and Theoretical Studies of the Colloidal Stability of Nanoparticles: A General Interpretation based on Stability Maps. ACS Nano. 2011, 5, 4658–4669.

[32] Piacenza E, Presentato A, Turner RJ. Stability of biogenic metal(loid) nanomaterial related to the colloidal stabilization theory of chemical nanostructures. Critl Rev Biotechnol 2018, 38, 1137–1156.

[33] Munusamy T, Settu K, Lee J-F, Short A. Review on Applications of Nanomaterials in Biotechnology and Pharmacology, Curr Bionanotechnol. 2016, 2(2).

[34] https://oecotextiles.blog/2010/09/01/silver-and-other-nanoparticles-in-fabrics/ accessed November 8, 2021

[35] Hofmann T, Lowry GV, Ghoshal S, et al. Technology readiness and overcoming barriers to sustainably implement nanotechnology-enabled plant agriculture. Nature Food. 2020, 1, 416–425.

[36] Paramo LA, Feregrino-Perez AA, Guevara R, Mendoza S, Esquivel K. Nanoparticles in Agroindustry: Applications, toxicity, challenges, and trends. Nanomaterials. 2020, 10, 1654.

[37] Sharma A, Madhunapantula SV, Robertson GP. Toxicological considerations when creating nanoparticle based drugs and drug delivery systems. Expert Opin Drug Metab Toxicol. 2012, 8, 47–69.

[38] Hansen SF, Hansen OFH, Nielsen MB. Advances and challenges towards consumerization of nanomaterials. Nat Nanotechnol. 2020, 15, 964–965.

[39] Salieri B, Turner DA, Nowack B, Hischier R. Life cycle assessment of manufactured nanomaterials: Where are we? NanoImpact. 2018, 10, 108–120.

[40] https://www.fda.gov/regulatory-information/search-fda-guidance-documents/considering-whether-fda-regulated-product-involves-application-nanotechnology (accessed October 27, 2021)

[41] https://www.oecd.org/officialdocuments/publicdisplaydocumentpdf/?cote=env/jm/mono (2019)12&doclanguage=en (accessed October 27, 2021)

[42] https://echa.europa.eu/documents/10162/987906/tgdpart2_2ed_en.pdf/138b7b71%E2%80%90a069%E2%80%90428e%E2%80%909036%E2%80%9062f4300b752f (accessed October 27, 2021)

[43] Moeta PJ, Wesley-Smith J, Maity A, Thwala M. Nano-enabled products in South Africa and the assessment of environmental exposure potential for engineered nanomaterials. SN Appl Sci. 2019.

[44] Geranio L, Heuberger M, Nowack B. The Behavior of Silver Nanotextiles during Washing. Environ Sci Technol. 2009, 43, 8113–8118.

[45] Lehutso RF, Tancu Y, Maity A, Thwala M. Aquatic toxicity of transformed and product-released engineered nanomaterials: An overview of the current state of knowledge. Process Saf Environ Prot. 2020, 138, 39–56.

[46] https://eur-lex.europa.eu/legal-content/EN/TXT/?uri=CELEX%3A52021DC0403&qid=1627289682807 (accessed October 27, 2021).

[47] Heilgeist S, Sekine R, Sahin O, Stewart RA. Finding nano: Challenges involved in monitoring the presence and fate of engineered titanium dioxide nanoparticles in aquatic environments. Water. 2021, 13, 734,

[48] Moloi MS, Lehutso RF, Erasmus M, Oberholster PJ, Thwala M. Aquatic Environment Exposure and Toxicity of Engineered Nanomaterials Released from Nano-Enabled Products: Current Status and Data Needs. Nanomaterials. 2021, 11, 2868.

[49] Lazareva A, Keller AA. Estimating potential life cycle releases of engineered nanomaterials from wastewater treatment plants. ACS Sustainable Chem Eng. 2014, 2, 1656–1665.

[50] Bakshi M, Liné C, Bedolla DE, Stein RJ, Kaegi R, Sarret G, Pradas Del Real AE, Castillo-Michel H, Abhilash PC, Larue C. Assessing the impacts of sewage sludge amendment containing nano-TiO_2 on tomato plants: A life cycle study. J Hazard Mater. 2019, 369, 191–198.

[51] Surette MC, Nason JA, Kaegi R. The influence of surface coating functionality on the aging of nanopartilces in wastewater. Environ Sci: Nano. 2019, 6, 2470–2483.

[52] Reed RB, Zaikova T, Barber A, et al. Potential Environmental Impacts and Antimicrobial Efficacy of Silver- and Nanosilver-Containing Textiles. Environ Sci Technol. 2016, 50, 4018–4026.

[53] Echavarri-Bravo V, Paterson L, Aspray TJ, Porter JS, Winson MK, Hartl MGJ. Natural marine bacteria as model organisms for the hazard-assessment of consumer products containing silver nanoparticles. Mar Environ Res 2017, 130, 293–302.

[54] Gottschalk F, Sun T, Nowack B. Environmental concentrations of engineered nanomaterials: Review of modeling and analytical studies. Environ Pollut. 2013, 181, 287–300.

[55] Iavicoli I, Leso V, Beezhold DH, Shvedova AA. Nanotechnology in agriculture: Opportunities, toxicological implications, and occupational risks. Toxicol Appl Pharmacol 2017, 329, 96–111.

[56] Shin SW, Song IH, Um SH. Role of physicochemical properties in nanoparticle toxicity. Nanomaterials. 2015, 5, 1351–1365.

[57] Sahu SC, Hayes AW. Toxicity of nanomaterials found in human environment: A literature review. Toxicol Res Appl. 2017, 1, 1–13.

[58] Zhao F, Zhao Y, Liu Y, Chan X, Chen C, Zhao Y. Cellular uptake, intracellular trafficking, and cytotoxicity of nanomaterials. Small. 2011, 7, 1322–1337.

[59] Sharifi S, Behzadi S, Laurent S, Forrest ML, Stroeve P, Mahmoudi M. Toxicity of nanomaterials. Chem Soc Rev, 2012, 41, 2323–2343.

[60] Sizochenko N, Mikolajczyk A, Jagiello K, Puzyn T, Leszczynski J, Rasulev B. How the toxicity of nanomaterials towards different species could be simultaneously evaluated: A novel multi-nano-read across approach. Nanoscale. 2018, 10, 582.

[61] Ma H, Williams PL, Diamond SA. Ecotoxicity of manufactured ZnO nanoparticles–a review. Environ Pollut. 2013, 172, 76–85.

[62] Gugala N, Lemire JA, Turner RJ. The efficacy of different anti-microbial metals at preventing the formation of, and eradicating bacterial biofilms of pathogenic indicator strains. J Antibiot (Tokyo). 2017, 7, 775–780.

[63] Strachan JB, Dyett BP, Nasa Z, Valery C, Conn CE. Toxicity and cellular uptake of lipid nanoparticles of different structure and composition. J Colloid Interface Sci. 2020, 576, 241–251.

[64] Imbraguglio D, Gikovannozzi AM, Roxxi A. Nanometrology. Proc Int School Phy. 2013, 185, 193–220.

[65] Praetorius A, Gundlach-Graham A, Goldberg E, et al., Single-particle multi-element fingerprinting (spMEF) using inductively-coupled plasma time-of-flight mass spectrometry (ICP-TOFMS) to identify engineered nanoparticles against the elevated natural background in soils. Environ Sci: Nano. 2017, 4, 307–314.

[66] Wegener J. (ed) Measuring Biological Impacts of Nanomaterials. Bioanalytical Reviews, 5. Springer, Cham. 2015.

[67] Forest V, Hochepied J-F, JPourchez J. Importance of Choosing Relevant Biological End Points To Predict Nanoparticle Toxicity with Computational Approaches for Human Health Risk Assessment. Chem Res Toxicol, 2019, 32, 1320–1326.

Som Niyogi, Kamran Shekh, and Solomon Amuno

Chapter 10
Toxicology of trace metals in the environment: a current perspective

10.1 Introduction

Metals are elements with good heat and electrical conduction properties. Due to their practical utility for making tools and their attractive appearance, metals are extensively used in practically every industrial sector. Since metals are elements, they can neither be created nor be degraded. Owing to their lack of degradability in the environment, metals essentially persist in the environment forever. Since metals occur in nature as a constituent of numerous minerals, rocks and geological formations, the biotic and abiotic weathering of these entities mobilizes metal ions to soil, water and sediment. Metal ions released from naturally occurring sources are also often referred to as trace metals. The origin of the term "trace metal" is not very clear but in the context of this chapter, it refers to the metals that naturally occur in the environment at background levels. Anthropogenic activities can increase the environmental levels of many metals significantly to create a situation of environmental concern. Therefore, the main sources of metal pollution are anthropogenic activities, such as mining and smelting or natural processes such as volcanic activity [1].

The world production of industrially important metal compounds and alloys has increased significantly during the last century [2]. This development prompted an interest in regulating metal levels in national waters and resulted in the establishment of the Environmental Protection Agency (EPA) in the United States of America in the 1970s. Around this time, information on the biochemical mechanisms of metal toxicity started to appear in the research literature. The first mechanistic studies utilized research tools and techniques which were common in physiological research, which expanded our understanding of how metals cause toxicity in aquatic organisms. These studies also revealed how the chemistry of water influences the chemical forms of metals (speciation of metals) and therefore their toxicity. The advent of sophisticated molecular biology techniques in the 1990s further expanded our knowledge about the uptake, the handling and the toxicity of metals in living organisms.

Som Niyogi, Department of Biology & Toxicology Centre, University of Saskatchewan, Saskatoon, SK, Canada, e-mail: som.niyogi@usask.ca
Kamran Shekh, Yordas Group, Hamilton, ON, Canada
Solomon Amuno, School of Environment and Sustainability, University of Saskatchewan, Saskatoon, SK, Canada

https://doi.org/10.1515/9783110626285-010

Although there are many ways to classify metals, in the field of toxicology, one of the major ways to classify them is to consider whether or not a metal has any essential function in an organism. Accordingly, metals such as copper (Cu), iron (Fe), zinc (Zn) and cobalt (Co), which are required in trace amounts for biological processes, are called "essential" metals. On the other hand, metals which are not currently known to possess any essential function in living organisms are called "non-essential" metals and some of the non-essential metals of major environmental concern are cadmium (Cd), silver (Ag), lead (Pb) and mercury (Hg). In the same context, metalloids that are often found at elevated concentrations in the environment can also be classified as essential (e.g., selenium (Se)) and non-essential (e.g., arsenic (As)). Since non-essential metals or metalloids are not required for any biological functions, exposure to these elements can result in severe adverse health effects in organisms even at rather low concentrations. On the other hand, extremely low concentrations of essential metals and metalloids in the body can also result in adverse health effects due to deficiency and high concentrations can result in toxic effects in organisms since 'the dose makes the poison'.

The uptake, the transport, the distribution and the excretion of metals in an organism are all physiologically regulated. Metals generally cause toxicity to organisms when these regulatory processes are overwhelmed by elevated exposure which can result in the disruption of homeostatic processes that are critical to maintain health and wellbeing. At the cellular level, essential metals are absorbed into cells by dedicated uptake mechanisms, which involve various protein channels and transporters. In contrast, non-essential metals do not have dedicated uptake mechanisms in cells and hence they rely on transport systems for essential elements to gain access. Newer developments in cellular techniques have shown that once inside the cells, metals accumulate into sensitive or insensitive compartments which determine the toxic potential of any given metal [3]. Moreover, there are many storage and detoxification mechanisms within cells that have evolved to maintain homeostatic control to tightly regulate metals [4]. Similar to intracellular handling, the transport of metals by the circulatory system to specific organs is also regulated by different carrier proteins that are present in the bloodstream [5].

The main objective of this chapter is to provide the reader with a solid foundation on the toxicology of metals and metalloids of major environmental concern. This chapter covers fundamental aspects that are involved in the molecular mechanisms of metal uptake, as well as the handling and the toxicity of metals in individual organisms along with a discussion of important environmental factors that influence metal uptake and bioaccumulation. Furthermore, this chapter also discusses the population level effects of metals using a case study that investigated the impacts of industrial metal pollution in sub-arctic Canada. We also highlight the usefulness of biochemical and histopathological biomarkers in assessing the population levels effects of metal pollution in resident species. The chapter ends with a brief review of current and emerging regulatory approaches that can be employed for the environmental risk assessment of metals in North America and the European Union.

10.2 Environmental factors that influence metal uptake and bioaccumulation

Up until the latter part of the twentieth century, the environmental toxicity of metals was considered to be a function of the total metals that were present in the exposure media. However, it was eventually recognized that only a certain fraction of metals that are present in the environment is actually available for biological uptake and therefore responsible for causing toxicity to organisms. In general, the most bioavailable and toxic form of most metals is believed to be the free metal ions (e.g., Cu^{2+}, Zn^{2+}, Cd^{2+}, Pb^{2+}, Ag^+) [6, 7].

Several different environmental factors, in both the aquatic and terrestrial environments, can influence the chemical speciation of metals in the exposure media and thereby alter the bioavailability and toxicity of metals to organisms. Therefore, it is critical to have a general understanding of the concept of bioavailability and the factors that modify metal bioavailability in the environment. The term "bioavailability"is used in two different contexts, which often leads to confusion. In the field of pharmacology, the term "bioavailability" is defined as the ratio of the amount of a substance (such as chemicals and pharmaceuticals) that reaches the circulation in its original chemical form relative to the amount administered. In environmental toxicology, the term "bioavailability" is interchangeably used with "environmental bioavailability", and it is defined as the amount of a substance or contaminant that is taken up by the organisms from the environment. In other words, the environmental bioavailability of a metal can be defined as the fraction of the total metal in the environment that is free for biological uptake (e.g., the fraction that is present in free ionic form).

In water, metals can exist in dissolved and insoluble particulate form. Technically, the dissolved metal is defined as the component of total metal in water that can pass through a 0.45 μM filter. Metals present in particulate form are generally not available for surface epithelial uptake, such as by the gills in fish and aquatic invertebrates. Dissolved metals in solution can be present as free cationic metal ions and as metal-ligand complexes. The predominant environmental ligands in natural waters are inorganic anions such as hydroxide, bicarbonate, carbonate, chloride, and sulfate, and dissolved organic matter (DOM; usually expressed in terms of dissolved organic carbon (DOC)), which can readily bind to free metal ions and form metal-ligand complexes which are typically not bioavailable to aquatic organisms. DOC is the result from the decomposition of dead plant and animal matter in the environment, and can be present in particulate and dissolved forms – both can form complexes with free metal ions and thereby reduce metal bioavailability [8]. Moreover, natural cations that are present in water, such as Ca^{2+}, Mg^{2+} and Na^+, can also compete with free metal ions for biological uptake by the epithelium at the surface of a biological tissue and thereby decrease metal bioavailability. This is why metal toxicity decreases with increasing hardness and salinity of the water.

Finally, the pH of the water can have profound implications on the metal speciation in water. At acidic pH (i.e., pH < 7), the free metal cations are generally more prevalent in freshwater, and thus metals tend to be more bioavailable in acidic water. Some metals, such as copper and silver, however, become less bioavailable at low pH than at neutral pH due to the increased competition between the monovalent forms of copper (Cu^+) or silver (Ag^+) and H^+ for uptake via gills or the body surface. On the other hand, the fraction of free metal ions decreases at high pH due to increased oxide and bicarbonate/carbonate complexation of free metal ions, and thus metals become less bioavailable in alkaline water [9].

The underlying principles which govern metal speciation and bioavailability in soil are similar to those in the aquatic environment; however, since the structure and environmental chemistry of soil is different from water, the process may appear very different, which is discussed below. Soil mass is a three-phase system, which consists of solid particles, water and air. The liquid phase is considered to be divided into two zones: the first zone is near the solid phase which controls the diffusion of the solute present in the water to the solid phase, and the second zone is the free water zone which controls the water flow and solute transport in soils. Together, these two zones are defined as 'pore water' [10] (Figure 10.1). For organisms living in soil, exposure to metals can occur either directly through the contact of their body surface with the pore water or through the ingestion of soil particles into their gut. Within the pore water, metals can exist either as free ions or as aquo-complexes with organic and inorganic ligands, and thus metal bioavailability in the soil can be modulated by the same factors that were already discussed for aquatic systems (e.g., pH, natural cations, inorganic and organic ligands). Moreover, the chemical composition and chemistry of soil (e.g., amount of metal adsorbing agents such as clay, organic matter, iron and manganese oxides, and pH) can also alter the transfer of metal ions from the solid phase to pore water and thereby influence the bioavailability of metals in the soil. For example, cadmium was found to be less bioavailable in soil with a higher clay content [11]. Similarly, lead was reported to be less bioavailable in soil which contained a higher content of iron and manganese oxides [12].

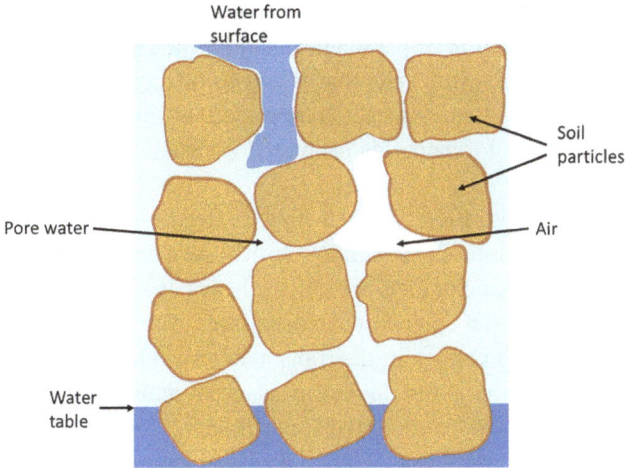

Figure 10.1: Different phases of soil, which is a three-phase mixture.

10.3 Exposure pathways and uptake mechanisms of metals

Organisms absorb metals from the environment via multiple exposure pathways depending on the environmental compartment that they live in. For example, aquatic organisms take up metals directly from the water via their gills or body surface (skin) as well as from the diet via their gastrointestinal tract. In contrast, the absorption of metals in most terrestrial organisms occurs predominantly from the diet across the gastrointestinal tract, although it can also involve absorption via the lung during inhalation or via dermal exposure to pore water in organisms that live in soil. Irrespective of the route of exposure or the organs involved in their uptake, the translocation of metals into organisms is frequently mediated by mechanisms that include facilitated diffusion and active transport. Below, the basic principles of these metal transport mechanisms will be discussed and the major transport proteins that mediate this transport will be identified.

Metals are predominantly absorbed by organisms from the environment as free ions which are polar or charged entities, and thus epithelial transport of metals across the hydrophobic plasma membranes, is not possible by simple diffusion except for some non-polar and lipophilic metal complexes, such as the organic forms of mercury (methyl- and ethyl-mercury) and inorganic forms such as mercuric chloride, which can pass through lipid bilayer membranes by simple diffusion [13, 14]. Metal ions, therefore, need the assistance of transmembrane proteins such as ion channels and carrier proteins to allow them to access the cytoplasm across the plasma membrane by shielding the metal ions from the hydrophobic core of the membrane. Facilitated diffusion is a form of passive transport which allows molecules to move from extracellular media into cells that is driven by a concentration or electro-chemical gradient without expending cellular energy [15]. One of the most common mechanisms of cellular metal transport is facilitated diffusion through protein ion channels.

In living organisms, various types of ion channels have evolved for the absorption and internal handling of essential ions, which are critical for various cellular and physiological functions. These ion channels are either always open or they are blocked by a "gate" which regulates the opening of the channels. The binding of a particular ion or other molecules or a change in the transmembrane electric potential often regulates the opening of the gate allowing passage of materials. Many metals gain entry into the cell by a process called ion mimicry, in which ionic charge and radius of a toxic metal closely resembles that of specific essential cations (e.g., Ca^{2+} or Na^+) and pass through the respective ion channels. Facilitated diffusion of metals into cells also occurs via transmembrane carrier proteins. In this transport process, binding of metal ions to the carrier protein triggers a change of its shape or structure allowing the movement of the bound metal ions across the plasma membrane into the cell or outside the cell depending on the gradient.

Active transport is a type of transport across the cell membrane that requires energy because this process transports ions against their concentration gradient. Primary active transport is a subtype of active transport that involves membrane proteins known as ion pumps (e.g., Ca pump or Ca-ATPase, Na/K pump or Na/K-ATPase) and directly utilizes energy released from the hydrolysis of ATP [16]. For example, Ca-ATPase utilizes energy released from the hydrolysis of ATP to move Ca^{2+} across the plasma membrane against its concentration gradient, but can also transport metal ions (e.g., Cd^{2+}) that are Ca ion analogues [17]. Another type of active transport is secondary active transport which couples the bidirectional movement of one ion with the movement of another ion. One of the two ions (known as the driving ion) in this type of transport is moved down its electrochemical gradient and the energy stored in the electrochemical gradient of the driving ion is used for the transport of the other ion against its electrochemical gradient. Some secondary active transporter proteins allow the movement of both ions in the same direction across the plasma membrane, and these transporters are known as symporters or co-transporters. One very well-characterized symporter is the natural-resistance-associated macrophage protein 2 (Nramp2; also known as divalent metal transporter 1 (DMT1)) which co-transports H^+ and essential (e.g., Fe^{2+} and Co^{2+}) or non-essential metal ions (e.g., Cd^{2+} and Pb^{2+}) [18]. On the other hand, other secondary active transporters allow the movement of two ions in opposite directions and thus are known as antiporters or exchangers. For example, Na/Ca exchanger (NCE) is an antiporter that usually transports Ca^{2+} against its concentration gradient, but it can also transport Cd^{2+} in exchange of Na^+ [19].

In contrast to metals, metalloids like Se and As are bioavailable in their oxyanion forms (selenite (SeO_3^{2-}) and selenate (SeO_4^{2-})) for Se, and arsenite ($As(OH)_3$) and arsenate (AsO_4^{3-}) for As. The uptake of these oxyanions occurs by multiple transport mechanisms including facilitated diffusion or secondary active transport pathways. For example, cellular absorption of oxyanions of Se (selenite and selenate) is mediated by anion transporters involved in the uptake of sulfite and sulfate and by water channels or aquaporins [20]. Similarly, oxyanions of As enter the cell adventitiously through transmembrane carrier proteins that transport nutrients such as phosphate (arsenate) and glucose (arsenite), and also via aquaporins (arsenite) [21]. Moreover, Se can also occur in the environment in form of organic forms, mainly as sulfur-substituted amino acids (selenocystine and selenomethionine), which are naturally produced from inorganic Se through microbial activity, especially in aquatic systems. These organoselenium compounds are often key micronutrients in the diet, which is the primary source of Se exposure to most aquatic organisms including fish living in Se-contaminated environments [22]. The uptake of these seleno-amino acids across the intestinal epithelium also occurs by facilitated diffusion pathways that transport cystine and methionine [23]. Free metal ions can also bind to sulfur-containing amino acids, such as cysteine and form metal-amino acid complexes, which can also be transported across the intestinal epithelium via specific amino acid transporters. For example, the intestinal absorption of Cd-cysteine complexes is believed to be mediated by cysteine transporters [24].

10.3.1 Major transporters and ion channels involved in epithelial metal uptake

A wide diversity of transmembrane proteins is involved in the cellular transport of metals, and their expression is often cell- or tissue-specific and regulated by the functions they carry out in respective cells or tissues. Therefore, the transport of some metals often involves different subsets or combinations of membrane transporters depending on the cell or tissue type. An individual metal can be transported by multiple transporters across the cell membrane, but which transporter would play the predominant role is governed by the affinity of the metal to different transporters and the concentration of the metal in the exposure medium. Similarly, a single transporter is often capable of transporting multiple metals albeit with different transport efficiency and/or transport kinetics. Moreover, the abundance of different metal transporters varies widely, with some transporters found in greater abundance relative to others. When the concentration of a metal is relatively low in the exposure media, it is predominantly transported by high-affinity transporters; however, with increasing concentration of the metal in the exposure media other low-affinity transporters come into play as the high-affinity transporters become saturated. Since a comprehensive overview of all the different types of metal transporters is beyond the scope of this chapter, the following section provides a description of well-characterized apical and basolateral transporters that are found in uptake epithelia in organisms (Figure 10.2). Epithelial transport of metals involves the uptake of metals across the apical membrane (facing the exposure media) followed by the extrusion of metals across the basolateral membrane of the cell (into the body fluid or the bloodstream).

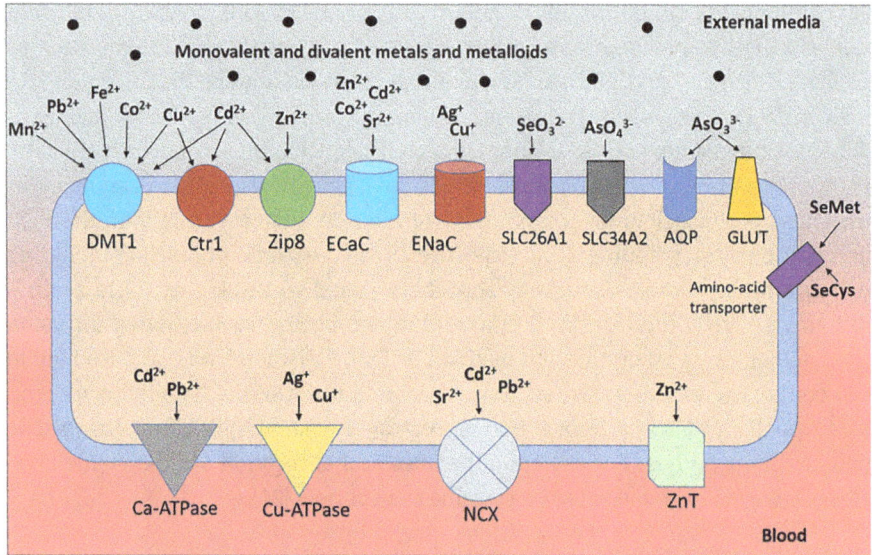

Figure 10.2: Major apical and basolateral transporters involved in the uptake of metals and metalloids in epithelial tissues such as fish gills and intestine. The expression of these transporters varies according to cell and tissue type. In addition, target cells may also contain transporters that excrete metals in the extracellular compartment. Metals in external media, blood and target cells are depicted as black solid circles. DMT1: divalent metal transporter 1; Ctr1: Copper transporter 1; Zip8: Zrt, Irt-like protein; ECaC: Epithelial Calcium Channel; ENaC: Epithelial sodium channel; SLC26A1: sulfate/anion exchanger; SLC34A2: sodium/phosphate cotransporter; AQP: aquaglyceroporin; GLUT: glucose transporter; Ca-ATPase: calcium ATPase; Cu-ATPase: P type copper ATPase; NCX: Na/Ca-Exchanger; ZnT: Zinc transporter. SeMet: Selenomethionine; SeCys: Selenocystine.

10.3.2 Apical ion channels and transporters

The transient receptor potential (TRP) transporter superfamily consists of several integral membrane proteins which function as ion channels. TRPs typically play an essential role in the uptake of Ca^{2+} [25, 26]. Among the members of the TRP superfamily, TRPV5 and TRPV6, which are also known as epithelial calcium channel-2 and -1 (ECaC2 and ECaC1) respectively, are the most relevant channels for the epithelial transport metals. In aquatic organisms like fish, only a single type of ECaC is found, and it is predominantly expressed in the mitochondria rich cells (MRCs – ion transporting cells) of the fish gill [27, 28]. Apart from the apical uptake of Ca^{2+}, ECaC in MRCs are likely involved in the uptake of several divalent metal ions, such as Cd^{2+}, Zn^{2+}, Co^{2+} and Sr^{2+} from the water based on mechanistic information that has been obtained from cellular studies and kinetic studies that were conducted in fish [29–32]. Similar to ECaC, another apical channel that is expressed on the apical membrane is the epithelial sodium channel (ENaC), which transports Na^+ and its analogues such as Cu^+ and Ag^+ from the water into the MRCs [33, 34].

The solute carrier (SLC) transporter superfamily is one of the largest families of membrane bound transporters. SLC transporters are responsible for the transport of a great variety of solutes including inorganic ions, amino acids, sugars, and other solutes [35]. Major SLC families which are involved in apical metal uptake are SLC11, SLC31 and SLC39. The SLC11 family, also known as natural resistance-associated macrophage protein (Nramp), and two paralogs (proteins with close sequence identity and similarities) Nramp1 and Nramp2 have been found in mammals. Both Nramp1 and Nramp2 transport Fe^{2+}; however, Nramp2 is designated as divalent metal transporter 1 (DMT1) because of its promiscuous transport properties for several divalent metals as Mn^{2+}, Fe^{2+}, Cu^{2+}, Co^{2+}, Cd^{2+} and Pb^{2+} [36]. DMT1 is expressed in a wide variety of transport epithelia including intestine and gills [37, 38]. The SLC39 family, also designated as the ZIP (Zrt-, Irt-like protein) family of transporters are usually involved in regulating the cellular uptake and the homeostasis of Zn^{2+} [39]. Fourteen members (ZIP1–ZIP14) have been identified to date, which are differentially expressed in various tissues including the gills and the intestinal epithelium of vertebrates. Among the ZIP family of transporters, ZIP8, which is an electroneutral $Zn^{2+}/[HCO_3^-]^2$ symporter, is also capable of apical transport of Cd^{2+} in fish and mammals [38–41]. The SLC31 family, also known as the family of copper transporters (Ctr), is dedicated for Cu uptake and its regulation in eukaryotes [42]. Currently two of the most well-recognized Ctr proteins are Ctr1 and Ctr2, which are high-affinity copper transporters. Ctr1 is expressed in both the gills and the intestinal epithelia in vertebrates and plays a major role in the uptake of Cu^+ from water and the diet, but is also capable of transporting other metals such as Cd^{2+} (43–45). Copper is reduced from Cu^{2+} to Cu^+ by the membrane reductase prior to its transport via Ctr1. A member of the SLC26 family, which is also known as the sulfate/anion exchanger (SLC26A1), has been proposed as a candidate which mediates the transport of the oxy-anionic form of selenium, selenite [46]. Another anionic transporter that is known as the sodium/phosphate cotransporter type IIb (SLC34A2) has been suggested to play an important role in the uptake of the pentavalent form of arsenic, known as arsenate (AsO_4^{3-}) [47]. The trivalent form of arsenic (arsenite; $As(OH)_3$), on the other hand, has been shown to be transported by a subclass of water channels known as aquaglyceroporins as well as by several members of SLC2A family of glucose transporters (GLUT), which have been reviewed elsewhere [48].

10.3.3 Basolateral transporters and channels

The transport mechanisms that are involved in the basolateral transport or the extrusion of metals from the cells are not as well understood as the apical metal uptake mechanisms. P-type ATPases are abundantly expressed in all transport epithelia and believed to mediate the basolateral transport of several metals. For example, Ca-ATPase (PMCA), which plays a critical role in homeostatic regulation and cellular transport of Ca^{2+} has been suggested to be involved in basolateral extrusion of metals such as Cd^{2+} and Pb^{2+} from the MRC of the fish gills into the bloodstream [49, 50]. Similarly, two P type Cu-ATPases, known as ATP7a and ATP7b, are primary basolateral Cu transporters [51]. Both ATP7a and ATP7b show a varied expression pattern in different tissues in mammals [52]. In fish, ATP7a appears to be expressed both in the gills and the intestine, and it likely also mediates the basolateral extrusion of Cu^+ in these epithelia [53, 54]. ATP7b has also been proposed to be capable of transporting Ag^+ [55].

The family of SLC8, also known as Na/Ca exchangers (NCXs), shows a wide distribution across various cell types. Along with Ca-ATPase, NCX plays a key role in basolateral extrusion of Ca^{2+} and maintaining the balance of cellular Ca^{2+} [56]. Several different isoforms of NCX genes have been identified in mammals and zebrafish [57]. A subset of NCX isoforms are expressed in the MRCs of fish gills and have been suggested to be potential sites of the basolateral extrusion of Cd^{2+}, Pb^{2+} and Sr^{2+} [19, 58, 59]. Another SLC family which is also known to be involved in the basolateral extrusion of Zn^{2+} is SLC30 (ZnT, zinc transporter) [60]. To date multiple isoforms of ZnT have been identified, which primarily function to maintain cellular Zn balance by exporting Zn^{2+} from cytoplasm to the extracellular fluid or by transporting excess cytoplasmic Zn^{2+} into intracellular sequestering compartments [61].

10.4 Transport and distribution of metals in organisms

10.4.1 Transport of metals in the bloodstream

Once absorbed from the environment by the epithelial tissue, metals are transported to different target organs via blood. Metals are most reactive and toxic when present in their free ionic form, and thus they do not typically occur in free forms after they enter the bloodstream. Several different high molecular serum proteins are known to act as the major carriers of metals in the systemic blood circulation (Figure 10.3).

Albumin, the most abundant plasma protein which is synthesized and released into the blood by the liver, acts as an important carrier of essential elements such as Cu, Zn, Ni and Se, in addition to other vital functions such as maintaining the

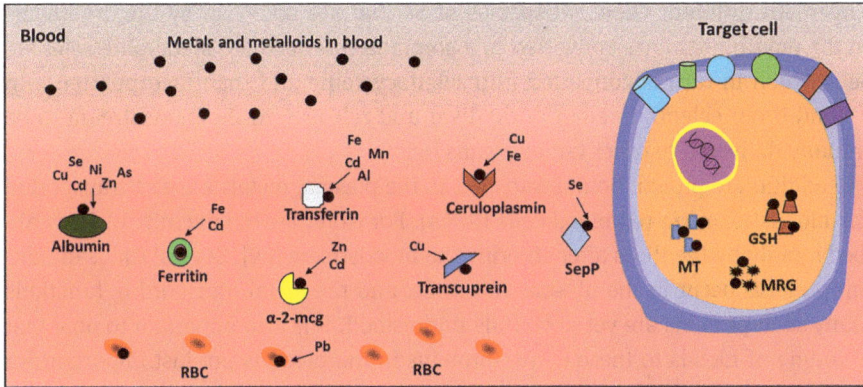

Figure 10.3: Major carrier proteins of metals and metalloids in the bloodstream and intracellular ligands involved in complexation of metals in target cells. α-2-mcg: alpha-2-macroglobulin; SelP: selenoprotein P; MT: metallothionein; GSH: glutathione; MRG: metal rich granules.

osmotic pressure of the blood and transport of fatty acids, hormones and vitamins to tissues. Albumin has also been found to bind and transport non-essential elements such as Cd and As^{III} in the bloodstream [62, 63].

There are also other major plasma proteins that play a critical role in transporting specific metals through the bloodstream. For example, transferrin is a plasma protein that transports most of the iron in the blood. Transferrin is also synthesized by the liver and released in the blood where it binds with ferric iron (Fe^{3+}). Although transferrin is predominantly a carrier protein for iron, it can also bind and transport multiple other metal ions such as Cd^{2+}, Al^{3+} and Mn^{2+} through the bloodstream [64]. Ferritin, which is mainly an intracellular iron storage protein, is also found in the plasma and transports iron in the blood, albeit in smaller amounts relative to transferrin. Ferritin is also known to bind and transport non-essential metals, such as Cd [65]. Similarly, alpha-2-macroglobulin, which is mainly a Zn transporting plasma protein, can also bind and transport Cd in the blood [66].

Transcuprein, a high-affinity Cu carrier in the plasma, plays a critical role along with albumin in the initial transport and distribution of Cu entering the systemic circulation to the target organs. However, more than 95% of Cu in the blood are bound with an evolutionarily conserved protein known as ceruloplasmin. Ceruloplasmin does not directly complex free Cu in the blood, instead Cu absorbed by the liver is incorporated into ceruloplasmin during its synthesis and then released into the bloodstream. In addition to Cu transport, ceruloplasmin acts as a ferroxidase by converting Fe^{2+} to Fe^{3+} and thereby allowing the binding and transport of iron by transferrin [67].

Selenoprotein P (SelP) is another important plasma protein which contains most of the Se present in the blood. Similar to ceruloplasmin for Cu, SelP does not bind Se directly in blood; rather, Se is incorporated into SelP during its synthesis in the liver. All different chemical species of Se that are taken up by the organisms from the environment are converted to a common intermediate metabolite selenide (HSe⁻), which in turn is converted into selenocysteine and then incorporated into SelP (and other selenoproteins) in the liver and released to the bloodstream to be eventually delivered to other target organs [68].

In addition to protein bound transport in the plasma, metals are also transported in the bloodstream *via* red blood cells (RBCs). For example, the majority of Pb is absorbed rapidly by the RBCs after entering the bloodstream [69]. Several thousand different proteins occur in the plasma and RBCs, and the metal-transporting functions of many such proteins are yet to be fully understood. It is also important to note that the binding of metals to these carrier proteins in the blood is not just important for their transport and distribution to the target tissues, but it can also have critical toxicological implications. For example, competitive interactions of essential and nonessential metals for binding to transporting proteins can disrupt the homeostasis of essential metals leading to adverse health consequences in organisms. Metals that are transported via the systemic blood circulation in the form of metal-protein complexes are delivered into target cells through receptor mediated uptake. For example, the Fe-transferrin complex or the Cd-transferrin complex can bind to transferrin receptors on the target cell membrane and enter the cell through clathrin-mediated endocytosis and Fe or Cd is then eventually released inside the cell [70]. Similarly, Cu bound to ceruloplasmin is released into the target cells through receptors; however, the identity of these receptors is not yet fully known [71]. The mechanism through which albumin, which is a systemic carrier of multiple metals, releases metals into the target cells is presently unknown.

10.4.2 Metal accumulation in tissues and subcellular distribution

When metal accumulation in target organs reaches a certain threshold, it can lead to structural or functional impairments therein and thus cause toxicity [72]. Metal accumulation in target organs in mammals and non-mammalian organisms can vary depending on the metal or metalloid in question, and include but is not limited to the liver, the kidneys, the gastrointestinal tract, the gills, bones as well as nervous and cardiovascular tissues. In general, the liver and the kidneys accumulate the highest concentrations of metals due to the central role they play in their metabolism and excretion. In aquatic organisms like fish, the route of exposure strongly influences the metal accumulation in target organs, as gills become a key site of accumulation and toxicity when metal exposure occurs primarily via water. Conversely, the gastrointestinal tract becomes one of the main sites of metal accumulation when exposure occurs predominantly *via* diet. Metals such as Cd and Pb can accumulate in high concentrations in bones during long-term chronic exposure because of their ability to displace Ca from bone tissues, which can lead to bone pathology like osteomalacia and osteoporosis [64]. Metalloids like As^{III} can also accumulate in keratin-dense tissues, such as body hair and nail, and thus these tissues can be used as non-invasive markers of chronic As exposure [73].

The mechanisms through which metals cause toxicity at the cellular level involve their interactions with various endogenous molecules found in the cells of target organs. The intracellular environment can be functionally divided into different compartments with distinct physicochemical and biochemical properties, and the distribution of metals to these different intracellular compartments is referred to as 'subcellular distribution'.

From a toxicological perspective, the inside of a cell can be functionally classified into two main compartments or pools: the biologically inactive metal pool (BIM) and the biologically active metal pool (BAM). Structurally, BAM consists of essential proteins, enzymes and organelles; therefore, metals distributed in the BAM can interact with critical proteins and enzymes required for normal functioning of cells and thus cause adverse effects. On the other hand, the BIM pool consists of a heat stable protein fraction and metal-rich granules, both of which are involved in the detoxification of metals and hence metals that accumulate in the BIM pool are considered to be not toxic to the cells [74] (Figure 10.3). The heat stable fraction of the BIM is not a single chemical entity, but actually a collection of sulfhydryl rich low-molecular weight proteins including metallothioneins and metallothionein-like proteins and other sulfhydryl molecules, such as glutathione [75]. Transition metals such as Cd and Cu have a high affinity toward sulfhydryl (-SH) groups; therefore, an interaction between metals and ligands with sulfhydryl groups in the heat stable fraction of the BIM leads to the formation of a stable complex, which is a key detoxification mechanism for metals.

Not much is known about the mechanisms that lead to the formation of metal-rich granules, but they mostly contain precipitates of metals complexed with ligands that contain functional groups such as phosphate, carbonate, and sulfide [76]. Overall, metals that accumulate in the BIM pool are not freely available to cause toxicity, whereas the metals accumulated in the BAM pool are toxicologically available. The pattern of subcellular distribution differs among metals, target organs and species [77]. Recent studies have demonstrated that the differences in species sensitivity to metals in fish can be explained by the difference in the relative distribution of metals between the BIM and BAM fractions in target organs, with more sensitive species having a greater fraction of metals in the BAM pool than in the BIM [3, 77, 78]. Nonetheless, the prediction of the toxicity based on subcellular distribution of metals must be treated with caution because the distribution of metals in the BIM and BAM pools is often a dynamic process as metals may move from one pool to the other depending on the exposure duration and metabolic state of the organism at a given point in time.

10.5 Mechanisms of metal toxicity

The environmental toxicity profile of metals and metalloid species can broadly be classified as acute and chronic toxicity. Acute toxicity refers to the type of toxicity that occurs over a short duration of exposure (24 to 96 h) to relatively high exposure concentrations. On the other hand, toxicity effects that occur following a long-term exposure such as weeks or months or even years to relatively lower exposure concentrations are considered as sub-chronic or chronic toxicity. In metal-contaminated natural environments, organisms mostly suffer from chronic toxicity, as the exposure concentrations of metals only reach the levels that can cause acute toxicity in rare and isolated situations. The mechanisms of acute metal toxicity are well characterized in aquatic organisms, especially fish. Moreover, the mechanistic underpinnings of chronic metal toxicity have also been extensively investigated in fish as well as model mammalian species such as mice and rats, which are commonly used as a surrogate to derive toxicity information of metals for humans. The following section provides an overview of cellular and physiological mechanisms by which metals cause acute and chronic toxicity.

10.5.1 Mechanisms of acute toxicity

In fish, the gills represent the primary target organ of metal toxicity during acute exposure. The gill is a multifunctional organ that regulates various critical physiological functions such as O_2 and CO_2 exchange, the uptake and the homeostasis of essential ions, and the excretion of nitrogenous waste. In freshwater fish, blood is hypertonic to the ambient water which leads to continual loss of essential ions from the body through passive diffusion, and the gills play a predominant role in compensating the loss of ions by actively absorbing ions from the water. At LC_{50} (concentration that causes lethality in 50% test population), many metals have been demonstrated to cause toxicity by inhibiting the uptake of essential ions, particularly Ca and Na in freshwater fish. For example, metals such as Cd^{2+}, Pb^{2+}, Zn^{2+}, Co^{2+} and Sr^{2+} are characterized as Ca antagonists because they compete with Ca for uptake via apical ECaC in the fish gill and thereby inhibit gill Ca absorption. In addition, Cd^{2+} and Pb^{2+} can further inhibit branchial Ca absorption by binding and inactivating the basolateral Ca-ATPase [79, 80]. The inhibition of branchial Ca absorption by these metals causes the disruption of Ca homeostasis and depletion of Ca from the body (hypocalcaemia), ultimately leading to cardiovascular collapse and death. Similarly, metals like Cu and Ag are characterized as Na antagonists because they inhibit gill Na^+ absorption by competing with it for uptake via apical ENaC and also by inactivating the basolateral Na-K-ATPase in the fish gills. Moreover, Cu and Ag can also inactivate the branchial (gill) carbonic anhydrase enzyme. Carbonic anhydrase catalyzes the production of bicarbonate ion (HCO_3^-) [81, 82]

and the branchial absorption of essential chloride ion (Cl^-) occurs through apical electroneutral exchange of HCO_3^- and Cl^-; therefore, the inactivation of carbonic anhydrase disrupts branchial Cl^- absorption. Therefore, Cu and Ag causes toxicity by inhibiting Na^+ and Cl^- uptake and thereby causing disruption of essential ion homeostasis in fish during acute exposure. Both Na^+ and Cl^- ions are critical for maintaining osmotic balance in the body and thus their depletion from the body often leads to eventual death. Thus, the disruption of ion regulation and homeostasis is the primary mechanism of metal toxicity during acute exposure in freshwater fish and this is also true for many freshwater invertebrate organisms such as crustaceans and gastropods. Nevertheless, metals can also cause acute toxicity by disrupting respiration and excretion of metabolic waste from the body. For example, Ni during acute exposure triggers inflammation and hypertrophy in the fish gills, which decreases the efficiency of branchial oxygen absorption and thereby causes hypoxia in the body and eventual death [83]. Moreover, Cu during acute exposure, in addition to disrupting ion homeostasis, also inhibits the branchial diffusion of ammonia (NH_3) from the body [84]. Most fish excrete nitrogenous waste primarily in the form of ammonia across the gills. Ammonia is a highly toxic molecule and thus disruption of branchial ammonia excretion by Cu causes ammonia build up in the body leading to adverse effects [85].

10.5.2 Mechanisms of chronic toxicity

In contrast to acute toxicity, chronic toxicity for metals is much more complex since long-term low-level exposure to metals often affects several vital cellular and physiological processes. These processes include the inactivation of vital cellular proteins and enzymes, the displacement of essential ions from biomolecules, the induction of oxidative stress, the disruption of endocrine and neuronal signaling pathways, histopathological damage, and immunosuppression. These effects may cumulatively affect the homeostasis of essential elements over time and thus adversely affect the survival, growth, behavior, reproduction and development of organisms. Due to the complex nature of chronic toxicity, manifestation of these effects often varies depending on the metals or species involved.

At the molecular level, one of the major mechanisms by which metals cause toxicity is by the inactivation of cellular proteins and enzymes. This typically occurs either due to the binding of metal ions directly to the active site of a protein or enzyme molecule or by the substitution of an essential metal from a metalloenzyme or metalloprotein by a non-essential metal.

Metalloproteins or metalloenzymes are a group of proteins that use metal cations as a cofactor which is essential for their functionality. Metals such as Cd and Pb inactivate cellular enzymes like ATPases and phosphatases by directly binding to the sulfhydryl groups on the active site of these enzymes [17, 80]. On the other hand, Cd is also known to inactivate Zn metalloproteins, such as DNA repair proteins by Zn substitution [86]. Similarly, Pb inhibits several enzymes involved in the hemoglobin synthesis pathway in RBC, mainly d-aminolaevulinic acid dehydratase (ALAD) by displacing Zn from this metalloenzyme, which is a key mechanism in Pb-induced anemia [87, 88].

Another important mechanism by which metals cause toxicity at the cellular level especially during chronic exposure is oxidative stress. Oxidative stress occurs when the production and accumulation of reactive oxygen species (ROS) is overwhelmed by the cellular scavenging capacity for ROS. The blanket term ROS is used to define the highly reactive derivatives of molecular oxygen including superoxide anion (O_2^{-}), hydrogen peroxide (H_2O_2) and hydroxyl radical (OH^{-}). ROS are produced normally as by-products during aerobic respiration, which can oxidize and damage vital cellular macromolecules such as proteins, lipids and nucleic acids unless neutralized. Hence, organisms have evolved antioxidative defense mechanisms to scavenge ROS and prevent their accumulation in cells. These defense mechanisms include enzymatic antioxidants such as superoxide dismutase (SOD), catalase (CAT), glutathione peroxidases (GPXs), and thioredoxin (Trx), as well as the non-enzymatic antioxidant molecules mainly glutathione (GSH) and ascorbic acid. Metals can induce oxidative stress by two different mechanisms. Redox active transition metals, such as Cu, Fe and Cr can enhance cellular ROS production via Fenton reaction because of their ability to change oxidation states by accepting and donating electrons [89]. In contrast, redox inactive transition metals like Cd, Hg and Pb can induce ROS accumulation and oxidative stress by directly binding with and inactivating the enzymatic and non-enzymatic antioxidants, which results in the depletion of cellular ROS scavenging capacity [89]. Oxidative stress may eventually lead to a wide array of adverse effects from cellular to physiological level including loss of structural and functional integrity of cells and tissues in critical organs, DNA damage, neurotoxicity, and immunosuppression.

Oxidative stress is also a key mechanism by which metalloids like Se^{IV} and As^{III} cause toxicity to organisms. Although Se protects against oxidative stress under normal physiological conditions because of selenoproteins with antioxidative properties, excess Se beyond the physiological threshold causes oxidative stress by inducing ROS production [90]. The metabolism of both inorganic and organic forms of Se in the presence of GSH produces metabolites that undergo redox cycling and generates ROS [91]. Similarly, As^{III} also contributes to ROS generation through targeting an enzyme known as NADPH oxidase as well as through its methylation during metabolism [92, 93]. Both Se^{IV} and As^{III} are potent reproductive and developmental toxicants to oviparous (egg laying) animals like fish, amphibians and birds [94–97]. These effects are mediated by the transfer of these elements from the mother to the eggs via yolk [98, 99]. Oxidative stress induced by maternally transferred Se and As decreases hatching success and causes morphological and skeletal deformities in the larvae and reduces their survival. Moreover, oxidative stress is also believed to be the main driver of the neurobehavioral toxicity of Se and As. Recent evidence from fish and mammalian studies indicates that chronic exposure to these metalloid species impairs social learning, anxiety response and cognitive performance, essentially by inducing oxidative stress in the brain and results in the subsequent disruption of important neurotransmission pathways such as dopaminergic, serotonergic, and glutamatergic pathways [100]. Interestingly, these neurobehavioral effects have been found to be transgenerational in nature, as the effects persist for multiple generations even with just maternal exposure to Se and As. This indicates that these heritable effects do not involve alterations in DNA sequence and thus involve epigenetic pathways that are yet to be characterized.

Metals can also act as endocrine disruptors, especially by interfering with the signaling and synthesis of steroid hormones. Steroid hormones act as male and female sex hormones, but they also mediate physiological responses to stress. Several elements, such as Cd, Cu, Pb, Ni, Co, Se (selenite) and As (arsenite) are characterized as metalloestrogens because of their ability to bind to cellular estrogen receptors and thus mimic the actions of physiological estrogens [101]. Estrogens are sex steroid hormones that play an important role in the normal sexual and reproductive development in females, and estrogens exert their cellular functions by binding to their receptors. The exact mechanism by which metalloestrogens bind to estrogen receptors is currently unknown; nonetheless, the binding of metalloestrogens to estrogen receptors alters the gene expression in cells responding to estrogen leading to adverse effects. Metalloestrogens have been linked to carcinogenic effects in reproductive organs in humans. In fish, the estrogen-mimicking activity of Cd has been linked to the decrease in spawning frequency and egg production [102]. In addition, several elements such as Cd^{2+}, Pb^{2+} and As^{III} are believed to disrupt steroidogenesis, which is a cellular process that converts cholesterol into biologically active steroid hormones [103–105]. Potential mechanisms underlying this effect may include the substitution of Zn or Ca from cellular motifs that regulate the steroidogenesis pathway or oxidative injury to the tissues where steroid hormones are synthesized.

The olfactory system in fish is also highly sensitive to metals. Fish depend on olfaction for a wide variety ecologically important behaviors such as finding mating partners, kin recognition, avoidance of predators and contaminants, and locating food, and thus impaired olfaction is a threat to their survival and reproductive success. The olfactory epithelium of fish is in direct contact with water and thus accumulates metals during waterborne exposure. Several metals including Cu, Ag, Cd and Pb cause olfactory toxicity in fish even at extremely low exposure levels [106]. The precise mechanisms of how these metals affect fish olfaction are not fully understood, but recent evidence suggests that metal ions disrupt the detection of odorants and transduction of sensory signals in olfactory neurons by interfering with ligand-gated or voltage-gated ion channels. Additional mechanisms may also involve metal induced inflammation or oxidative injury to the olfactory epithelium [106].

10.5.3 Summary

It must be re-emphasized here that only a certain fraction of metals that are present in the environment is actually available for biological uptake and therefore responsible for causing toxicity. In general, the most bioavailable and toxic form of most metals are their free metal ions. Environmental factors, in both the aquatic and terrestrial environments, greatly influence the bioavailability and toxicity of metals to organisms. Therefore, environmental factors must always be taken into consideration whenever metal toxicity is evaluated. We also learned that organisms absorb metals from the environment via multiple exposure pathways depending on the environmental compartment that they live in. Since free metal ions are polar or charged entities, their epithelial transport across the hydrophobic plasma membranes is not possible by simple diffusion except for some non-polar and lipophilic metal complexes. Therefore, metal ions need the assistance of transmembrane proteins such as ion channels and carrier proteins to allow them to enter into the cytoplasm across the plasma membrane.

Once absorbed from the environment, metals do not typically remain in free ionic forms after they enter the bloodstream. Several different high molecular serum proteins are known to act as the major carriers of metals in the systemic circulation. When metal accumulation in target organs reaches a certain threshold, it can lead to structural or functional impairments therein and thus cause toxicity. The inside of a cell can be functionally classified into two main compartments or pools: the biologically inactive metal pool (BIM) and the biologically active metal pool (BAM). Metals that accumulate in the BIM pool are in detoxified form and thus are not available to interact with critical cellular macromolecules and cause toxicity, whereas the metals accumulated in the BAM pool are toxicologically available.

The primary mechanisms by which many metals cause acute toxicity is by inhibiting the uptake of essential ions, particularly Ca and Na in freshwater organisms like fish. Other notable mechanisms of acute toxicity are disruption of respiration and excretion of metabolic waste from the body. In contrast, chronic toxicity of metals is much more complex and involves the alteration of several vital cellular and physiological processes. These processes can include the inactivation of vital cellular proteins and enzymes, the displacement of essential ions from biomolecules, the induction of oxidative stress, the disruption of endocrine and neuronal signaling pathways, histopathological damage and immunosuppression.

10.6 Biomarkers and their application in assessing the effects of environmental metal pollution: case study example

The previous sections described how chronic environmental exposures of organisms to metals or metalloid species can induce a continuum of biological responses. They can have effects across different levels of biological organizations, ranging from molecular and cellular alterations, to tissue and organ damage, which can compromise the health of individuals and ultimately put populations at risk. Such concerns have led to numerous clinical research and field studies that have focused on identifying "early warning" indicators, or "biomarkers" which can be used to monitor exposure levels and predict adverse biological responses that may occur if the pollution continues unabated in the environment [107, 108]. Due to their sensitivity and ease of detection, the use of biomarkers continues to gain much scientific attention due to their ability to characterize both point-source and non-point-source pollution including exposure-related effects in exposed wildlife [109, 110]. The purpose of this section is to present a relevant case study with examples of different types of biomarkers that can be used to assess the exposure-related effects in wildlife species breeding in an area of Canadian sub-arctic that has been impacted by industrial metal pollution over several decades.

10.6.1 Case study: assessing the impacts of chronic arsenic and cadmium exposure in small mammals inhabiting Yellowknife area in Northwest Territories of Canada

Previous gold mining activities and arsenopyrite ore roasting activities at the Giant mine site (1948 to 2004) was responsible for the contamination of As and other trace metals in the terrestrial and aquatic ecosystems of Yellowknife [111–116]. For many decades, elevated levels of As have been consistently reported in the vicinity of the Giant mine area and in surrounding locations, which has prompted a lot of concerns regarding the current health status of wildlife species inhabiting the area. This case study describes how a suite of biomarkers was used to evaluate the chronic As and Cd exposure levels and the health status of wild populations of muskrat (*Ondatra zibethicus*) and red squirrel (*Tamiasciurus hudsonicus*) living in and around Yellowknife.

Representative animals were sampled from three different field sites (Figure 10.4): [1] an area surrounding the 2 km radius of Giant Mine (Site 1 – contaminated site), [2] intermediate area located approximately 20 km away from the mine area (Site 2), and [3] a reference location spanning 52–105 km from the mine area and outside of the City of Yellowknife (Site 3). The small mammal species chosen for this study was based on the small perimeter of their home ranges and their sensitivity to contaminant exposure from their habitats and different feeding habits (muskrat – omnivore vs. red squirrel – predominantly herbivore) [111].

In this study, the total As and Cd concentrations were measured in nails, brain, livers, kidneys, bones, and the stomach content of muskrats and red squirrels. In addition, a histopathological evaluation of tissues was conducted and eye lesions were analyzed. A total of 30 animals, 10 from each site, were surveyed for this study. Different biomarkers were used for evaluating the status of chronic As and Cd exposure and other toxicological effects in the animals [111–114]:

1. Biomarker of exposure: Total As levels and Cd levels were measured in specific body tissues of animals from the contaminated site which were then compared with those from intermediate and reference sites to determine whether there was a significant difference in metal exposure across three populations of animals.

2. Biomarkers of effects: Markers of oxidative stress (lipid peroxidation and antioxidant/enzymatic activities such as superoxide dismutase (SOD), glutathione peroxidase (GP_x) and catalase (CAT)) were measured in the brain tissues to evaluate the biochemical toxicity at the organ level. In addition, critical neurotransmitters in the brain such as dopamine and serotonin were also measured to monitor neurological dysfunction in response to contaminant exposure. Histopathological effects were assessed to understand the extent and severity of morphological tissue changes in eye tissues caused by contaminant exposure. Optical coherence tomography (OCT) was employed to evaluate eye lesions and changes of the retina layer. Finally, magnetic resonance imaging (MRI) was used for evaluating structural changes in the brain and bone radiography analysis was performed to evaluate the patterns and severities of bone lesions and abnormalities induced by contaminant exposure.

10.6.1.1 Contaminant levels in gut content, nails, brains and bones

The As concentration in the gut contents of muskrats from the contaminated site was 5 to 49 times higher than those from the reference site suggesting that the habitats and the diets of the animals are contaminated. The maximum concentration of As in the gut content of squirrels from the contaminated site was ~ 40 times higher than those from the reference site. The As concentration in the nails of muskrats collected from Site 1 and 2 was in the range of 11 to 48 times higher than those from the Site 3 (reference site). An elevated As accumulation was also observed in the nails of squirrels from the sites closest to the Giant mine, but the As concentration was below the detection limit in the nails of squirrels from the reference location. Cadmium was not detected in the nails of muskrats and squirrels from all study sites. The cadmium concentrations in the guts of muskrats from Site 1 were almost double than those from the reference site. Arsenic was also detected in the brains of

a)

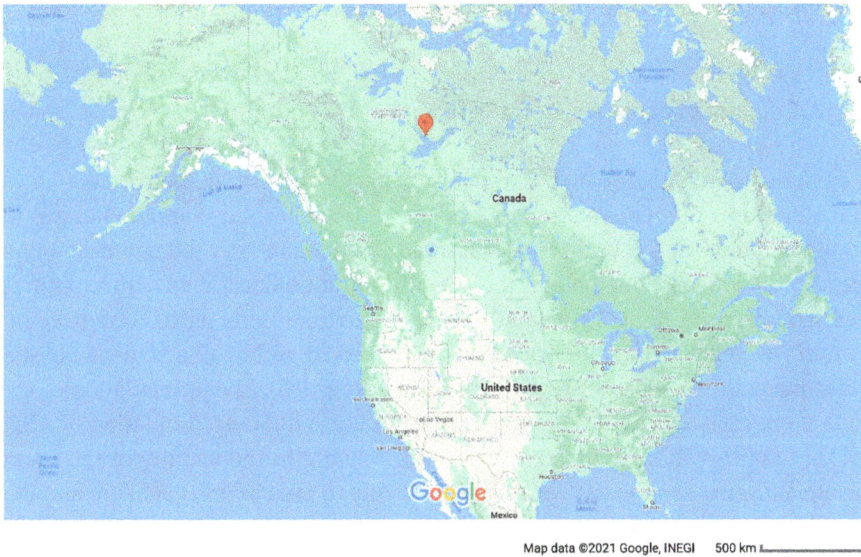

Map data ©2021 Google, INEGI   500 km

b)

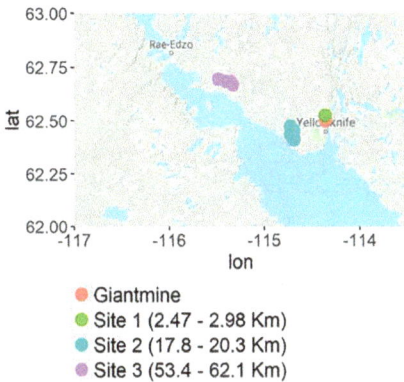

lat
lon

● Giantmine
● Site 1 (2.47 - 2.98 Km)
● Site 2 (17.8 - 20.3 Km)
● Site 3 (53.4 - 62.1 Km)

c)

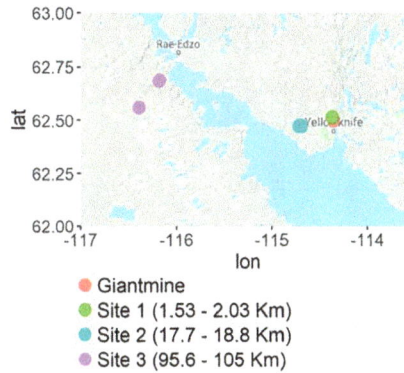

lat
lon

● Giantmine
● Site 1 (1.53 - 2.03 Km)
● Site 2 (17.7 - 18.8 Km)
● Site 3 (95.6 - 105 Km)

Figure 10.4: Map of study area showing sampling areas relative to the mining area: (a) location of Yellowknife city on the map of Canada, (b) muskrat collection sites, (c) red squirrel collection sites.

squirrels from Sites 1 and 2 but was not detected in the squirrel brain samples from the reference site. Cadmium accumulation was also recorded in all the brain samples of squirrels from all three sites. Arsenic levels were markedly higher in the bones of muskrats captured from Sites 1 and 2 compared to Site 3. Cadmium was not detected in the bones of muskrats, but both As and Cd levels in the bones of squirrels collected from Sites 1 and 2 were significantly higher compared to that in squirrels from Site 3. These observations established a clear pattern of chronic exposure to elevated levels of As and Cd to small mammals living in the vicinity of the Giant mine.

10.6.1.2 Evidence of oxidative stress in the brain

The oxidation of membrane lipids, which is characterized as lipid peroxidation, is a reliable marker of oxidative damage in tissues. However, no marked changes in lipid peroxidation in the brain were observed in muskrats and red squirrels in all three sampling sites. Nonetheless, the activities of antioxidative enzymes, such as superoxide dismutase (SOD), catalase (CAT), and glutathione peroxidase (GP_x) in the brain were elevated at Site 1 relative to Site 2 and 3, although a statistically significant increase was only observed with GP_x (Figure 10.5). The upregulation of antioxidant enzymes usually occurs in response to the cellular accumulation of ROS; thus, these observations indicated that animals living in the vicinity of the Giant mine are suffering from oxidative stress, likely due to the chronic exposure and the accumulation of As and Cd in the brain.

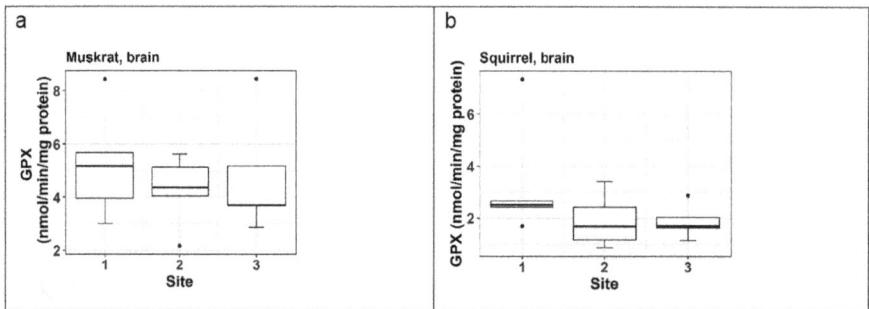

Figure 10.5: Glutathione peroxidase (GP_x) activity in the brain of muskrats (a) and red squirrels (b) across different sampling sites used in the case study (111; Reproduced with copyright permission from Elsevier).

10.6.1.3 Evidence of cataract and changes in retinal layers of the eye

Significant eye lesions were histologically confirmed in all the muskrats collected from Site 1 with evidence of moderate to severe eye inflammation, keratitis and subcapsular cataracts (Figure 10.6a). Inner retinal degeneration was also prevalent in all the muskrats collected from Site 1, while muskrats from the reference area showed predominantly normal eye morphology. No significant eye lesions were detected in the squirrels from the three population groups except for incidental mild scars in the cornea which were observed in the squirrels from all the three sampling sites. In addition,

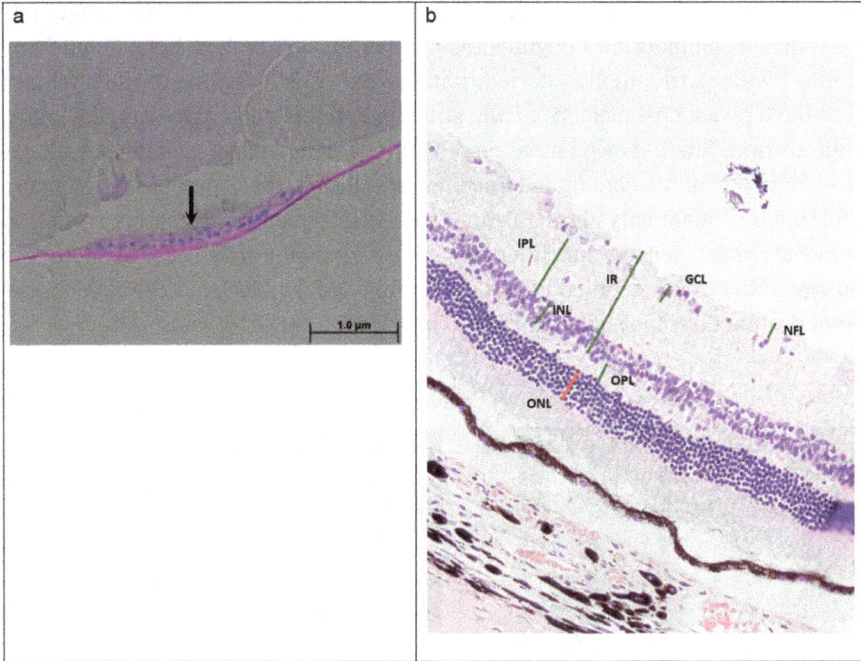

Figure 10.6: Representative structural eye histopathology in muskrats from the Giant mine site which is indicative of cataract formation. (a) Metaplastic lens epithelium of the eye has migrated under the liquefied cortex along the posterior lens capsule from intermediate location (black arrow). (b) Histopathology of the retinal layers in the eye. ONL: outer nuclear layer, OPL: outer plexiform layer, IR: inner retina, INL: inner nuclear layer, IPL: inner plexiform layer, GCL: ganglion cell layer, NFL: nerve fiber layer (112; Reproduced with copyright permission from Elsevier).

the thicknesses of ganglion cell layer (GCL), the retina nerve fiber layer (NFL), and the inner retina layer (IR) of the eye were reduced in the muskrats living near the Giant Mine (Figure 10.6b). Oxidative stress induced by As and Cd accumulation in the eye has been found to cause structural and functional damage to mammalian eye [117, 118]. Thus, elevated As and Cd exposure is likely to be a major contributing factor for the observed eye damage in muskrats living near the Giant mine.

10.6.1.4 Alterations of neurotransmission and brain structures

The neurotransmitters dopamine and serotonin regulate several critical physiological and behavioral functions in mammals including motor control, cognition, motivation, reproductive behaviors, appetite, mood and feelings. However, no marked changes in either dopamine or serotonin levels in the brain were observed in muskrats and red squirrels across three different sampling sites. Similarly, an MRI of the brain did not demonstrate any major differences in the brain volumes of muskrats and red squirrels across different sampling sites (Figures 10.7). However, core brain regions in muskrats from Site 1 were substantially affected, notably the critical regions of the forebrain such as the hippocampal memory circuit, striatum and thalamus. Moreover, the brains of squirrels from Site 1 showed more prominent neuroanatomical alterations with extensive shrinkage of the core brain structures, specifically the cortex in the forebrain. The forebrain in mammals plays a central role in information processing related to cognitive activities, sensory functions and voluntary motor activities. Therefore, the pathology of the brain recorded in this study provided evidence indicating neurobehavioral dysfunctions in small mammals living close to the Giant mine.

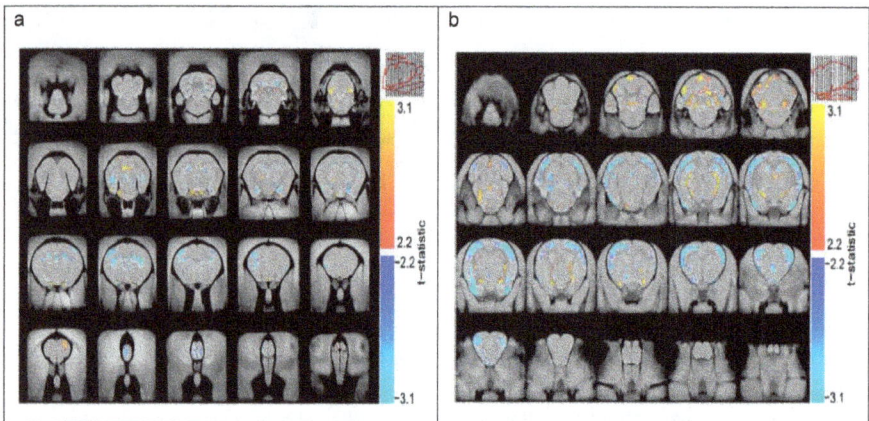

Figure 10.7: MRI coronal views showing structural alterations of animal brains from Site 1 (Giant mine; top 2 panels) *vs.* Site 3 (reference; bottom 2 panels): (a) muskrats and (b) red squirrels. The uncorrected t-statistic values indicate relative brain volume differences (113; Reproduced with permission from Elsevier).

10.6.1.5 Radiographic evidence of bone pathology

Radiographic analysis of bones of squirrels demonstrated abnormalities such as bone rarefaction (decreased density of the bones or osteoporosis), osteopenia (loss of bone mass) and thinning of the femoral shafts with significant lesions and bowing. These skeletal pathologies were observed in squirrels collected from all three sampling sites but were more common in Site 1 (Figure 10.8a). In muskrats, the femur which is the longest and one of the main load-bearing bones in the body showed bone sclerosis (Figure 10.8b), which is caused by a disturbance of the balance between bone formation and bone resorption leading to abnormal accumulation of bone minerals and density. These bone pathologies are indicative of bones that are weaker and more susceptible to fracture and are likely caused by chronic exposure to Cd because of its capacity to substitute Ca from bones.

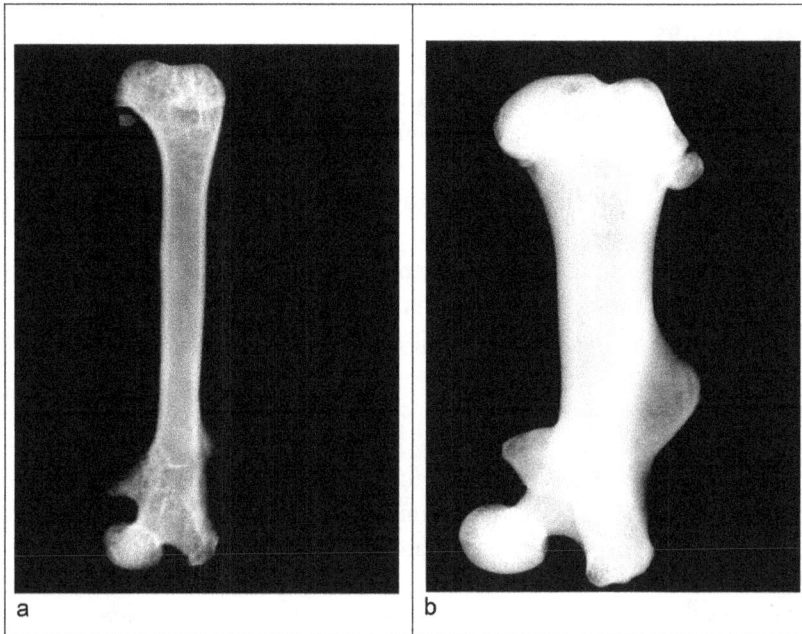

Figure 10.8: Skeletal pathology in red squirrels of small mammals from the Giant mine site: (a) decreased density of femur (osteoporosis) in red squirrels, (b) sclerosis (i.e., irregular bone calcification) in femur of muskrats (114; Reproduced with copyright permission from Elsevier).

10.6.2 Summary

Overall, the case study demonstrated how different types of biochemical and physiological assessments (biomarkers) can be useful in monitoring and understanding the real-world implications of industrial metal pollution for the resident wildlife. Specifically, the study provided strong evidence of elevated chronic exposure to As and Cd in small mammals near the Giant mine in Yellowknife, leading to oxidative stress and pathology of brain, eye, and bones. Such adverse effects can ultimately reduce their survival scope and reproductive fitness, and thereby affect their long-term sustenance in the impacted area.

10.7 Basic framework of environmental risk assessment and regulation of metals

10.7.1 Environmental risk assessment

Due to the hazardous nature of metals, their release into the environment must be controlled through public policy. Environmental regulation approaches, however, vary greatly depending on the jurisdiction and need to ascertain to protect the environment without hampering the economy. To achieve this goal, regulators rely on the process of risk assessment, which is a process of determining whether a hazardous substance will actually cause harm to exposed individuals or populations. The basic framework of risk assessment employs a four-step process:

a) Hazard identification: The purpose of this step is to simply determine whether a substance is hazardous. Metals, as discussed in this chapter, are hazardous to the receptors in the environment irrespective of whether they are essential or non-essential.

b) Hazard characterization (dose–response assessment): Once a hazardous substance is identified, it is essential to determine the relationship between the dose or concentration of the substance and the magnitude of adverse health effects. Hazard characterization also establishes useful values for risk assessment such as (i) the threshold dose or concentration at or below which adverse effects are not observed in a toxicity assessment test, which are reported as NOAEC (no observed adverse effects concentration) or NOAEL (no observed adverse effects level) values, (ii) the concentration or dose which causes death in 50% of the test organisms which is expressed as LD_{50} (lethal dose 50) and LC_{50} (lethal concentration 50), and (iii) the concentration which causes an adverse health effect in test organisms that is half of the maximum effect possible, also known as EC_{50} (effects concentration 50). During risk assessment, usually, all the available toxicity data across different species are collected for a substance and organized according to genus, trophic level for the environmental compartments they inhabit. Since these collected values generally come from laboratory testing and are mostly available for a very selective number of species, they are further processed mathematically such as by calculating the cumulative probability of 0.05 in acute toxicity values for the genera or by dividing the lowest reported EC_{50} or NOAEC with an appropriate assessment factor. The purpose of further processing is to make the assessment more environmentally realistic as well as representative of all species. This finally leads to values that can then be used for regulations, such as Criterion Maximum Concentration (CMC) or Criterion Continuous Concentration (CCC) in the USA and Predicted No Effect Concentration (PNEC) in the European Union.

c) Exposure assessment: This step deals with determining the frequency, the duration and the levels of contaminants to which an organism of interest is in contact with. Exposure assessment of a metal is performed through computer modelling and/or through direct measurement of the hazardous substance in the environment. As discussed in Section 2 of this chapter, the dissolved fraction, speciation, bioavailable fraction and environmental chemistry are all very critical factors for the assessment of metal exposure, and as such these factors require very careful consideration in the exposure assessment.

d) Risk characterization: In the final step, the exposure level is compared with the hazardous potential of the contaminant to determine whether harmful effects are expected in the natural environment. Approach for risk characterization may differ among different regulatory regimes but most common approach is to simply calculate the ratio of hazard data to exposure data and a ratio of less than 1 indicates adequate protection of the population of interest against the hazard in question.

10.7.2 Risk assessment framework in different jurisdictions

The risk assessment framework is used directly to determine the safe use of a hazardous substance. It is also used by the regulators to set an environmental release limit or maximum allowable concentration of a hazardous substance in a receiving environmental compartment. Although the fundamentals of environmental risk assessment are consistent, the methodology of risk assessment often differs depending on jurisdictions and the regulatory goal. A brief overview of the environmental regulations of metals in Canada, the USA and European Union is provided here.

10.7.2.1 Canada

In Canada, a major federal statute to manage pollution is the Fisheries Act, which prohibits the deposition of hazardous substances into waters that are frequented by fish. One of the key provisions of the Fisheries Act is the pollution prevention provision (also known as Section 36), which is administered by the Department of Environment and Climate Change Canada (ECCC). ECCC administers several effluent regulations to exercise the Section 36 of the Fisheries Act, one of which is relevant to metal mining effluents and it is known as Metal and Diamond Mining Effluent Regulations (MDMER). The MDMER authorizes the metal mining facilities to discharge the mine effluent containing specified deleterious substances into waters frequented by fish and to deposit mine waste into water bodies according to prescribed conditions. Under MDMER, mining facilities are required to manage all their effluent and the discharge of effluent is only permitted through an authorized final discharge point (FDP). All effluent must meet the concentration-based limits for the metals and other compounds listed in the regulation, the allowable pH limit and must not be acutely lethal to rainbow trout. In addition, this regulation also specifies effluent testing, reporting and environmental effects monitoring (EEM) requirements for facilities. EEM is a regulatory tool that is used in the MDMER as a condition governing the authority to deposit effluent. The purpose of EEM studies is to detect and measure changes in aquatic ecosystems (i.e., receiving environments) and it includes studies characterizing effluents, their sublethal toxicity, water quality monitoring and biological monitoring studies in the aquatic receiving environment. In addition to the federal regulations for the metal effluents, there are also Provincial and Territorial regulations that further enhance the regulatory framework for metal mining that affects aquatic organisms.

Within Canada, the Canadian Soil Quality Guidelines and Canadian Water Quality Guidelines are important scientific tools to assess the in-place contaminants including various metals in soil and water. These soil and water quality guidelines establish a numerical concentration for a contaminant including several metals that are recommended as levels expected to result in negligible risk to the health of the ecosystems. The quality guidelines are published by the Canadian Council of Ministers of the Environment (CCME) and these guideline values are not legally enforceable. Moreover, the CCME quality guidelines are generic and are not reflective of the unique conditions of a local site; therefore, the quality guidelines are used further by provincial and territorial resource managers and regulators to develop broad ecosystem management goals such as screening and assessment of contaminated sites and development of site-specific water and soil quality objectives.

Another federal act that protects the Canadian environment and human health from harmful chemicals including several metals is the Canadian Environmental Protection Act, 1999 (CEPA). Within Canada, the Domestic Substances List (DSL) is an inventory of substances that are manufactured in the country or imported into Canada. The DSL currently has 28,373 substances and it is updated regularly. Substances that are not on the DSL are considered new substances to Canada. If these 'new substances' are imported or manufactured over a certain threshold, they must be notified and their toxicity must be assessed. All major metals and metalloids discussed in this chapter are listed on the DSL. Substances (including metals and metalloids) that are toxic to humans or non-human organisms or persistent or bioaccumulative or may present to individuals in Canada the greatest potential for exposure are subjected to a screening level risk assessment. Following a screening level risk assessment, if it is concluded that a more comprehensive assessment is required for a substance, it can be added to the Priority Substances List. Finally, under the CEPA 1999, there are also provisions for developing codes of practice. One such code is the Environmental Code of Practice for Base Metals Smelters and Refineries. This code provides descriptions of processes used in the metal industry sector and environmental concerns associated with it. Moreover, it also recommends environmental performance standards for resolving these concerns. The Code provides guidance for environmental management systems and guidelines for environmental releases to air, water, and land, with particular focus on pollution prevention and control.

10.7.2.2 USA

As in Canada, the environmental regulation of metals is a complex combination of federal and state laws. All the current mining operations and the clean-up of abandoned mining sites are regulated in the USA. Two major acts that play a significant role in the regulation of metals and mining are National Environmental Policy Act (NEPA) and Clean Water Act (CWA), both of which are discussed briefly below. Other acts such as the Clean Air Act (CAA), the Toxic Substances Control Act (TSCA), and the Comprehensive Environmental Response, Compensation, and Liability Act (CERCLA) are also the part of the regulatory framework for metals, but a detailed discussion of these additional acts is beyond the scope of this chapter.

NEPA is considered as the backbone of environmental policies in the USA. Federal approvals of mining operations are granted in the USA through the application of NEPA. Essentially, NEPA establishes processes for evaluating the consequences of proposed mining on federal lands through impact and risk assessments. CWA also covers significant territory in metal regulation. The discharge of toxic pollutants including metals into the surface waters is covered by CWA in order to make surface water bodies safe. This is achieved by enforcing a system that limits the discharges to surface waters. In addition, the CWA requires the US Environmental Protection Agency (US EPA) to develop criteria for determining safe water for humans and wildlife. Accordingly, the US EPA develops Aquatic Life Criteria to determine how much of a particular chemical can be present in surface water before it harms wildlife (plants and aquatic animals) and human health. It is important to note that these water quality criteria are recommendations and as such are not legally binding; however, the States and authorized Tribes can use these criteria for developing their own water quality standards, which are legally binding and can be used to control discharges of pollutants into waters. Water quality criteria are set separately for both acute exposure as well as chronic exposure. First of all, toxicity data are collected for all available species and the species level data are then arranged based on the genus. Subsequently, different genus mean acute values (GMAV) are calculated by taking the geometric mean of species level data within each genus, and arranged in the order of high to low. Cumulative probability for each genus is then calculated, and finally a cumulative probability of 0.05 of toxicity values is selected and labelled as the final value. The final value, when divided by a factor of 2, provides the water quality criteria. CWA also indirectly regulates metals in soils by regulating the use and disposal of sewage sludge. Sewage sludge is a by-product produced from the wastewater treatment plants which is then commonly applied to land as a fertilizer. The concentration limit of several hazardous metals has been set in the sewage sludge by the US EPA and the concentration of metals in the sludge should always be below a threshold concentration limit.

Another tool used by USEPA to derive a risk-based screening levels of various soil contaminants (predominantly metals) are called ecological soil screening levels (Eco-SSLs). Similar to water quality criteria, Eco-SSLs are concentrations of contaminants in soil that are protective of plants, soil invertebrates, birds, and mammals that routinely come into contact with soil or consume biota that live in soil. There are 24 Eco-SSL contaminants in total, of which 17 are metals. It should be noted that Eco-SSL is only a screening level tool which can be used to identify contaminants of potential concern in soils requiring further characterization in the ecological risk assessment. In other words, any federal or state environmental assessment program can use Eco-SSL values to get preliminary information on soil contaminants and sites and conclude whether further ecological risk assessment is warranted or not.

10.7.2.3 European Union (EU)

Three regulations in the EU contribute significantly to the control of hazards to environmental and human health posed by metals. These regulations are: Registration, Evaluation, Authorisation and Restriction of Chemicals (REACH), Industrial Emission Directive (IED), and Water Framework Directive (WFD). In addition, there are several sector-specific regulations for controlling chemical (including metals) exposure such as the Toy Safety Directive, Regulation (EC) 1223/2009 on Cosmetics Products, and Regulation (EC) 1935/2004 on Food Contact Materials.

REACH applies, in principle, to all chemical substances that are used in the EU market, irrespective of the type of products they are used in. Manufacturers, producers and importers of the chemicals must demonstrate the safety of their chemicals for all the proposed uses of the said substance. To achieve this, the companies must register their substances by submitting a dossier for each substance to the European Chemicals Agency (ECHA) and these dossiers must include information about the quantity produced or manufactured or imported per year, hazard, exposure, risk assessment and risk management of the substance. Once a dossier is submitted, the substance may be evaluated, restricted or outright banned depending on the risk posed by the chemical. Many metal and metal compounds are currently on the Restricted Substance List in the REACH, which means that the manufacture, marketing or use of these metals and metal compounds is limited to the conditions set in the restriction or banned in the European Union.

Emissions from industrial activities are a major source of environmental pollution in Europe. One of the main tools for regulating pollutant emissions from industrial installations is the Industrial Emission Directive (IED). The list of polluting substances within the IED includes metals and their compounds including many other classes of chemicals. The goal of IED is to reduce the harmful industrial emissions across the EU and it is to be achieved mainly through a better application of Best Available Techniques (BAT). Industrial installations undertaking the industrial activities listed in the IED are required to operate in accordance with a permit which contains certain conditions set in accordance with the principles of the IED.

The third regulatory regime that contributes significantly in controlling metal levels in the environment is the Water Framework Directive (WFD). The main purpose of WFD is to restore the quality of surface and ground waters, and the responsibilities to achieve the goal have been put on member states. As part of the implementation of WFD, first a list of dangerous substances known as "priority substances" was created. Some of the substances within this list were further classified as 'priority hazardous substances'. The goal of WFD is to progressively reduce the discharge of priority substances and phase-out of the priority hazardous substances within a defined time frame. Among many other chemical groups, metals such as Cd, Hg, Pb and Ni and their compounds are also included in this list. The WFD sets up environmental quality standards (EQSs) for these priority substances, which are essentially the threshold concentration of these substances which must not be exceeded. EQSs were determined through the risk assessment framework described above, and it is meant to ensure protection of the aquatic environment from the exposure to pollutants including metals.

10.7.3 Summary

Risk assessment is a process of determining whether a hazardous substance (e.g., a metal or a metalloid) will actually cause harm to exposed individuals or populations. Risk assessment requires characterization of both exposure as well as adverse effects. The most common approach of risk assessment is generally to calculate the ratio of hazard data to exposure data and a ratio of less than 1 indicates adequate protection of the population of interest against the hazard in question. Although the fundamentals of environmental risk assessment are consistent, the methodology of risk assessment often differs depending on jurisdictions and the regulatory goal. In general, policies that regulate metals are complex and involve federal, state (or provincial) and local or municipal authorities. Furthermore, there are industry-specific regulations to control the metals in different products such as food, toys and consumer products.

Suggested readings

Tchounwou PB, Yedjou CG, Patlolla AK, Sutton DJ. Heavy Metal Toxicity and the Environment. In: Molecular, Clinical and Environmental Toxicology (Volume 3: Environmental Toxicology). 2012, Springer, Basel, Switzerland.

Mercurio SD. Hazard, Exposure, and Risk Modeling. In: Understanding Toxicology – A Biological Approach. 2017. Jones & Bartlett Learning, Burlington, MA, USA.

Adams W, Blust R, Dwyer R, Mount D, Nordheim E, Rodriguez H, Spry D. Bioavailability Assessment of Metals in Freshwater Environments: A Historical Review. Environl Toxicol Chem, 2019, 39, 48–59.

Gilbert SG. Metal toxicology. In: Information Resources in Toxicology (5th Edition), Volume 1: Background, Resources, and Tools. 2020, Academic Press, New York, USA.

Sengul AB, Asmatulu E. Toxicity of metal and metal oxide nanoparticles: A review. Environ Chem Lett, 2020, volume 18, 1659–1683.

Jørgensen SE. Ecotoxicology- A Derivative of Encyclopaedia of Ecology. 2010. Academic Press, Amsterdam, The Netherlands.

References

[1]   Tchounwou PB, Yedjou CG, Patlolla AK, Sutton DJ. Heavy metal toxicity and the environment. Experientia Suppl. 2012, 101, 133–164.

[2]   Han FX, Banin A, Su Y, Monts DL, Plodinec MJ, Kingery WL, et al. Industrial age anthropogenic inputs of heavy metals into the pedosphere. Naturwissenschaften. 2002, 89, 497–504.

[3]   Le Croizier G, Lacroix C, Artigaud S, Le Floch S, Munaron J-M, Raffray J, et al. Metal subcellular partitioning determines excretion pathways and sensitivity to cadmium toxicity in two marine fish species. Chemosphere. 2019, 217, 754–762.

[4]   Tamas MJ, Martinoia E, editors. Molecular Biology of Metal Homeostasis and Detoxification: From Microbes to Man. Springer, Berlin, Heidelberg, 2006.

[5]   Friedberg F. Albumin as the major metal transport agent in blood. FEBS Lett. 1975, 59, 140–141.

[6]   Niyogi S, Wood CM. Biotic ligand model, a flexible tool for developing site-specific water quality guidelines for metals. Environ Sci Technol. 2004, 38, 6177–6192.

[7]   Hogstrand C, Galvez F, Wood CM. Toxicity, silver accumulation and metallothionein induction in freshwater rainbow trout during exposure to different silver salts. Environ Toxicol Chem. 1996, 15, 1102–1108.

[8]   Playle RC, Dixon DG, Burnison K. Copper and cadmium binding to fish gills: Modification by dissolved organic carbon and synthetic ligands. Can J Fish Aquat Sci. 1993 Dec 1,50(12), 2667–2677.

[9]   de Paiva Magalhães D, Da Costa Marques MR, Baptista DF, Buss DF. Metal bioavailability and toxicity in freshwaters. Environ Chem Lett. 2015, 13, 69–87.

[10]   Bonito MD, Di Bonito M, Breward N, Crout N, Smith B, Young S. Overview of selected soil pore water extraction methods for the determination of potentially toxic elements in contaminated soils: Operational and technical aspects. Environ Geochem. 2008. p. 213–249.

[11]   García-Miragaya J, Cardenas R, Page A. Surface loading effect on Cd and Zn sorption by kaolinite and montmorillonite from low concentration solutions. Water Air Soil Pollut. 1986, 27, 181–190.

[12]   McKenzie RM. The adsorption of lead and other heavy metals on oxides of manganese and iron. Soil Res. 1980, 18, 61–73.

[13]   Gutknecht J. Inorganic mercury (Hg2+) transport through lipid bilayer membranes. J Membr Biol. 1981, 61, 61–66.

[14]   Olson KR, Fromm PO. Mercury uptake and ion distribution in gills of rainbow trout (Salmo gairdneri): Tissue scans with an electron microprobe. J Fish Res Board Can. 1973, 30, 1575–1578.

[15]   Cooper GM. Transport of Small Molecules. In: The Cell: A Molecular Approach 2nd edition. Sinauer Associates, 2000.

[16]   Reuss L. Mechanisms of Ion Transport Across Cell Membranes and Epithelia. In: Alpern RJ, Hebert SC editors,. The Kidney. Elsevier, 2008. p. 35–56.

[17]   Verbost PM, Flik G, Lock RA, Wendelaar Bonga SE. Cadmium inhibits plasma membrane calcium transport. J Membr Biol. 1988, 102, 97–104.

[18]   Illing AC, Shawki A, Cunningham CL, Mackenzie B. Substrate profile and metal-ion selectivity of human divalent metal-ion transporter-1. J Biol Chem. 2012, 287, 30485–30496.

[19]   Verbost PM, Van Rooij J, Flik G, Lock RAC, Bonga SEW. The movement of cadmium through freshwater trout branchial epithelium and its interference with calcium transport. J Exp Biol. 1989, 145, 185–197.

[20]   Misra S, Kwong RWM, Niyogi S. Transport of selenium across the plasma membrane of primary hepatocytes and enterocytes of rainbow trout. J Exp Biol. 2012, 215, 1491–1501.

[21]   Garbinski LD, Rosen BP, Chen J. Pathways of arsenic uptake and efflux. Environ Int. 2019, 126, 585–597.

[22]   Luoma SN, Presser TS. Emerging opportunities in management of selenium contamination. Environ Sci Technol. 2009, 43, 8483–8487.

[23]   Thiry C, Ruttens A, Pussemier L, Schneider Y-J. An in vitro investigation of species-dependent intestinal transport of selenium and the impact of this process on selenium bioavailability. Br J Nutr. 2013, 109, 2126–2134.

[24]   Thévenod F, Fels J, Lee W-K, Zarbock R. Channels, transporters and receptors for cadmium and cadmium complexes in eukaryotic cells: Myths and facts. Biometals. 2019, 32, 469–489.

[25]   Nilius B, Vennekens R, Prenen J, Hoenderop JG, Droogmans G, Bindels RJ. The single pore residue Asp542 determines Ca2+ permeation and Mg2+ block of the epithelial Ca2+ channel. J Biol Chem. 2001, 276, 1020–1025.

[26]   Kim BJ, Hong C. Role of transient receptor potential melastatin type 7 channel in gastric cancer. Integr Med Res. 2016, 5, 124–130.

[27]   Shahsavarani A, McNeill B, Galvez F, Wood CM, Goss GG, Hwang -P-P, et al. Characterization of a branchial epithelial calcium channel (ECaC) in freshwater rainbow trout (Oncorhynchus mykiss). J Exp Biol. 2006, 209, 1928–1943.

[28]   Pan T-C, Liao B-K, Huang C-J, Lin L-Y, Hwang -P-P. Epithelial Ca2+ channel expression and Ca2+ uptake in developing zebrafish. Am J Physiol Regul Integr Comp Physiol. 2005, 289, R1202–11.

[29]   Galvez F, Wong D, Wood CM. Cadmium and calcium uptake in isolated mitochondria-rich cell populations from the gills of the freshwater rainbow trout. Am J Physiol Regul Integr Comp Physiol. 2006, 291, R170–6.

[30]   Kovacs G, Danko T, Bergeron MJ, Balazs B, Suzuki Y, Zsembery A, et al. Heavy metal cations permeate the TRPV6 epithelial cation channel. Cell Calcium. 2011, 49, 43–55.

[31]   Monteilh-Zoller MK, Hermosura MC, Nadler MJS, Scharenberg AM, Penner R, Fleig A. TRPM7 provides an ion channel mechanism for cellular entry of trace metal ions. J Gen Physiol. 2003, 121, 49–60.

[32]   Varanasi U, Gmur DJ. Influence of water-borne and dietary calcium on uptake and retention of lead by coho salmon (Oncorhynchus kisutch). Toxicol Appl Pharmacol. 1978, 46, 65–75.

[33]   Bury NR, Wood CM. Mechanism of branchial apical silver uptake by rainbow trout is via the proton-coupled Na+channel. Am J of Physiology-Regulatory, Integrative and Comparative Physiol. 1999, 277(5), R1385–91.

[34]   Grosell M, Wood CM. Copper uptake across rainbow trout gills: Mechanisms of apical entry. J Exp Biol. 2002, 205, 1179–1188.

[35]   He L, Vasiliou K, Nebert DW. Analysis and update of the human solute carrier (SLC) gene superfamily. Hum Genomics. 2009, 3, 195–206.

[36]   Garrick MD, Singleton ST, Vargas F, Kuo HC, Zhao L, Knöpfel M, et al. DMT1: Which metals does it transport? Biol Res. 2006, 39, 79–85.

[37]   Kwong RWM, Andrés JA, Niyogi S. Molecular evidence and physiological characterization of iron absorption in isolated enterocytes of rainbow trout (Oncorhynchus mykiss): Implications for dietary cadmium and lead absorption. Aquat Toxicol. 2010, 99, 343–350.

[38]   Komjarova I, Bury NR. Evidence of common cadmium and copper uptake routes in zebrafish (Danio rerio). Environ Sci Technol. 2014, 48, 12946–12951.

[39]   Qiu A, Shayeghi M, Hogstrand C. Molecular cloning and functional characterization of a high-affinity zinc importer (DrZIP1) from zebrafish (Danio rerio). Biochem J. 2005, 388, 745–754.

[40]   Dalton TP, He L, Wang B, Miller ML, Jin L, Stringer KF, et al. Identification of mouse SLC39A8 as the transporter responsible for cadmium-induced toxicity in the testis. Proc Natl Acad Sci. 2005, 102, 3401–3406.

[41]   Kwong RWM, Niyogi S. Cadmium transport in isolated enterocytes of freshwater rainbow trout: Interactions with zinc and iron, effects of complexation with cysteine, and an ATPase-coupled efflux. Comp Biochem Physiol. 2012, 155C, 238–46.

[42]   Eisses JF, Kaplan JH. The Mechanism of Copper Uptake Mediated by Human CTR1. Vol. 280, J of Biol Chem. 2005, 280, 37159–37168.

[44]   Minghetti M, Leaver MJ, Carpenè E, George SG. Copper transporter 1, metallothionein and glutathione reductase genes are differentially expressed in tissues of sea bream (Sparus aurata) after exposure to dietary or waterborne copper. Comp Biochem. 2008, 147C, 450–459.

[45]   Lee J, Peña MMO, Nose Y, Thiele DJ. Biochemical characterization of the human copper transporter Ctr1. J Biol Chem. 2002, 277, 4380–4387.

[46]   Fairweather-Tait SJ, Bao Y, Broadley MR, Collings R, Ford D, Hesketh JE, et al. Selenium in human health and disease. Antioxid Redox Signal. 2011, 14, 1337–1383.

[47]   Villa-Bellosta R, Sorribas V. Arsenate transport by sodium/phosphate cotransporter type IIb. Toxicol Appl Pharmacol. 2010, 247, 36–40.

[48]   Roggenbeck BA, Banerjee M, Leslie EM. Cellular arsenic transport pathways in mammals. J Environ Sci. 2016, 49, 38–58.

[49]   Schoenmakers TJ, Klaren PH, Flik G, Lock RA, Pang PK, Bonga SE. Actions of cadmium on basolateral plasma membrane proteins involved in calcium uptake by fish intestine. J Membr Biol. 1992, 127, 161–172.

[50]   Marshall WS. Na+, Cl-, Ca2+ and Zn2+ transport by fish gills: Retrospective review and prospective synthesis. J Exp Zool. 2002, 293, 264–283.

[51]   Mercer JFB, Barnes N, Stevenson J, Strausak D, Llanos RM. Copper-induced trafficking of the Cu-ATPases: A key mechanism for copper homeostasis. Biometals. 2003, 16, 175–184.

[52]   Linz R, Lutsenko S. Copper-transporting ATPases ATP7A and ATP7B: Cousins, not twins. J Bioenerg Biomembr. 2007, 39, 403–407.

[53] Mendelsohn BA, Yin C, Johnson SL, Wilm TP, Solnica-Krezel L, Gitlin JD. Atp7a determines a hierarchy of copper metabolism essential for notochord development. Cell Metab. 2006, 4, 155–162.

[54] Minghetti M, Leaver MJ, George SG. Multiple Cu-ATPase genes are differentially expressed and transcriptionally regulated by Cu exposure in sea bream, Sparus aurata. Aquat Toxicol. 2010, 97, 23–33.

[55] Ibricevic A, Brody SL, Youngs WJ, Cannon CL. ATP7B detoxifies silver in ciliated airway epithelial cells. Toxicol Appl Pharmacol. 2010, 243, 315–322.

[56] Brini M, Carafoli E. The Plasma Membrane Ca2+ ATPase and the Plasma Membrane Sodium Calcium Exchanger Cooperate in the Regulation of Cell Calcium. Cold Spring Harb Perspect Biol. 2011, 3.

[57] Liao B-K, Deng A-N, Chen S-C, Chou M-Y, Hwang -P-P. Expression and water calcium dependence of calcium transporter isoforms in zebrafish gill mitochondrion-rich cells. BMC Genomics. 2007, 8, 354.

[58] Mager EM Lead. In: Wood CM, Farrell AP, Brauner CJ, editors. Homeostasis and Toxicology of Non-essential Metals. Academic Press, In: Fish Physiology, vol. 31B. 2012. p. 185–236.

[59] Chowdhury MJ, Blust R. Strontium. In: Wood CM, Farrell AP, Brauner CJ editors,. Fish Physiology. Academic Press, In: Fish Physiology, 31B. 2012, 351–390.

[60] Bafaro E, Liu Y, Xu Y, Dempski RE. The emerging role of zinc transporters in cellular homeostasis and cancer. Signal Transduct Target Ther. 2017, 2. Available from: http://dx. doi.org/10.1038/sigtrans.2017.29

[61] Lopez V, Kelleher SL. Zinc transporter-2 (ZnT2) variants are localized to distinct subcellular compartments and functionally transport zinc. Biochem J. 2009, 422, 43–52.

[62] Bal W, Sokołowska M, Kurowska E, Faller P. Binding of transition metal ions to albumin: Sites, affinities and rates. Biochim Biophys Acta. 2013, 1830, 5444–5455.

[63] Shooshtary S, Behtash S, Nafisi S. Arsenic trioxide binding to serum proteins. J Photochem Photobiol B. 2015, 148, 31–36.

[64] Liu J, Goyer RA, Waalkes MP. Toxic Effects of Metals. Casarett and Doull's Toxicology: The Basic Science of Poisons 7th ed New York, McGraw-Hill. 2008, 931–979.

[65] Li C, Li Z, Li Y, Zhou J, Zhang C, Su X, et al. A ferritin from Dendrorhynchus zhejiangensis with heavy metals detoxification activity. PLoS One. 2012, 7, e51428.

[66] Carson SD. Cadmium binding to human α2-macroglobulin. Biochim Et Biophys Acta (BBA) – Protein Structure and Molecular Enzymology. 1984, 791, 370–374.

[67] Vulpe CD, Kuo YM, Murphy TL, Cowley L, Askwith C, Libina N. Hephaestin, a ceruloplasmin homologue implicated in intestinal iron transport, is defective in the sla mouse. Nat Genet. 1999, 21, 195–199.

[68] Burk RF, Hill KE. Regulation of Selenium Metabolism and Transport. Annu Rev Nutr. 2015, 35, 109–134.

[69] Calderón-Salinas JV, Quintanar-Escorcia MA, González-Martínez MT, Hernández-Luna CE. Lead and calcium transport in human erythrocyte. Hum Exp Toxicol. 1999, 18, 327–332.

[70] Andrews NC, Schmidt PJ. Iron homeostasis. Annu Rev Physiol. 2007, 69, 69–85.

[71] Ramos D, Mar D, Ishida M, Vargas R, Gaite M, Montgomery A, et al. Mechanism of Copper Uptake from Blood Plasma Ceruloplasmin by Mammalian Cells. PLoS One. 2016, 11, e0149516.

[72] Foulkes EC. The concept of critical levels of toxic heavy metals in target tissues. Crit Rev Toxicol. 1990, 20, 327–339.

[73] Rakib MA, Huda ME, Hossain SM, Naher K, Khan R, Sultana MS, et al. Arsenic Content in Inactive Tissue: Human Hair and Nail. J Sci Res. Rep. 2013,522–535.

[74] Rainbow PS. Trace metal concentrations in aquatic invertebrates: Why and so what? Environ Pollut. 2002, 120, 497–507.

[75] Rosabal M, Pierron F, Couture P, Baudrimont M, Hare L, Campbell PGC. Subcellular partitioning of non-essential trace metals (Ag, As, Cd, Ni, Pb, and Tl) in livers of American (Anguilla rostrata) and European (Anguilla anguilla) yellow eels. Aquat Toxicol 2015. 160, 128–141.

[76] Brown BE. The form and function of metal-containing "granules" in invertebrate tissues. Biol Rev Camb Philos Soc. 1982, 57, 621–667.

[77] Shekh K, Saeed H, Kodzhahinchev V, Brinkmann M, Hecker M, Niyogi S. Differences in the subcellular distribution of cadmium and copper in the gills and liver of white sturgeon (Acipenser transmontanus) and rainbow trout (Oncorhynchus mykiss). Chemosphere. 2021, 265, 129142.

[78] Eyckmans M, Blust R, De Boeck G. Subcellular differences in handling Cu excess in three freshwater fish species contributes greatly to their differences in sensitivity to Cu. Aquat Toxicol. 2012, 118–119, 97–107.

[79] Verbost PM, Flik G, Lock RA, Wendelaar Bonga SE. Cadmium inhibition of Ca2 uptake in rainbow trout gills. Am J of Physiol-Regulatory, Integrative and Comparative Physiology. 1987. 253, R216–21.

[80] Rogers JT, Wood CM. Characterization of branchial lead-calcium interaction in the freshwater rainbow trout Oncorhynchus mykiss. J Exp Biol. 2004, 207, 813–825.

[81] Morgan TP, Grosell M, Gilmour KM, Playle RC, Wood CM. Time course analysis of the mechanism by which silver inhibits active Na+ and Cl– uptake in gills of rainbow trout. Am J of Physiol-Regulatory, Integrative and Comparative Physiology. 2004, 287, R234–42.

[82] Nogueira LS, Chen CC, Wood CM, Kelly SP. Effects of copper on a reconstructed freshwater rainbow trout gill epithelium: Paracellular and intracellular aspects. Comp Biochem Physiol. 2020, 230C, 108705.

[83] Al-Attar AM. The Influences of Nickel Exposure on Selected Physiological Parameters and Gill Structure in the Teleost Fish, Oreochromis niloticus. J Biol Sci. 2006. 7, 77–85.

[84] Lim MY-T, Zimmer AM, Wood CM. Acute exposure to waterborne copper inhibits both the excretion and uptake of ammonia in freshwater rainbow trout (Oncorhynchus mykiss). Comp Biochem Physiol. 2015, 168C, 48–54.

[85] Zimmer AM, Barcarolli IF, Wood CM, Bianchini A. Waterborne copper exposure inhibits ammonia excretion and branchial carbonic anhydrase activity in euryhaline guppies acclimated to both fresh water and sea water. Aquat Toxicol. 2012, 122–123, 172–180.

[86] Braga MM, Dick T, de Oliveira DL, Scopel-Guerra A, Mussulini BHM, Souza DO. Evaluation of zinc effect on cadmium action in lipid peroxidation and metallothionein levels in the brain. Toxicol Reports. 2015, 2, 858–863.

[87] Jackim E. Influence of Lead and other Metals on Fish δ-Aminolevulinate Dehydrase Activity. J Fish Res Board of Canada. 1973, 30, 560–562.

[88] Magyar JS, Weng T-C, Stern CM, Dye DF, Rous BW, Payne JC, et al. Re-examination of lead(II) coordination preferences in sulfur-rich sites: Implications for a critical mechanism of lead poisoning. J Am Chem Soc. 2005, 127, 9495–9505.

[89] Ercal N, Gurer-Orhan H, Aykin-Burns N. Toxic metals and oxidative stress part I: Mechanisms involved in metal-induced oxidative damage. Curr Top Med Chem. 2001, 1, 529–539.

[90] Misra S, Niyogi S. Selenite causes cytotoxicity in rainbow trout (Oncorhynchus mykiss) hepatocytes by inducing oxidative stress. Toxicol In Vitro. 2009, 23, 1249–1258.

[91] Misra S, Peak D, Niyogi S. Application of XANES spectroscopy in understanding the metabolism of selenium in isolated rainbow trout hepatocytes: Insights into selenium toxicity. Metallomics. 2010, 2, 710–717.

[92]   Chou WC, Jie C, Kenedy AA, Jones RJ, Trush MA, Dang CV. Role of NADPH oxidase in arsenic-induced reactive oxygen species formation and cytotoxicity in myeloid leukemia cells. Proceed Nat Acad Sci. 2004, 101, 4578–4583.

[93]   Xu Y, Wang Y, Zheng Q, Li X, Li B, Jin Y, et al. Association of oxidative stress with arsenic methylation in chronic arsenic-exposed children and adults. Toxicol Appl Pharmacol. 2008, 232, 142–149.

[94]   McDonald BG, deBruyn AMH, Elphick JRF, Davies M, Bustard D, Chapman PM. Developmental toxicity of selenium to Dolly Varden char (Salvelinus malma). Environ Toxicol Chem. 2010, 29, 2800–2805.

[95]   Gaworecki KM, Chapman RW, Neely MG, D'Amico AR, Bain LJ. Arsenic exposure to killifish during embryogenesis alters muscle development. Toxicol Sci. 2012, 125, 522–531.

[96]   Spallholz JE, Hoffman DJ. Selenium toxicity: Cause and effects in aquatic birds. Aquat Toxicol. 2002, 57, 27–37.

[97]   Fernández AJG, Virosta PS, Espín S, Eeva T. A review on exposure and effects of arsenic in passerine birds. Toxicol Lett. 2016, 258, S203.

[98]   Janz DM, DeForest DK, Brooks ML, Chapman PM, Gilron G, Hoff D. Selenium toxicity to aquatic organisms. In: Chapman PM, Adams WJ, Brooks ML, Delos CG, Luoma SN, Maher WA, et al., editors. Ecological Assessment of Selenium in the Aquatic Environment. CRC Press Boca Raton, FL, 2010. p. 141–231.

[99]   Gonzalez HO, Roling JA, Baldwin WS, Bain LJ. Physiological changes and differential gene expression in mummichogs (Fundulus heteroclitus) exposed to arsenic. Aquat Toxicol. 2006, 77, 43–52.

[100] Naderi M, Puar P, Zonouzi-Marand M, Chivers DP, Niyogi S, Kwong RWM. A comprehensive review on the neuropathophysiology of selenium. Sci Total Environ. 2021, 767, 144329.

[101] Darbre PD. Metalloestrogens: An emerging class of inorganic xenoestrogens with potential to add to the oestrogenic burden of the human breast. J Appl Toxicol. 2006, 26, 191–197.

[102] Sellin MK, Kolok AS. Cd exposures in fathead minnows: Effects on adult spawning success and reproductive physiology. Arch Environ Contam Toxicol. 2006, 51, 594–599.

[103] Knazicka Z, Forgacs Z, Lukacova J, Roychoudhury S, Massanyi P, Lukac N. Endocrine disruptive effects of cadmium on steroidogenesis: Human adrenocortical carcinoma cell line NCI-H295R as a cellular model for reproductive toxicity testing. J Environ Sci Health A. 2015, 50, 348–356.

[104] Wang H, Ji Y-L, Wang Q, Zhao X-F, Ning H, Liu P. Maternal lead exposure during lactation persistently impairs testicular development and steroidogenesis in male offspring. J Appl Toxicol. 2013, 33, 1384–1394.

[105] Tian M, Wang Y-X, Wang X, Wang H, Liu L, Zhang J, et al. Environmental doses of arsenic exposure are associated with increased reproductive-age male urinary hormone excretion and in vitro Leydig cell steroidogenesis. J Hazard Mater. 2021, 408, 124904.

[106] Green AJ, Planchart A. The neurological toxicity of heavy metals: A fish perspective. Comp Biochem Physiol 2018, 208C, 12–19.

[107] Eason C, O'Halloran K. Biomarkers in toxicology versus ecological risk assessment. Toxicology. 2002, 181–182, 517–521.

[108] Peakall DB. The role of biomarkers in environmental assessment. Ecotoxicology. 1994, 3, 157–160.

[109] Bernard A. Biomarkers of metal toxicity in population studies: Research potential and interpretation issues. J Toxicol Environ Health A. 2008, 71, 1259–1265.

[110] Duffy LK, Oehler MW Sr, Bowyer RT, Bleich VC. Others. Mountain sheep: An environmental epidemiological survey of variation in metal exposure and physiological biomarkers following mine development. Am J Environ Sci. 2009, 5, 295–302.

[111] Amuno S, Shekh K, Kodzhahinchev V, Niyogi S. Neuropathological changes in wild muskrats (Ondatra zibethicus) and red squirrels (Tamiasciurus hudsonicus) breeding in arsenic endemic areas of Yellowknife, Northwest Territories (Canada): Arsenic and cadmium accumulation in the brain and biomarkers of oxidative stress. Sci Total Environ. 2020, 704, 135426.

[112] Amuno S, Bedos L, Kodzhahinchev V, Shekh K, Niyogi S, Grahn B. Comparative study of arsenic toxicosis and ocular pathology in wild muskrats (Ondatra zibethicus) and red squirrels (Tamiasciurus hudsonicus) breeding in arsenic contaminated areas of Yellowknife, Northwest Territories (Canada). Chemosphere. 2020, 248, 126011.

[113] Amuno S, Rudko DA, Gallino D, Tuznik M, Shekh K, Kodzhahinchev V, et al. Altered neurotransmission and neuroimaging biomarkers of chronic arsenic poisoning in wild muskrats (Ondatra zibethicus) and red squirrels (Tamiasciurus hudsonicus) breeding near the City of Yellowknife, Northwest Territories (Canada). Sci Total Environ. 2020, 707, 135556.

[114] Amuno S, Shekh K, Kodzhahinchev V, Niyogi S, Al Kaissi A. Skeletal pathology and bone mineral density changes in wild muskrats (Ondatra zibethicus) and red squirrels (Tamiasciurus hudsonicus) inhabiting arsenic polluted areas of Yellowknife, Northwest Territories (Canada): A radiographic densitometry study. Ecotoxicol Environ Saf. 2021, 208, 111721.

[115] Amuno S, Jamwal A, Grahn B, Niyogi S. Chronic arsenicosis and cadmium exposure in wild snowshoe hares (Lepus americanus) breeding near Yellowknife, Northwest Territories (Canada), part 1: Evaluation of oxidative stress, antioxidant activities and hepatic damage. Sci Total Environ. 2018, 618, 916–926.

[116] Cott PA, Zajdlik BA, Palmer MJ, McPherson MD. Arsenic and mercury in lake whitefish and burbot near the abandoned Giant Mine on Great Slave Lake. J Great Lakes Res. 2016, 42, 223–232.

[117] Kleiman NJ, Quinn AM, Fields KG, Slavkovich, Graziano JH. Arsenite accumulation in the mouse eye. J Toxicol Environ Health A. 2016, 79, 339–341.

[118] Wills NK, Sadagopa Ramanujam VM, Chang J, Kalariya N, Lewis JR, Weng T-X, van Kuijk FJGM. Cadmium accumulation in the human retina: Effects of age, gender, and cellular toxicity. Exp Eye Res. 2008, 86, 41–51.

Feiyue Wang and Robie W. Macdonald

Chapter 11
Socioeconomic, political, and legal ramifications of environmental and biochemical toxicology: the complicated story of mercury

11.1 Introduction

Many issues addressed by environmental and biochemical toxicology have a broad societal relevance in that they require coordinated human action at local, national and international scales. As has been witnessed with global climate change, what commences as a scientific challenge to collect, interpret and model data sufficiently to provide a scientific foundation for the issue ends up as a social and political challenge to mount an appropriate response. For such global issues, the science must be sufficiently robust to survive the doubt generated, often by vested interests, to thwart action (e.g., see [1]).

In this chapter we will explore socioeconomic, political, and legal ramifications of environmental and biochemical toxicology by following the story of mercury (Hg), an element that has both fascinated and intimidated humans for over 4,000 years. In particular, we wish to highlight the winding scientific path that has led eventually to the modern understanding of how Hg presents a global threat, and why international action is necessary.

Prior to any scientific evidence, Hg was regarded as the elixir of life and became the magic pill for a wide spectrum of diseases. This application arguably harmed or killed more people than it ever cured, and it was the use and misuse of Hg in medical practices that laid the foundation of modern toxicology. This initial medical misuse of Hg was largely founded on a mythology inspired by Hg's magical appearance:

Acknowledgments: The authors would like to acknowledge funding from the Canada Research Chairs program, the Natural Sciences and Engineering Research Council of Canada, ArcticNet, the Canadian Arctic GEOTRACES program, and Fisheries and Oceans Canada. F.W. would like to dedicate this chapter to the fond memory of co-author Dr. Robie W. Macdonald, who passed away during the book production. Robie was an intellectual giant in environmental, marine, and Arctic sciences, and a great mentor to me and many of my graduate students over the past two decades. His curiosity, story-telling, and wit and wisdom will be forever missed.

Feiyue Wang, Centre for Earth Observation Science, University of Manitoba, Winnipeg, MB, Canada, e-mail: feiyue.wang@umanitoba.ca
Robie W. Macdonald, Institute of Ocean Sciences, Department of Fisheries and Oceans, Sydney, BC, Canada

https://doi.org/10.1515/9783110626285-011

uniquely a shiny, silvery metal liquid at room temperature. Early on, Hg was found to be technically useful for extracting gold (Au) and silver (Ag) out of Earth's crust and this widespread application initiated global contamination and pollution (see Textbox 1) by human exploitation of Hg over at least the past 500 years [2, 3]. As a basis to accumulate the wealth associated with Au and Ag, Hg thus became a foundation for global trade and the economy. The red pigment of Hg, cinnabar, has long been used in royal seals to ink political and legal documents. It therefore seems fitting that the solution to the millennium-old conundrum has finally been "inked" in the Minamata Convention, which went into force in 2017 and presents a legally binding road map to global political action. But what is essential to note here is that it was irrefutable scientific evidence that led to this convention.

Here, we will use the Arctic as a case study. The Arctic has provided to us during the past two decades the strongest evidence of how fugitive, semi-volatile substances such as Hg widely released in temperate and tropical regions may threaten the food safety of remote Indigenous populations. We will also discuss how ecosystem recovery from Hg contamination under the Minamata Convention may be delayed by climate change.

Textbox 1: terminology matters!
Many terms used in environmental and biochemical toxicology have very distinctive scientific meanings but are commonly misused. It is important to clearly understand their differences and use them properly to avoid ambiguous, confusing, misleading, or even completely wrong statements in scientific communications, which can be rapidly propagated and misinterpreted by the media and public. Below are a few such terms that are especially relevant to the Hg story. More discussions can be found in [4, 5].

Contamination or pollution?
The distinction between these two terms is important because it permits discrimination between instances in which disturbance by human activities can be detected (contamination) and those in which adverse effects occur as a result of these disturbances (pollution). Difference between these terms implies that there exists a capacity for change in the environment without adverse effects occurring.

Contamination refers to detectable disturbances by human activities but without inference of adverse effects. Our ability to recognize contamination depends on our ability to analytically measure a substance in the environment: contamination occurs when the concentration of a naturally occurring substance (e.g., a metal) is clearly above the normal background level, or the concentration of an anthropogenically produced substance (e.g., polychlorinated biphenyls or PCBs) is readily detectable. Such a substance is referred to as a **contaminant**. In either case, contamination may or may not result in harm, and mitigation may not be required.

Pollution refers to more severe disturbances by human activities resulting in adverse effects (e.g., harm to living organisms including humans). Our ability to recognize pollution depends on our ability to toxicologically identify associated harm. Typically, pollution occurs when a substance (naturally occurring or anthropogenically produced) is present in the environment at high enough concentrations to cause harm to living organisms that are exposed to it. Such a substance is referred to as a **pollutant**. In other words, pollution is caused by more severe contamination, and mitigation is typically required under most jurisdictions.

Hazard or risk?

Hazard refers to the inherent capacity or possibility of a stressor (e.g., a contaminant) causing adverse effects to a receptor (e.g., an organism). Theoretically, any substance can be hazardous; the difference lies in the dosage needed to result in adverse effects (recall Paracelsus dictum).

Risk refers to the likelihood or probability of such adverse effects occurring to a receptor. Adverse effects occur only when an organism is exposed (directly or through food) to a substance after a certain duration. Thus, the risk of a substance is determined not only by its presence in the environment, but also by the organism's **exposure** to it.

Bioaccumulation, bioconcentration, or biomagnification?

These three terms describe processes by which an organism accumulates a substance from its environment.

Bioaccumulation refers to the general process by which an organism takes up a substance from its environment via any pathways (e.g., from water, soil, or dietary sources).

Bioconcentration refers to the specific process by which an aquatic organism takes up a substance directly from water.

Biomagnification refers to the specific process by which the tissue concentrations of a bioaccumulated substance increase as it passes up the food chain through two or more trophic levels. Biomagnification results in organisms at higher trophic levels containing higher concentrations of the substance.

Whereas many substances are bioaccumulated in organisms, biomagnification occurs to a much smaller number of substances such as some persistent organic pollutants (POPs). Methylmercury is one of very few metal contaminants known to biomagnify in aquatic ecosystems.

11.2 The global Hg cycle

We start with a brief overview of how Hg cycles through the global environment. As a naturally occurring element in the Earth's crust (~0.056 mg/kg in the upper continental crust [6]), Hg is continuously released to the surface environment by natural processes such as rock weathering, volcanism, and geothermal activity. The natural release rate, however, has been greatly exceeded in modern times due to human activities (known as *anthropogenic processes*) including mining and production of Hg, the use of Hg in various applications including Au and Ag refining, as well as from processes where Hg is released as an impurity or by-product such as fossil fuel combustion and other metal production [3]. In total, human activities have resulted in the release of more than 1.5 million tonnes (Mt; 1 Mt = 10^9 kg) of Hg throughout human history, 95% of which has occurred in the past 500 years [3] (Figure 11.1a).

Once released, Hg is free to cycle through atmospheric, terrestrial, and marine systems. Over the long term, the release of Hg is balanced by removal processes, such as the burial of Hg in accumulating marine sediments (Figure 11.2a). During the cycling, Hg may undergo chemical reactions, some of which are photochemically or microbially enhanced; these reactions change Hg's chemical *speciation* (i.e., the composition of various chemical forms of Hg), which has major implications for its volatility, transport pathways, *bioavailability* (i.e., the degree to which a chemical can be taken up by an organism) and toxicity of Hg (Figure 11.2b).

Mercury emitted to the atmosphere from natural and anthropogenic sources is almost exclusively inorganic and in two oxidation states: elemental (Hg^0) and divalent (Hg^{II}); the monovalent form (Hg^I) is thought to be less stable. Reactions between these oxidation states (i.e., redox reactions) in the atmosphere are primarily photochemically driven, and the bulk of the atmospheric Hg deposited onto the terrestrial and marine environments is in the form of inorganic Hg^{II}, which is readily scavenged by particles. The marine environment also receives inorganic Hg^{II} from rivers, catchment runoff, coastal erosion, submarine groundwater discharge, and hydrothermal vents (Figure 11.2). Some of the inorganic Hg^{II} in terrestrial and marine environments may be converted into organic Hg^{II}, the most important form being methylmercury (MeHg, CH_3Hg^+ compounds). The conversion process, known as *methylation*, is typically favored in reducing environments (e.g., sediments) by anaerobic bacteria that carry a two-gene cluster, *hgcA* and *hgcB* [8]. Methylmercury *bioaccumulates* and *biomagnifies* (see Textbox 1) through aquatic food chains, resulting in elevated MeHg concentrations in organisms at higher trophic levels such as fish, marine mammals and humans, potentially causing neurological damage. As we will see throughout this chapter, MeHg is the primary culprit responsible for Hg's reputation as one of the most potent global pollutants. The natural organic carbon cycle plays a special role in the transportation of Hg and, especially, in the conversion of inorganic Hg^{II} to MeHg. It does this by providing the energy and setting (hypoxia) required by microbial Hg methylators. This intertwinement of Hg and organic carbon cycles adds complexity to determining the risk at any given location, but also an enhanced sensitivity to climate change, which affects both the Hg and organic carbon cycles.

11.3 Historical context

Now let us turn our clock back to the early days of human civilization. Mercury is among the first elements to be discovered and used by humans. This should not come as a surprise, as the bright red color of cinnabar (α-HgS, an inorganic Hg^{II} compound), the most common Hg ore, makes it an attractive pigment (known as vermillion) for use in paintings, ointments, make-up, ink, seals, and scripts. The Chinese are known to have used cinnabar ("*dan*" or "*dansha*") well before 2000 BC, and vermillion-based pigments can still be seen on structures and paintings in ancient ruins and palaces throughout China. Elemental Hg (metallic Hg, or liquid Hg^0) may also occur naturally on cinnabar. Tubes filled with liquid Hg were found in Egyptian tombs dated to 1500 B.C.

Extracting elemental Hg from cinnabar may be accomplished simply by roasting the crushed ore and collecting and cooling the vapor (gaseous Hg^0). This process must have been known to the Chinese well before 200 BC, as the Mausoleum of the

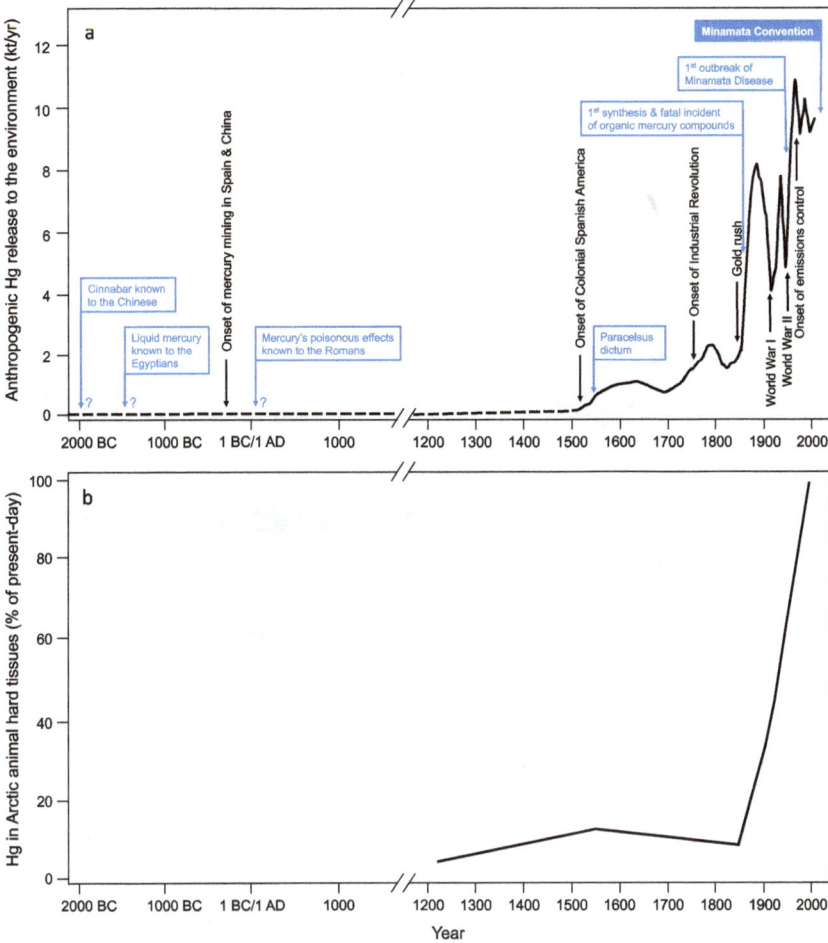

Figure 11.1: a) Global anthropogenic mercury (Hg) releases to the surface environment over time (1 kt = 1 kilotonne = 10^6 kg). The data for the period 1510–2010 are based on [3], and those prior to that (dashed line) are for illustration only. Also shown are major forces or events (black arrows) influencing the Hg releases, and some milestones (blue arrows) in the human relationship with Hg. b) Historical trends of Hg concentrations in hard tissues of Arctic animals, expressed as percentage of present-day maximum annual average concentrations. The trend line is based on data for seabird feather, ringed seal teeth, beluga teeth, polar bear hair, and human teeth [7].

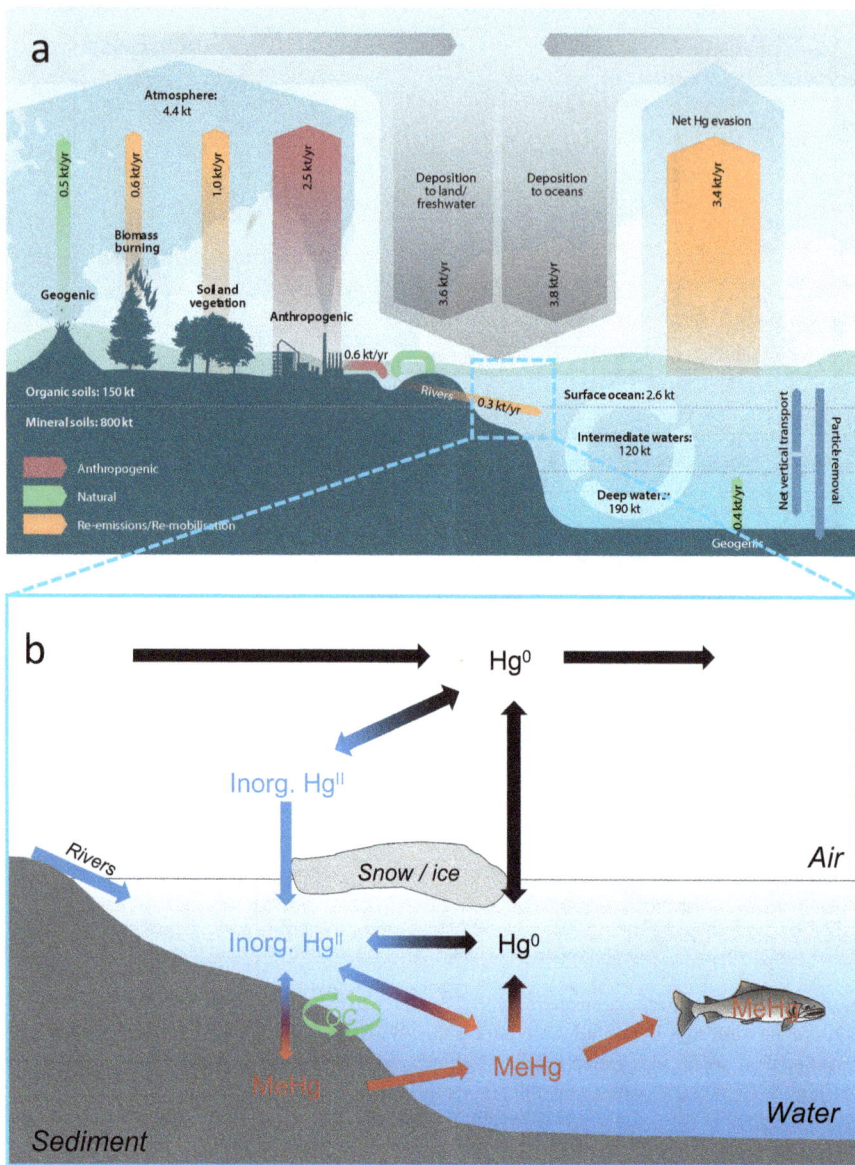

Figure 11.2: An illustration of global mercury (Hg) cycling, showing a) the best estimate values of the present-day masses of Hg (in kilotonnes (kt)) in the surface environment (the atmosphere, soils and ocean waters), and major fluxes among them (in kt/yr) (adapted with permission from [9]. Copyright 2018 American Chemical Society; also [10]), and b) major changes in chemical forms of Hg during its cycling between the atmosphere and ocean. Hg⁰: elemental Hg; inorg. Hg^{II}: inorganic divalent Hg; MeHg: methylmercury; OC: organic carbon.

First Qin Emperor was said to have used large quantities of liquid Hg to simulate flowing rivers and the sea. Similar traditions of using liquid Hg to recreate underworld rivers appear to have been prevalent in other cultures such as during the pre-Aztec era in Mexico. At the Almadén mine, Spain, the largest Hg mine in the world, mining of Hg, first for cinnabar and later for liquid Hg, dated back to at least 300 B.C. and continued until its closure in 2003 (Figure 11.1a). Southwest China had also seen extensive Hg mining for over two millennia until the early 2000s. Other large Hg mines include those in Idrija of Slovenia, Huancavelica of Peru, and San Jose of California, USA; all of them have been closed.

It is rather intriguing that Hg held not only a prominent role, but essentially the same role in the Arabic, Chinese and European philosophies of alchemy. Alchemists from these different regions and cultures somehow arrived at the same belief that Hg was the "philosopher's stone" that would transubstantiate base metals into precious ones, and the "elixir of life" that would make people immortal. Whether this belief was developed independently around the same time or was the result of cultural cross-pollination remains an interesting subject of debate.

No alchemist succeeded in using Hg to transubstantiate base metals into Au or Ag; however, Hg did transform dull, Au- and Ag-bearing ores to the shining precious metals. As the only metal that is liquid at room temperature, Hg is capable of dissolving (or more accurately, *amalgamating*) metals such as Au, Ag, tin (Sn), and lead (Pb). This fact was known to the Chinese some 2,000 years ago and was used to make Au powders and polish bronze mirrors. For thousands of years Hg amalgamation was the most common technique to extract Au and Ag from the Earth's crust; this practice peaked from the 1500s to 1900s during the Spanish colonial era in the Americas and the Gold Rush, making it one of the largest sources of anthropogenic Hg releases to the environment (Figure 11.1a). Mercury amalgam has also been commonly used as dental fillings and can still be found in some dental practices worldwide.

Mercury's other properties, such as high density, low viscosity, high rate of thermal expansion, and excellent electrical conductance, have been exploited for many other industrial applications, including in barometers and blood pressure monitors, thermometers, electrical switches, fluorescent lighting, and as cathodes in batteries and in the chlor-alkali process to manufacture chlorine and sodium hydroxide. Inorganic Hg compounds are also used as catalysts in organic synthesis in research laboratories and in manufacturing of industrial products such as acetaldehyde and polyvinyl chloride (PVC) plastic polymer. Organic Hg compounds, such as MeHg and ethylmercury (EtHg), were first synthesized in the 1860s. Their potent antifungal properties resulted in widespread global application in agriculture, especially to seed grains. The use of Hg in most of these applications has been phased out since the 1970s, although small-scale practices, such as the use of Hg in artisanal Au and Ag mining, remains in many places around the world. Very low levels of organic Hg (thimerosal) are also used in some critical products such as multidose vaccines and eyedrops due to their exceptional disinfecting property.

11.4 Early signs of trouble

As opposed to being the elixir of life, Hg and many of its compounds have turned out to be exterminators of life, especially effective against bacteria and fungi. This could explain Hg's early medicinal use to treat skin infections, which dates back to ancient Egypt. Mercury vapor, inorganic Hg compounds, and later synthetic organic Hg compounds remained a popular treatment for syphilis from the fifteenth to the mid-twentieth centuries. Calomel (Hg_2Cl_2, a Hg^I compound) was also used to treat malaria, yellow fever, and helminths ("worm candy").

Paracelsus (1492–1541) is widely recognized as the founder of modern toxicology, and his famous dictum "*the dose makes the poison*" is a basic principle of toxicology to this day. His original quote was in German, as he wrote in his *Die Dritte Defension* (Third Defence), which literally translates to "*solely the dose determines that a thing is not a poison*" and is probably more accurate as it implies a no-effect-level rather than an effect-level of a dose. Paracelsus arrived at this dictum largely from his extensive practice with Hg therapy. He strongly advocated the use of Hg to treat various diseases including syphilis, but also realized a little too much of it could harm the patient.

We now know that Hg is a non-essential element and is not required by any known essential biochemical functions of any life form. The medicinal benefits of Hg, if any, are due to its poisonous effect. At lower doses, below a threshold, it kills bacteria or fungi and hence benefits the patient, but when applied above that threshold, it would harm and even kill the patient. Paracelsus was among the first to realize that the threshold dosage for Hg was rather small. Despite the lack of statistics, it is almost certain that Hg harmed or killed more people than it ever cured throughout the human history.

Mercury's poisonous effects (*mercurialism*) were known long before Paracelsus' work. They are typically associated with tremor, abnormal mental excitement or irritation, unusual timidity, and personality change (*erethism*), and movement and coordination disorder (*ataxia*). Noting that many workers at the Almadén mine lost their reason and some their lives, Roman scholars attributed the poisoning to the inhalation of Hg vapor. The use of $Hg(NO_3)_2$ in the seventeenth century in carroting felt hats gave rise to the "mad hat disease", popularized by the Mad Hatter in *Alice's Adventures in Wonderland*. Mercurous (Hg^I) compounds such as calomel were thought to be less toxic and appeared to be the most common form of Hg used by Paracelsus. The widespread use of calomel in treating teething (teething powders) in the early twentieth century caused outbreaks of "pink disease" (*acrodynia*) in children.

Different from inorganic Hg, poisoning by organic Hg compounds selectively targets the central nervous system. Occupational poisoning by organomercuric compounds occurred literally at the same time they were first synthesized. Two of the three chemists involved in the synthesis of dimethylmercury (($CH_3)_2Hg$, DMHg) in a London laboratory in 1864 died a few weeks afterward [11]. History unfortunately repeated itself more than a century later, when another two scientists died from DMHg poisoning, one from chronic exposure during its synthesis, and the other from a single acute exposure while handling the compound [12]. Four cases of non-fatal poisoning by inhalation of MeHg were reported in 1940 in a London factory that manufactured MeHg as a fungicide that was used a seed dressing, resulting in severe ataxia, speech impediment (*dysarthria*), and constriction of the visual fields [13]. Despite the researchers' stern warning of risk associated with the manufacture and use of such seed dressings, serious incidents of mass poisoning occurred in the 1950s to early 1970s in several countries, where MeHg or EtHg-treated grains were accidentally consumed by people.

11.5 The turning point: Minamata disease

Although the poisonous effects of Hg and its compounds had been known for thousands of years in many cultures, it remained a "strange disease" when first reported in Minamata, Japan, in the 1950s. This single event placed Hg in the global spotlight. At Minamata, Hg-containing industrial waste was released directly to a semi-enclosed ocean bay out of which local residents harvested seafood. In hindsight it seems obvious that Hg poisoning was at play but, nevertheless, it took about a decade from the first incident to identify Hg, specifically MeHg, as the cause of the disease. This discovery marked the beginning of a new realization of the severity and complexity of environmental pollution at local and regional scales.

11.5.1 First outbreak of Minamata disease

Located on the south shore of the Yatsushiro Sea in southwestern Japan, Minamata is a small town that had a population of about 50,000 in the mid-twentieth century. A mysterious nervous illness first struck the community around 1953, with an outbreak occurring by 1956. The earliest victims were mostly young children. Referred to as Minamata disease, typical symptoms included ataxia, dysarthria, tremor, sensory disorders, and visual and hearing impairment, with damage to the central nervous system confirmed by pathological examinations [14]. In severe cases, victims lost consciousness, and some died. Congenital Minamata disease with cerebral palsy was also observed in infants born from mothers with no or mild symptoms. As of 2016, a total of 2,280 patients from Minamata and the nearby region were officially recognized to have suffered from Minamata disease [15], although the actual number of victims is believed to be much larger.

Minamata was the only industrialized city in the surrounding area, and Chisso Co. Ltd. operated a large factory in Minamata producing fertilizers, acetaldehyde, synthetic resins, among other chemicals, using Hg as a catalyst. It did not take long for scientists to suspect the illness was caused by consumption of fish and shellfish that were contaminated by the effluent from the factory released to Minamata Bay [16]. Even local cats, which ate the same fish, exhibited similar nervous disorders and many died. Several candidate contaminants were proposed [16], but the focus rapidly narrowed down to MeHg, as the clinical symptoms of Minamata disease were very similar to those of MeHg poisoning [13, 14, 17]. Soon after, MeHg was detected in shellfish from Minamata Bay [18], in the Chisso plant [19] and waste sludge [20], and in tissues of deceased patients and in umbilical cord samples from affected infants [14].

With mounting evidence, the Government of Japan in 1968 officially announced that Minamata disease was caused by consumption of fish and shellfish contaminated by MeHg due to Chisso's operation. Chisso stopped manufacturing acetaldehyde in May 1968. A series of pollution control and reclamation measures ensued for the next three decades, with nearly 0.1 km^2 of the sludge area being covered with soil and 1.5 km^2 of the water area being dredged to remove contaminated sediments [14]. By 1997, Minamata Bay was declared safe and reopened for fishing.

A key missing link in the early investigation was the origin of MeHg. Chisso's plant had used Hg as a catalyst for the synthesis of acetaldehyde since 1932, but in the form of inorganic $HgSO_4$. The synthesis reactions were known to involve organo-mercuric intermediates [21], which were thought to produce small amounts of MeHg as a by-product [15]. Chisso's increased production of acetaldehyde with a modified process in 1951 [22], notably the replacement of the co-catalyst from MnO_2 to $FeSO_4$, was thought to have enhanced the production of MeHg [15], which could have explained the timing of the outbreak. The pathway of MeHg formation from the acetaldehyde synthesis, however, remains a subject of debate to the present day [23]. Given what we now know about MeHg (see Section 11.2), at least some of the MeHg in Minamata must have been formed from the large amount of inorganic Hg contained in the sludge and the Bay's sediments. This MeHg was then bioaccumulated and biomagnified in fish and shellfish.

11.5.2 A new type of disease

Minamata disease is a nervous disorder in humans caused by indirect poisoning by MeHg through the food chain as a result of environmental pollution [24]. It was a new type of disease that had been rarely seen up until the 1950s, and the determination of its cause has had far reaching implications.

Different from most poisoning incidents, which occur via **direct** occupational exposure, medical use, or inadvertent consumption, Minamata Disease is caused by **indirect** poisoning through ecosystem exposure as a result of environmental pollution [24]. The conversion of inorganic Hg to MeHg in the receiving aquatic environment together with the ability of MeHg to biomagnify through the foodwebs make it especially challenging for pollution control and ecosystem recovery.

Moreover, unlike most other diseases, Minamata disease may be passed from mothers who exhibit no or mild symptoms to infants and children through the placenta [24]. The ability of MeHg to cross the blood–brain barrier and placental barriers makes it one of only a handful of known developmental neurotoxins to humans [25].

11.5.3 Beyond Minamata

In addition to Minamata and the surrounding region, Japan had a second Minamata disease outbreak in the mid-1960s along the lower reaches of the Agano River and its estuary at the Sea of Japan in Niigata of Honshu. Once again, mercuric catalyst used for acetaldehyde synthesis was identified as the source of pollution. As of 2016, 704 patients in this region were officially recognized to have suffered from Minamata disease, bringing the total official number in Japan close to 3,000 [15].

While so far only the Minamata and Niigata cases in Japan are officially recognized as Minamata disease, outbreaks may have occurred elsewhere, as a definitive diagnosis can be challenging especially when the symptoms are relatively mild. One of the best-known examples is in the Asubpeeschoseewagong (also known as Grassy Narrows) First Nation in northwest Ontario, Canada. Mercury poisoning symptoms were observed in local residents in the early 1970s. The source of pollution was a chlor-alkali plant in the town of Dryden, which used metallic Hg in the process to produce chlorine and sodium hydroxide. Large quantities of Hg were dumped into the Wabigoon-English River system, resulting in highly elevated MeHg concentrations in fish that the community consumed. There is no evidence that MeHg was produced in the chlor-alkali process; therefore, the MeHg in the fish must have been accumulated from the receiving aquatic environment where the dumped inorganic Hg was converted to MeHg through microbial methylation.

Several Japanese scientists who extensively studied Minamata disease visited the Canadian community first in 1975 with follow-up studies in 2002, 2004, and 2010. They concluded that the community had experienced an outbreak of Minamata disease based on the Japanese criteria [26–28]. However, North American researchers came up with diverging opinions without reaching a definitive diagnosis [29], adding an "international twist" to the complex situation due to "differences in interpretation of clinical findings" [30]. The Canadian government has not recognized the Asubpeeschoseewagong case as Minamata disease, although negotiations between the community, the company, and the provincial and federal governments on compensation, patient care, and cleanup continue to this day.

11.6 The Arctic connection

11.6.1 A pivotal discovery from the least likely place

So far, the confirmed and suspected incidents of Minamata disease have been regional with a clear connectivity to a local, "point" source of industrial pollution. Once the immediate pollution source was identified and placed under control, the situation improved. Following that logic, the Arctic would be among the last places on Earth to expect any serious Hg contamination, as it is located far away from any major industrial activities and features no known use of Hg by the sparse Indigenous populations that live there. Yet what was first observed in the 1970s in the Canadian Arctic has played a pivotal role in the call for global action on Hg pollution control.

The first hint of a Hg problem in the Arctic appeared in a 1972 study by a consulting company [31]. The study reported that, among people from western and northwestern Canada, Inuit from coastal communities such as Tuktoyaktuk and Ulukhaktok (known also as Holman) in the Canadian Arctic had the highest blood Hg concentrations (> 30 µg/L; for reference, Health Canada's guideline for blood Hg concentration is 20 µg/L for the general adult population and 8 µg/L for children, pregnant women and women of childbearing age). In that same year, high Hg concentrations were also detected in the livers of ringed seals collected from the same region [32]. This prompted a study in 1973 to include more marine and terrestrial animals [32]. While very low in fish (Arctic char) and terrestrial mammals (caribou, Arctic foxes, and wolves), Hg concentrations were found to be exceptionally high in seals: averaging 27.5 µg/g in the liver (where the majority is inorganic Hg) and 0.72 µg/g in the muscle (where almost all Hg is MeHg) of ringed seals, and 143 µg/g in the liver and 0.53 µg/g in the muscle of bearded seals. One individual bearded seal had an astonishingly high Hg concentration of 420 µg/g in the liver and 88.7 µg/g in the muscle. These numbers were and remain today among the highest ever reported for pinnipeds and cetaceans anywhere in the world, exceeded only by those from areas with known industrial sources of Hg [32]. A follow-up study in 1978 confirmed that such elevated Hg concentrations in ringed seals and bearded seals constituted a widespread phenomenon for Arctic marine mammals across the Canadian Arctic, from the Beaufort Sea in the west to the Baffin Bay in the east, and to Hudson Bay in the south [33].

Since then, Hg concentrations in seals, whales and polar bears have been continuously or intermittently monitored in the Canada Arctic, as well as in many other parts of the Arctic, making it one of the longest time-series of biological Hg records in the world [10, 34, 35]. Mercury concentrations have also been measured in Inuit and northern peoples in many parts of the Arctic (Figure 11.3). We now know that Hg concentrations in most of the Arctic marine mammals and humans who rely on these animals as traditional food have remained stubbornly high, and that regional and temporal variations are confusingly complex. A recent review of Hg concentrations in hard tissues of high trophic animals in the Arctic (e.g., seabird feather, ringed seal teeth, beluga teeth,

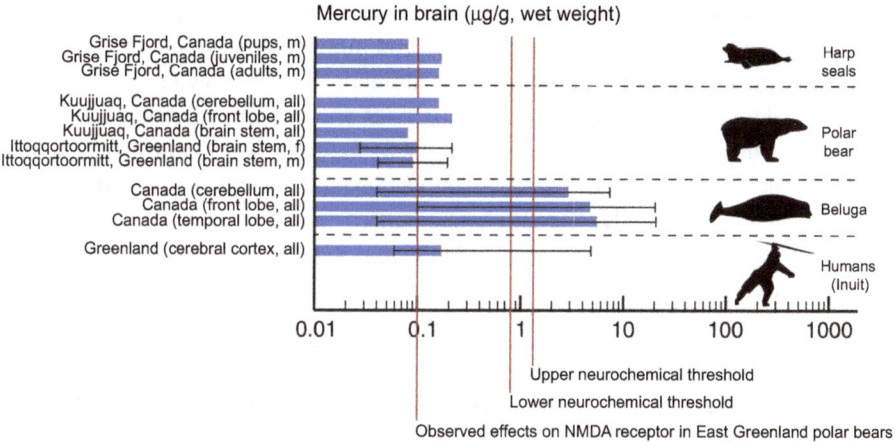

Figure 11.3: Mercury concentrations in brain of marine mammals and humans in the Arctic. Red lines indicate the mean mercury concentrations in East Greenland polar bear brain stem that were associated with lower N-methyl-D-aspartate (NMDA) receptor levels and the mercury-associated neurochemical effect thresholds based on studies on fish-eating mammals. Adapted from [38], Copyright 2013, with permission from Elsevier.

polar bear hair and human teeth) showed that on average Hg concentrations increased sharply, by about 10-fold, from the mid- to late-nineteenth century to the present-day (Figure 11.1b), and that ~90% of the present-day Hg concentrations in these tissues can be attributed to anthropogenic processes [7]. Strong correlations often seen between Hg and selenium (Se) in the liver of Arctic marine mammals suggest that high levels of Se may have offered a protective mechanism against the Hg toxicity (e.g., via the formation of HgSe(s)) [36, 37]. However, epidemiological studies are too scarce to assess how elevated levels of Hg may have affected the health of marine mammals and humans in the Arctic.

11.6.2 A major dilemma

Nearly four million people live in the Arctic, about 10% of whom are Indigenous peoples [39]. The total population in the Canadian Arctic is about 120,000, of which nearly 50% are Indigenous peoples. Inuit and many other Arctic Indigenous peoples have always relied on *country food* (i.e., foods that originate from the population's local habitat, such as plants, berries, fish, waterfowl and seabirds, and land and marine mammals [39]) not just for their energy and nutritional needs, but also as an integral part of their social, cultural, spiritual, and economic well-being. Marine mammals (e.g., seals, beluga, narwhal, walrus, and polar bears) are important sources of lipids (including omega fatty acids), protein, vitamins, and essential elements for Inuit. The observation that traditional food items such as marine mammals contain elevated concentrations of Hg (and other contaminants) thus presents a major dilemma to Inuit and other Arctic Indigenous peoples who now require a basis to make informed choices considering risks and benefits of country foods.

There is no human consumption guideline for Hg in marine mammals. However, a simple calculation, using fish as an example (Textbox 2), suggests the safe limit of MeHg intake may be readily exceeded by many Arctic Indigenous peoples. This could explain elevated Hg levels often seen in blood and brain samples of Inuit across the Arctic (Figure 11.3).

Textbox 2: How much fish should you eat, from a mercury perspective?
Health Canada has established a maximum level of Hg of 0.5 µg/g (wet weight) in the edible portion of commercially sold fish (except for shark, swordfish, and tuna) for human consumption. It has also established a provisional tolerable daily intake (pTDI) of MeHg of 0.47 µg/kg body weight (bw)/day for adults in general, and 0.2 µg/kg bw/day for pregnant women, women of child-bearing age and young children [40].

If a fish muscle contains 0.2 µg/g (wet weight) of Hg and 95% of the Hg is present as MeHg, how much of that fish muscle can one consume daily?

Assuming a bw of 60 kg, the pTDI of MeHg would be:

$$0.47\,µg/kg\,bw/day \times 60\,kg\,bw = 28.2\,µg/day$$

This would be equivalent to

$$28.2\,µg/day / (0.2\,µg/g \times 95\%) = 148\,g\text{ of fish muscle/day}$$

which is about one serving of the fish per day.

For a pregnant woman or a woman of child-bearing age, the amount would be reduced to 63 g of the fish per day, or about one serving every two days.

To reduce the exposure to Hg, an obvious option seems to be for Inuit and other Arctic Indigenous People to reduce their dietary dependence on marine mammals. This is, however, neither possible nor advisable. Although there is other country food sources (e.g., terrestrial animals), they may not be available to some communities and only available at certain times of a year for others. For instance, marine mammals may be the only foods available for coastal communities during late spring months. Replacing country food with imported food is even more problematic economically, nutritionally and socioculturally. Not only are the costs of nutritious imported food prohibitively high in many northern communities, a shift away from country food diets to imported foods has also been shown to increase the risk of cancers, cardiovascular diseases, diabetes, excess weight and obesity [41]. Country foods and related activities such as hunting, fishing, collecting, distributing, sharing, preparing, and consuming food are also central to the way of life for Inuit and many Arctic Indigenous peoples. Therefore, the contamination of Arctic marine country food by Hg raises problems that cannot be resolved simply by risk-based health advisories or food substitutions alone [41].

11.7 What makes Hg a global contaminant?

The basic understanding of the role of MeHg in producing Minamata disease took decades to develop. The mechanisms by which MeHg may be produced in quantities that present risks to residents of remote regions like the Arctic is an even harder scientific problem to solve.

11.7.1 The organochlorine background story: a brief diversion

We now turn briefly to semi-volatile organochlorines (SVOCs, such as many pesticides and PCBs), because research on these compounds provided the roadmap to understand Hg as a global pollutant.

Unlike Hg, the evidence for SVOCs is not obfuscated by an underlying natural cycle. When SVOCs are detected, we know they have escaped from human activities. More than five decades ago, Holden and Marsden [42] inadvertently discovered high SVOCs of industrial origin in the fatty tissue of seals collected near the Arctic. These authors, who had collected remote samples to determine detection limits and test blanks, immediately drew the conclusion that distance was not a protection from semi-volatile toxic chemicals that were being applied widely in temperate regions (considering this finding, we might well have had concerns about Hg at the time; more on this later).

The scientific evidence from the Arctic clearly showed that contaminants released in large amounts (~Mt quantities) by industry or agriculture could be transported to anywhere on the globe. Further research, much of it conducted under the auspice of the Arctic Monitoring and Assessment Programme (AMAP) [35], refined the understanding of SVOC pathways and also determined the importance of concentrating processes within the environment. A widely held view in the 1960s was that *the solution to pollution was dilution*. Science conducted by the AMAP international community produced clear evidence that not only could SVOCs arrive in the Arctic, but that the cryosphere in particular was a favored global sink due to thermodynamic forcing of these chemicals from warm to cold regions, known as *global distillation* or *the grasshopper effect* [43]. Furthermore, a number of concentrating processes occur within the Arctic, an important one comprising aquatic foodwebs [44]. Thus, these studies showed unequivocally that the Arctic and its foodwebs were contaminated by SVOC, but to make the case for pollution is more difficult and requires toxicological work (see Textbox 1). The AMAP program continued to produce extensive science to demonstrate that SVOCs were producing harm through biomagnifying processes, making a solid case for pollution of the Arctic. This AMAP work was foundational and led to the inclusion of more SVOCs under the Stockholm Convention which entered into force in 2004. Subsequent time series monitoring has provided further evidence of the response of Arctic ecosystems to the curtailing of the release of SVOCs.

Out of the SVOC research we can state with high confidence that an anthropogenic chemical likely to lead to harm (i.e., pollution) in remote regions has these defining characteristics: (1) it is released in large quantities (e.g., in Mt) over long periods of time; (2) it is persistent in the environment (i.e., high resistance to degradation; this characteristic is exceptionally important); (3) it is bioaccumulative and biomagnifying, and prone to other magnifying processes; and (4) it is toxic in some way to biota when it exceeds certain thresholds.

11.7.2 Back to Hg

Mercury is not an SVOC, yet it has all four of the "bellwether" properties listed above. It is, therefore, a strong candidate for producing harm in remote regions and especially the Arctic [4].

Let us first highlight some fundamental differences between Hg and SVOCs. First and foremost, Hg has a natural background cycle, which makes it harder to establish an accurate anthropogenic loading into that cycle (Figure 11.2a). This is further complicated by the fact that Hg has been released by human activities for over two thousand years compared to the ~ seven decades for SVOCs. The locations of Hg release also differ from the SVOCs and have migrated and changed in importance with time (e.g., from cinnabar/elemental Hg mining in Roman times, to Au and Ag extraction in the 1600s, to emissions from coal burning over the last century; Figure 11.1a). Second, while Hg may be considered "permanently" persistent as an element, it exhibits several chemical forms (e.g., Hg^0, inorganic Hg^{II}, organic Hg^{II} including MeHg), which affect its volatility, transport pathways and toxicity (see Section 11.2). Furthermore, as mentioned earlier, whereas the cycles of Hg and SVOCs in aquatic systems are both closely intertwined with that of organic carbon, the conversion of inorganic Hg^{II} to MeHg is fueled by the organic carbon cycle (Figure 11.2). This process is crucial to understanding the toxic risks of Hg and for providing an accurate assessment of harm to ecosystems and humans who depend on these for food. The important point here is that unlike the case for SVOCs, the toxicity of Hg is to a large extent determined by the organic carbon cycle of the receiving environment. This by itself has presented the greatest challenge to understanding the complex patterns of risk from Hg toxicity observed in the Arctic.

The effort to understand Hg cycles in the Arctic, especially in seawater, is exceptionally demanding because the minute concentrations observed (typically < 1 ng/L) are easily contaminated during sampling and laboratory processing. There have also been diversions along the road. The discovery of Arctic Mercury Depletion Events

(MDEs) in the 1990s [45] led to a vigorous research activity to understand why atmospheric Hg was removed to the ocean surface during polar sunrise in the Arctic. This phenomenon was well documented due to the excellent time series available from modern sampling instrumentation, and the mechanism has been extensively studied. But at the end of the day these so-called Hg showers were followed by a large proportion of the Hg being rapidly re-emitted to the atmosphere due to photochemical reduction in snow and sea ice. On the other hand, the conversion of inorganic Hg to MeHg in the ocean and other aquatic environments was less emphasized in studies and far more crucial to the risks Hg presents to foodwebs.

A crucial advance in the science of global Hg cycling during the past decade has, accordingly, been the establishment of modern analytical standards for trace element research, especially clean sampling techniques and sophisticated instrumentation that now permit ship-based high-resolution measurements of water-column profiles of Hg and its various chemical species including MeHg. This game changer for ocean work has produced the first reliable profiles for Hg in world oceans (e.g., [46, 47]), and therefore the first accurate understanding of the ocean's Hg cycle.

Figure 11.4 shows one such recent development in the Arctic that solved a long-time mystery as to why marine animals in the western Canadian Arctic are more contaminated by Hg than their counterparts in the east [48]. By measuring the vertical distributions of Hg in seawater along a 5,200-km transect in the Canadian Arctic, we discovered the presence of a subsurface MeHg enrichment layer: MeHg is lowest at the sea surface, increases to a maximum at depths between 100 and 300 m, and then decreases toward the bottom of the ocean. The peak MeHg concentration in this enrichment layer is highest in the western Canadian Arctic and lowest in the east, mirroring the Hg trends observed in marine animals. This makes sense as the shallow MeHg-enrichment layer lies within the habitat of zooplankton and other organisms near the bottom of the foodweb, which allows MeHg to be readily taken up by these animals, and subsequently biomagnified in marine mammals. This discovery would not have been possible if we were only focusing on Hg loadings – comprised primarily of inorganic Hg – from the atmosphere or rivers. As seen in Figure 11.4, Hg trends in marine animals bear no obvious relationship with the seawater distribution of total Hg (the sum of inorganic Hg and MeHg).

With these recent advancements in Hg analysis, global and regional budgets have started to emerge and continue to undergo refinement. These, together with research on Hg pathways in the ocean, especially the production and dynamics of MeHg, will provide a much better basis to understand sources, sinks and fluxes in the Arctic and beyond.

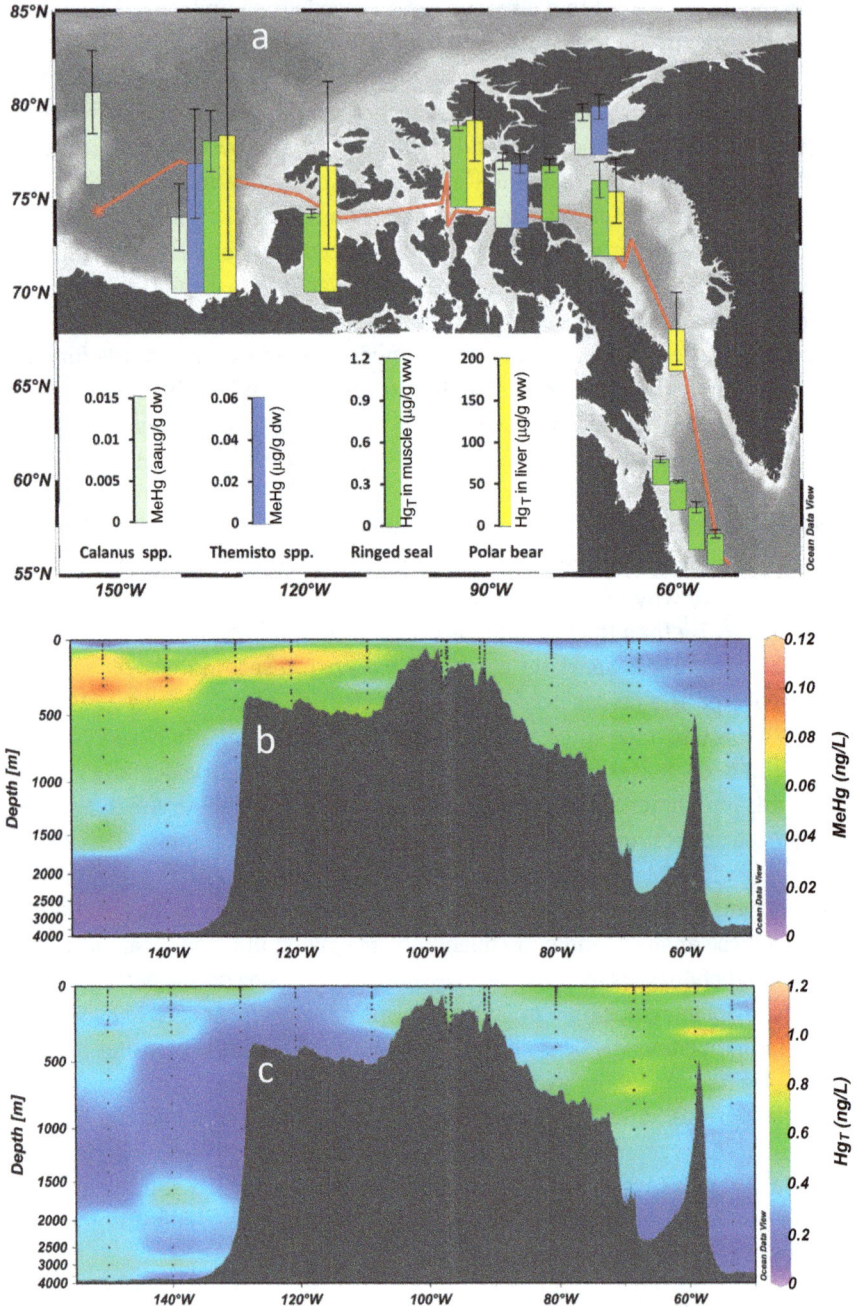

Figure 11.4: Mercury (Hg) concentrations in marine animals (a) and seawater (b and c) from the Canadian Arctic and Labrador Sea. The red line in (a) indicates the transect along which the seawater Hg concentrations (b and c) are measured. Hg_T: total Hg; MeHg: methylmercury. Adapted from [48].

11.8 The future: Minamata Convention and climate change

The severity of local to regional Hg pollution, as exemplified by Minamata disease, and widespread global pollution, as exemplified in the Arctic, dictate that global action is needed to protect human health and the environment from anthropogenic emissions and releases of Hg and Hg-containing compounds. The overwhelming scientific evidence presented in the 2003 Global Mercury Assessment by the United Nations Environment Programme (UNEP) [49] prompted initial discussions on a comprehensive legally binding treaty on Hg. Formal negotiations started in 2009, and the resulting Minamata Convention on Mercury was adopted in 2013 and entered into force in 2017. Canada played an active role in the negotiations of the treaty and was among the first nations to ratify it. The Convention noted specifically the vulnerability of Arctic ecosystems and Indigenous communities "because of the biomagnification of mercury and contamination of traditional foods".

The Minamata Convention sets out a range of measures to control the supply and trade of Hg including Hg mining and to control the use of Hg in products and processes including ASGM. As anthropogenic emissions of Hg are being placed under control globally, the Hg contaminated ecosystems including the Arctic marine ecosystems are expected to move toward recovery. But this recovery process will likely take a long time, on the order of decades or longer, before meaningful reduction is seen in food-web Hg levels and in human exposure. The timeline and extent of the recovery will depend on the nature of the system that determines the residence time of Hg (i.e., the inventory and in- and out-fluxes) and processes and pathways that control the Hg cycle, especially the production of MeHg and its subsequent uptake in the foodwebs [50]. As most of these processes are sensitive to climate change, the effectiveness of the Minamata Convention is going to be affected by a changing climate .

The impact of climate change on ecosystem recovery from Hg contamination is perhaps most profoundly felt in the Arctic, where rapid climate warming has resulted in dramatic changes in many biogeochemical and ecological processes that drive the Hg cycle [51, 52]. For example, the rapid decline in aerial coverage and thickness of Arctic sea ice and the replacement of multiyear sea ice by first-year sea ice have been shown to influence Hg distribution and transport across the ocean–sea ice–atmosphere interface, as ice prevents Hg^0 evasion and leads to its accumulation under the ice. This in turn can alter Hg methylation and demethylation rates, promote changes in primary productivity, and shift foodweb structures ("bottom-up processes"). Changes in animal social behavior associated with changing sea-ice regimes can also affect dietary exposure to Hg ("top-down processes") [51, 52].

A separate feature of climate change in the Arctic is the effect of warming on the permafrost, which covers much of the Arctic's drainage basins. During the past, extending back even to Roman times, inorganic Hg deposition has accumulated in surface soils and permanent snow and ice in the cryosphere, forming a large reservoir upstream from the Arctic Ocean. With warming, coasts have begun to erode, the permafrost is thawing and some, if not all, of the "permanent" snow and ice will melt: these processes will release some portion of the stored legacy inorganic Hg, allowing it to re-enter active cycling [53]. The release of legacy Hg will be accompanied by the release of dissolved and particulate organic carbon, which will mediate the transport of Hg and contribute labile carbon to support methylation. Given that the Arctic's drainage basins are large (twice the area of the Arctic Ocean) and span a wide range of climate conditions, we would expect a protracted release of soil Hg, as yet unquantified, to be released. This process will likely oppose the reduction in direct deposition as a consequence of curtailing modern sources of Hg to the atmosphere as prescribed in the Minamata Convention.

Presently, we lack the capability of projecting more quantitatively the effects of reducing Hg emissions in a changing climate, due to a lack of reliable models that capture the kinetics of the Hg cycle, the emerging effects of climate change on the carbon cycle, and the connectivity between the mercury and organic carbon cycles. Using a hypothetic aquatic ecosystem as an example, one way to qualitatively envision how climate change will affect ecosystem recovery from Hg contamination under the Minamata Convention is via a multi-phased diagram shown in Figure 11.5.

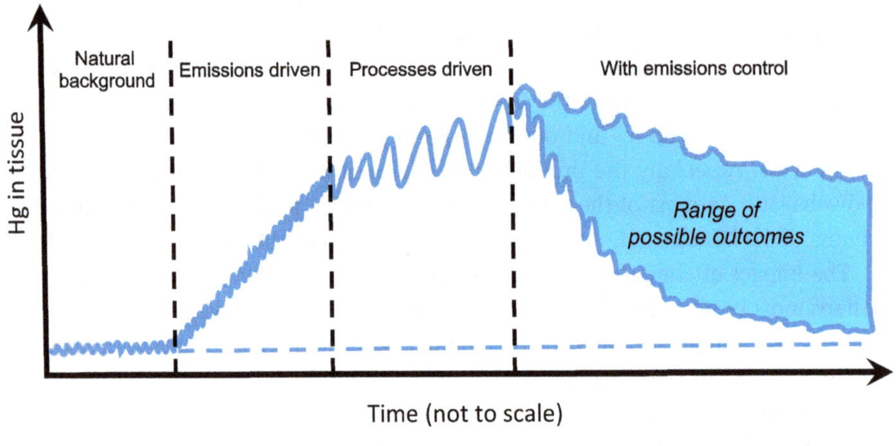

Figure 11.5: A schematic representation of evolution in mercury (Hg) concentrations in aquatic biota showing changes over time in the principal drivers of Hg bioaccumulation. Adapted from [51], Copyright 2019, with permission from Elsevier.

Prior to major human perturbations, Hg loadings to the ecosystem were very low, as were the concentration in the foodwebs (Phase I – "Natural background") (Figure 11.1a,b). As anthropogenic Hg releases to the environment increased rapidly around 1500s–1800s (see Figure 11.1a), foodweb Hg concentrations would have responded rapidly due to increasing Hg deposition, exposure and uptake from a small but growing Hg inventory (Phase II – "Emissions-driven"). This phase is clearly seen in long-term retrospective studies of Hg concentrations in Arctic biota (Figure 11.1b). Once the ecosystem has accumulated sufficient Hg, additional increases in Hg influx become secondary to the amount that has been accumulated in the system after decades to centuries of loading. Bioaccumulation then draws on both this legacy and newly released Hg, which is operated on by internal biogeochemical processes (Phase III – "Processes-driven"). Throughout all these three phases, biogeochemical and ecological processes (shown as sine-wave "noise" in Figure 11.5) determine the transport and transformation of Hg from the abiotic part of the ecosystem to biota. But it is in Phase III that these processes emerge to create a variability that is large enough to obscure the effect of the concurrent external Hg inputs, and hence produce the divergence between biotic and atmospheric Hg trends, which has been observed in many places around the world over the past decades including the Arctic [51].

Now that the Minamata Convention is being enforced, we can envision a new phase, Phase IV ("With emissions control") that describes the ecosystem recovery process. As anthropogenic Hg releases decrease, the preceding "processes-driven" bioaccumulation (Phase III) dictates that it will take much longer to establish a new steady-state in foodweb Hg concentrations because of the continued presence and re-release of legacy Hg stored in large environmental reservoirs (e.g., soils and oceans). In the shorter term, foodweb Hg concentrations in many systems, especially in some of the larger, deeper marine ecosystems, are likely to continue to increase despite recent emission controls. Biota in smaller waterbodies, such as lakes and coastal marine systems with restricted water mass turnover, are more likely to respond relatively rapidly to emissions controls because of the smaller mass of legacy Hg contained there relative to the decreased anthropogenic Hg loadings.

The fact that effective Hg emission controls are expected to be followed by long delays, in some cases, before an ensuing reduction is seen in food web Hg concentrations makes it all the more pressing to control and reduce Hg emissions as early as possible. Nonetheless, we should be prepared for the fact that the reductions in foodweb Hg concentrations and human Hg exposure are likely to be uneven and predictable only in the most general way, both in terms of their timing and in the degree of reduction.

11.9 Connecting science to policy

The step of developing legal and institutional frameworks to mitigate or curtail activities that can broadcast harm widely through air and water requires a sustained, globally organized approach. Examples include the Montreal Protocol (on ozone-depleting chemicals such as chlorofluorocarbons (CFCs), entered into force in 1989), and the Basel (hazardous waste, 1992), Rotterdam (hazardous chemicals, 2004), Stockholm (persistent organic pollutants, 2004) and most recently Minamata Conventions. Success with these international agreements has shown that for clearly defined, single issue threats (e.g., CFCs, organochlorines, Hg) the international community can together arrive at practical solutions to mitigate harm. But this process is slow and the scientific basis for action usually leads the international response by decades.

In science, laboratory methodologies have long been used to determine toxicity of individual compounds or their mixtures. The problem environmental scientists face is that laboratory toxicity is not the same as environmental toxicity: the latter is prone to surprises. For example, we discovered the hard way that CFCs, which are inert compounds with practically no toxic effect in the laboratory studies, present a serious "toxicity" to stratospheric ozone and a non-trivial contribution to global warming. We were fortunate to discover this fairly early in the 1970s because there was no organized effort to monitor this substance. Out of the four "bellwether" properties listed above (see Section 11.7.1), CFCs possess two – released in large quantities and persistent. A lesson here is that having just those two properties should put the use of most industrial products under intense scrutiny. Plastics come to mind as does e-waste.

Another problem we face in communicating science is the general confusion in the media and public about the nature of risk. Risk is a probability term consisting of two parts: the inherent hazard (toxicity) of a substance, and the extent to which the substance is exposed to a receptor (Textbox 1) (see also [54]). When dealing with our complex and dynamic environment, the probability component is the most difficult to assess, and it is especially difficult for policymakers and elected officials to devise preventative steps for risks that have a high hazard accompanied by a low probability of occurrence. In the case of climate science, getting to the point of establishing a high probability of occurrence is too late to mitigate catastrophic consequences.

In the political and social arenas, the challenges are perhaps harder to surmount than they are in science. In particular, the media plays a crucial role in communicating science to the public, and the modern electronic world has given large impetus to misinformation and disinformation, the rampant spread of which is protected by freedom of speech laws. This is not likely to change. This makes it all the more important that early school education must include development of critical thinking in young people. Equally important is that international organizations need to be formed and, among other tasks, they should become a prime source of accurate, transparent, and up-to-date scientific information. In terms of environmental pollution, one solution might be to form a sister organization to the Intergovernmental Panel on Climate Change (IPCC) that would provide a science-policy platform for the protection of biological diversity, ecosystem services and human health [55]. Such a platform would facilitate peer-reviewed science to be more rapidly engaged in policy development.

11.10 Summary

In this chapter we have briefly followed the evolution of the human relationship with Hg over the past 4000 years. We showed how a naturally occurring fugitive element like Hg that once fascinated humanity can lead to a global problem threatening food safety and human health in even the most remote regions of our planet. We further showed that the science to develop the modern understanding of the Hg cycle has been a long journey undertaken by a dedicated international community who have been at times stymied by complexity and inadequate trace-level measurements, but how peer-reviewed scientific evidence ultimately has triumphed. The relentless research on environmental Hg has, ultimately, led to an international agreement to reduce global Hg emissions under the Minamata Convention.

Together with previous environmental emergencies involving stratospheric ozone-depleting chemicals, POPs, lake eutrophication and acidification, the Hg story serves as one of many examples of how science may influence and inform policy; it does take time and there are surprises along the way. In the case of Hg, the time from inception to action took over six decades.

Perhaps there is a lesson in this. The process of science can be slow due to the complexity of the environment both in its processes and in the media that it requires us to analyze: soil, air, water, and biological tissue. But, big or small, many of our cases are similar to *crime scene investigations*. We identify a "crime", for example a "hole" in stratospheric ozone, high organochlorine and Hg concentrations in the Arctic's country food, shield lakes of North America that are acidifying. We then collect evidence and interpret that evidence to a high enough standard to identify appropriate action. The science community is tasked with finding out the who, what, where, when, why of the problem and, in the above examples, an international community then develops an understanding out of which solutions emerge. But this is of course reactive science, and as with crime scene work, we would feel far better if we could somehow do the science required to prevent the crime. In the cases mentioned above, we have not gone past a tipping point – that is, we should be able with appropriate action to back out of the problem by curtailing the cause. So, for the greatest existential threat facing us – global climate change – can we afford to wait for the crime to be completed? When is the evidence strong enough to provoke appropriate action?

References

[1] Oreskes N, Conway EM. Merchants of Doubt: How a Handful of Scientists Obscured the Truth on Issues from Tobacco Smoke to Global Warming. New York, Bloomsbury Press, 2010.

[2] Cooke CA, Hintelmann H, Ague JJ. et al., Use and legacy of mercury in the Andes. Environ Sci Technol. 2013, 47, 4181–4188.

[3] Streets DG, Horowitz HM, Lu Z, Levin L, Thackray CP, Sunderland EM. Five hundred years of anthropogenic mercury: Spatial and temporal release profiles. Environ Res Lett. 2019. 14, 084004, https://doi.org/10.1088/1748-9326/ab281f.

[4] Macdonald R, Bewers JM. Contaminants in the Arctic marine environment: Priorities for protection. ICES J Mar Sci. 1996, 53, 537–563.

[5] Chapman PM, Wang F. Issues in ecological risk assessment of inorganic metals and metalloids. Human Ecol Risk Assess. 2000, 6, 965–988.

[6] Wedepohl KH. The composition of the continental crust. Geochim Cosmochim Acta. 1995, 59, 1217–1232.

[7] Dietz R, Outridge PM, Hobson KA. Anthropogenic contributions to mercury levels in present-day Arctic animals – A review. Sci Total Environ. 2009, 407, 6120–6131.

[8] Parks JM, Johs A, Podar M. et al., The genetic basis for bacterial mercury methylation. Science. 2013, 339, 1332–1335.

[9] Outridge P, Mason R, Wang F, Guerrero S, Heimburger-Boavida L-E. Updated global and oceanic mercury budgets for the United Nations Global Mercury Assessment 2018. Environ Sci Technol. 2018, 52, 11466–11477.

[10] AMAP/UNEP. Technical Background Report for the Global Mercury Assessment 2018. Arctic Monitoring and Assessment Programme, Oslo, Norway / UN Environment Programme, Chemicals and Health Branch, Geneva, Switzerland: 2019.

[11] Edwards GN. Two cases of poisoning by mercuric methide. Saint Bartholomew's Hosp Rep. 1865, 1, 141–150.

[12] Clarkson TW, Magos L. The toxicology of mercury and its chemical compounds. Crit Rev Toxicol. 2006, 36, 609–662.

[13] Hunter D, Bomford RR, Russell DS. Poisoning by methyl mercry compounds. Quart J Med. 1940, 33, 193–213.

[14] Harada M. Minamata disease: Methylmercury poisoning in Japan caused by environmental pollution. Crit Rev Toxicol. 1995, 25, 1–24.

[15] Yokoyama H. Mercury Pollution in Minamata. Singapore, Springer, 2018.

[16] McAlpine D, Araki S. Minamata disease. An unusual neurological disorder caused by contaminated fish. Lancet. 1958, 2, 629–631.

[17] Takeuchi T, Morikawa N, Matsumoto H, Shiraishi Y. A pathological study of Minamata Disease in Japan. Acta Neuropathol. 1962, 2, 40–57.

[18] Uchida M, Hirakawa K, Inoue T. Biochemical studies on Minamata disease. IV. Isolation and chemical identification of the mercury compound in the toxic shellfish with special reference to the causal agent of the disease. Kumamoto Med J. 1961, 14, 181–187.

[19] Irukayama K, Kondo T, Kai F, Fujiki M. An organomercury compound extracted from the sludge in the acetaldehyde plant of Minamata-factory. Japan J Med Prog. 1962, 49, 536–541.

[20] Irukayama K, Kai F, Fujiki M, Kondo T. Studies on the origin of the causative agent of Minamata disease. III. Industrial wastes containing mercury compounds from Minamata Factory. Kumamoto Med J. 1962, 15, 57–68.

[21] Vogt RR, Nieuwland JA. The role of mercury salts in the catalytic transformation of acetylene into acetaldehyde, and a new commercial process for the manufacture of paraldehyde. J Am Chem Soc. 1921, 43, 2071–2081.

[22] Othmer DF, Kon K, Igarashi T. Acetaldehyde by the Chisso Process. Ind Eng Chem. 1956, 48, 1258–1262.
[23] James AK, Nehzati S, Dolgova NV. et al., Rethinking the Minamata tragedy: What mercury species was really responsible? Environ Sci Technol. 2020, 54, 2726–2733.
[24] Harada M. The Global Lessons of Minamata Disease: An Introduction to Minamata Studies. Takahashi T. ed. Taking Life and Death Seriously – Bioethics from Japan (Advances in Bioethics, Vol. 8). Bingley, UK, Emerald Group Publishing Limited, 2005, 299–335.
[25] Grandjean P, Landrigan PJ. Developmental neurotoxicity of industrial chemicals. Lancet. 2006, 368, 2167–2178.
[26] Harada M, Fujino T, Akagi T, Nishigaki S. Epidemiological and clinical study and historical background of mercury pollution on Indian Reservations in Northwestern Ontario, Canada. Bull Inst Const Med. 1976, 26, 169–184.
[27] Takaoka S, Fujino T, Hotta N. et al., Signs and symptoms of methylmercury contamination in a First Nations community in Northwestern Ontario, Canada. Sci Total Environ. 2014, 468–469, 950–957.
[28] Harada M, Fujino T, Oorui T. et al., Followup study of mercury pollution in Indigenous Tribe Reservations in the Province of Ontario, Canada, 1975–2002. Bull Environ Contam Toxicol. 2005, 74, 689–697. 10.1007/s00128-005-0638-7.
[29] Wheatley B, Barbeau A, Clarkson TW, Lapham LW. Methylmercury poisoning in Canadian Indians – The elusive diagnosis. Can J Neurol Sci. 1979. 6, 417–422, 10.1017/S0317167100023817.
[30] Shephard DAE. Methyl mercury poisoning in Canada. Can Med Assoc J. 1976, 114, 463–472.
[31] ERC. Levels of Mercury in the Blood of Persons Living in Selected Communities in Alberta, British Columbia, the Yukon and Northwest Territories. North Vancouver, BC, Environment Research Consultants Ltd (ERC), 1972.
[32] Smith TG, Armstrong FAJ. Mercury in seals, terrestrial carnivores and principal food items of the Inuit from Holman, NWT. J Fish Res Bd Can. 1975, 32, 795–801.
[33] Smith TG, Armstrong FAJ. Mercury and selenium in ringed and bearded seal tissues from Arctic Canada. Arctic. 1978, 31, 75–84.
[34] AMAP, AMAP Assessment 2011: Mercury in the Arctic. Oslo, Norway. Arctic Monitoring and Assessment Program. 2011.
[35] AMAP. AMAP Assessment Report: Arctic Pollution Issues. Oslo, Norway, Arctic Monitoring and Assessment Programme, 1998.
[36] Khan MAK, Wang F. Mercury-selenium compounds and their toxicological significance: Toward a molecular understanding of the mercury-selenium antagonism. Environ Toxicol Chem. 2009, 28, 1567–1577.
[37] Wagemann R, Trebacz E, Boila G, Lockhart WL. Methylmercury and total mercury in tissues of Arctic marine mammals. Sci Total Environ. 1998, 218, 19–31.
[38] Dietz R, Sonne C, Basu N. et al., What are the toxicological effects of mercury in Arctic biota? Sci Total Environ. 2013, 443, 775–790.
[39] AMAP. AMAP Assessment 2009: Human Health in the Arctic. Oslo, Norway, Arctic Monitoring and Assessment Program, 2009.
[40] Health Canada. Human Health Risk Assessment of Mercury in Fish and Health Benefits of Fish Consumption. 2007.
[41] Van Oostdam J, Donaldson SG, Feeley M. et al., Human health implications of environmental contaminants in Arctic Canada: A review. Sci Total Environ. 2005, 351–352, 165–246.
[42] Holden AV, Marsden K. Organochlorine pesticides in seals and porpoises. Nature. 1967, 216, 1274–1276.

[43] Wania F, Mackay D. A global distribution model for persistent organic chemicals. Sci Total Environ. 1995, 160–161, 211–232.

[44] Macdonald RW, Mackay D, Hickie B. Contaminant amplification in the environment: Revealing the fundamental mechanisms. Environ Sci Technol. 2002, 36, 457A–462A.

[45] Schroeder WH, Anlauf KG, Barrie LA. et al., Arctic springtime depletion of mercury. Nature. 1998, 394, 331–332.

[46] Sunderland EM, Krabbenhoft DP, Moreau JW, Strode SA, Landing WM. Mercury sources, distribution, and bioavailability in the North Pacific Ocean: Insights from data and models. Global Biogeochem Cycles. 2009. 23, GB2010, 10.1029/2008GB003425.

[47] Cossa D, Averty B, Pirrone N. The origin of methylmercury in open Mediterranean waters. Limnol Oceanogr. 2009, 54, 837–844.

[48] Wang K, Munson KM, Beaupré-Laperrière A, Mucci A, Macdonald RW, Wang F. Subsurface seawater methylmercury maximum explains biotic mercury concentrations in the Canadian Arctic. Sci Rep. 2018. 8, 14465, 10.1038/s41598-018-32760-0.

[49] UNEP. Global Mercury Assessment. Geneva, Switzerland, United Nations Environment Programme (UNEP, 2002.

[50] Wang F, Macdonald RW, Stern GA, Outridge PM. When noise becomes the signal: Chemical contamination of aquatic ecosystems under a changing climate. Mar Pollut Bull. 2010, 60, 1633–1635.

[51] Wang F, Outridge PM, Feng X-B, Meng B, Heimbürger-Boavida L-E, Mason RP. How closely do mercury trends in fish and other aquatic wildlife track those in the atmosphere? – Implications for evaluating the effectiveness of the Minamata Convention. Sci Total Environ. 2019, 674, 58–70.

[52] Stern GA, Macdonald RW, Outridge PM. et al., How does climate change influence Arctic mercury?. Sci Total Environ. 2012, 414, 22–42.

[53] Schuster PF, Schaefer KM, Aiken GR. et al., Permafrost stores a globally significant amount of mercury. Geophys Res Lett. 2018, 45, 10.1002/2017GL075571.

[54] Gilbert D. Buried by bad decisions. Nature. 2011, 474, 275–277.

[55] Lim XZ. Scientists call for IPCC-like group on chemical pollution. Chem Eng News. 2021, 99, 18.

Jürgen Gailer and Raymond J. Turner

Chapter 12
Suggested problems, case studies, assignments

12.1 Learner goals and evaluation

Environmental toxicants are ubiquitous and pervasive in aquatic environments and have the potential to impact biota at multiple stages of biological organization. As environmental stewards, we have the responsibility to monitor and mitigate risks of environmental contamination to help sustain healthy ecosystems. Effective risk assessment requires an understanding of what forms and concentrations of toxicants occur in the environment and how they interact with aquatic life to manifest toxicity. The goals of this textbook are to expose the new learner to environmental toxicology defined as the biochemical consequences of their toxicity. Below follows a listing of problems that are aimed to have the learners explore a specific toxin/ toxin class or a specific toxicological problem.

12.2 Group problems

Here we suggest topics/problems that groups of students can explore as a term assignment for oral and written presentations. The overall emphasis is for the learner to explore the sources of pollutants in the environment all the way through to the mechanism of their biochemical toxicity in relevant receptor organisms. Additionally, learners should aim to understand what led to the toxin being present in the environment in the first place.

12.2.1 Nanomaterial-enhanced product manufacturing

The use of nanomaterials (NMs) in the production of computer/TV screens can result in toxic effects to workers during the manufacturing process. In 2018, the management of Samsung Electronics apologized to workers and provided compensation for serious illnesses, including a number of cancers, miscarriages, and children with congenital

Jürgen Gailer, Department of Chemistry, 2500 University Drive NW, University of Calgary, Calgary, AB, Canada, e-mail: jgailer@ucalgary.ca
Raymond J. Turner, Department of Biological Sciences, University of Calgary, Calgary, Alberta, Canada, e-mail: turnerr@ucalgary.ca

https://doi.org/10.1515/9783110626285-012

diseases. These work-related health problems were acquired by young workers (20s and early 30s) building liquid crystal displays and semiconductors during the 1990s and 2000s. Briefly discuss all relevant processes that contribute to the exposure of workers to NMs. Your discussion should include which exposure pathways need to be considered as well as the exposure dose per day. In addition, briefly describe potential biochemical processes that contribute to the mechanisms underlying chronic toxicity in workers.

12.2.2 The use of nano silver in athletic apparel

Use the internet to identify which kinds of textiles contain Ag nanomaterials. Odor on clothing and body odor comes from bacteria metabolizing compounds in our sweat that leads to volatilization of noxious and putrid smelling molecules. Killing the microbes by nano silver formulations prevents this and thus helps clothing remain "fresh" longer as well as not disgusting your workout partner and preventing one from bringing one's dirty athletic gear into the house. One can visit the EPA site for a report on "NANOMA-TERIAL CASE STUDY WORKSHOP: DEVELOPING A COMPREHENSIVE ENVIRONMEN-TAL ASSESSMENT RESEARCH STRATEGY FOR NANOSCALE SILVER – 2011". Now consider the use of Ag NMs in wound bandages to control infectious bacteria. Consider the constant exposure from the textiles and what this means in terms of the efficacy in infection control. Are there any policies regulating the use of nano silver?

12.2.3 Do quantum dots present an environmental toxicity risk?

Even though Cd-based quantum dots are frequently integrated into electronic displays, there is still conflicting information as to how toxic QDs are to people that are involved in the manufacturing process. The stability of QDs is similar to other nanomaterials. To comprehensively assess their toxicity, one therefore needs to consider the contribution of the coating as well as the contribution of the crystalline metal core. QDs may be capped by an organic polymer or inorganic shells, such as ZnS. Environmental factors such as pH and oxygen levels can decrease the stability of the materials. Upon release of the QDs coatings the particles may aggregate which can lead to a decreased bioavailability. There are suggestions in the literature that dissolved metal ions are less toxic than pristine QDs, but less than aggregated QDs. Evaluate the magnitude of the potential exposure risk from computer/TV displays during their use and after their disposal into a landfill. Consider the product life cycle, the manufacturing their use, disposal, and recycling for exposure and waste and subsequent toxicity. (Lead paper: Bechu A, Liao J, Huang C, Ahn C, McKeague M, Ghoshal S, Moores A. Cadmium-Containing Quantum Dots Used in

Electronic Displays: Implications for Toxicity and Environmental Transformations ACS Appl. Nano Mater. 2021, 4, 8417–8428).

12.2.4 Zinc oxide nanoparticles

Zinc oxide nanoparticles (ZnO NP) are likely the most prevalently used in a variety of industries electronics as well as cosmetic preservatives. Because of their wide use, they are becoming more prevalent as a pollutant in a variety of aquatic and marine environments. Describe the antimicrobial mechanisms of toxicity of Zinc ions vs ZnO NPs to microbes. Is there any resistance to this metal? What is the toxicity potential of ZnO NPs to humans? Finally, provide an argument for their continued use or their elimination from use in products (i.e., if you were the EPA or FDA would you approve ZnO NPs for use).

12.2.5 Microplastics

Microplastics are a global environmental problem as they have been detected in tissues of fish and higher organisms. They have also been suspected to be a reservoir for pollutant sorption, loading up with both hydrocarbon and heavy metals. These plastic particles also turn out to be good supports for marine microbes to attach and form a biofilm on periphyton to effectively become mini ecosystems. Additionally, these microplastic microbiomes may also be a reservoir for antimicrobial resistance, as they share resistance genes between them easily in the biofilm. Write an opinion review where you A: Provide an overview of important issues that are associated with the exposure of various organisms to microplastics. B: Identify the variety of toxins that are frequently associated with microplastic particles. C: Choose 2 particular toxins that are associated with microplastics and describe conceptual processes by which each of these contribute to adverse effects on a microbial community (i.e., the biofilm associated with the microplastics). Consider if the microbes could potentially biotransform the toxins or even contribute to the degradation of the microplastic particles themselves. Finally, how would you use this information toward policies in use of single-use plastics.

12.2.6 Perfluorinated compounds

Non polymer per- and polyfluoroalkyl substances (**PFAS**) may be the most persistent industrial chemicals contaminants that man has introduced into in the environment that has prompted some people to refer to them as "forever molecules". Their ubiquitous presence in many environmental compartments has contributed to them receiving increased attention from regulatory bodies over the past five years and has prompted them to be banned from use in some regions of the world. Describe the issues that surround PFAS pollution. What use significantly contributed to their ubiquity in environmental compartments and how can they exert toxic effects in organisms? How would they affect marine and aquatic microbes? Could bacteria be exploited for degrading these persistent environmental pollutants?

12.2.7 Heavy oil mining operations (Canadian – Alberta – oil sands)

12.2.7.1 Toxins of the OSPW to microbes

The activities that are associated with the mining of Alberta oil sands has resulted in considerable volumes of tailings waste which comprises a complex "soup" that is referred to as "oil sands processed water" (OSPW). The toxin class that has received the most attention are the naphthenic acids. However, the tailings ponds contain other toxins, both organic and inorganic. Briefly describe the types of compounds that are present and of toxicity concern in the OSPW. You will find that hydrocarbon naphthenic acids show up as a major group of chemicals of concern. But non-naphthenic acid toxins are also present. Choose two different OSPW toxins (one inorganic and one organic). Describe how each of these may be toxic to microbes. Consider that the tailings ponds have a robust microbial community growing on the trace hydrocarbons released. How are these bacteria able to survive the two toxins you are interested in (i.e., how could they have evolved to deal with this harsh environment)? Also, finally in 2020, the companies have admitted these ponds are leaking into the environment. What if any effect would your two toxins have upon release to the surrounding rivers and streams to the bacteria in the periphyton? Provide information about the concentrations of toxins that are released into the environment and how they may be toxic to higher life forms (i.e., consider the food chain). Given that the government established a zero-release policy, why did it take so long to admit to the release? Why are the companies still operating?

12.2.7.2 Bird deaths and regulations

The activities that are associated with the mining of Alberta oil sands have resulted in considerable volumes of tailings waste that are collected in tailings ponds. These tailings ponds (lakes) are now >200 km² and are composed of a mixture of sand, clay water, residual bitumen, naphthenic acids, poly aromatic hydrocarbons, and heavy metals. These ponds lie in waterfowl migratory paths as well as in nesting grounds; thus, they are a natural attracted place to set down. Although there are some controls in place, several mass bird killings have been reported over the years: 1,600 birds in 2009; 230 ducks in 2010, 122 birds in 2014; 123 birds in 2017. The worst case was 30 blue herons, an endangered species, were killed in 2015. This led to a fine of CDN$ 2.57 M in 2019 to one of the operators. Although this sounds like a large fine, it amounts to only about 2 h of profit loss from a year's operations. Evaluate the pond waters' contents for the toxins most likely responsible for the acute death of the birds. What biochemical mechanisms are involved? Evaluate the political-economic-social environment and why would the status quo operations be allowed to continue. Consider the arguments for and against continuing operations and the consequences.

12.2.8 Methylmercury

Methylmercury is one of the most neurotoxic agents that are naturally present in the environment. Discuss four important aspects that pertain to this compound. First, provide some background information about its physicochemical properties and highlight those which make it particularly dangerous. Provide insight into the mechanisms as to how this compound is formed, where it is degraded and what concentration levels are found in surface waters, sediments, and organisms (**part 1**). Which human populations are predominantly exposed to methylmercury and which human population is most susceptible to its adverse health effects? What is known about the adverse health effects that this compound exerts in humans including pregnant women? How big of a problem is the human exposure to this chemical on a worldwide basis? What regulations are in place to protect humans from the chronic toxicity of methylmercury (e.g., consider that this mercury compound is contained in tuna)? (**part 2**). Discuss what is known about the biomolecular mechanisms by which this toxin exerts toxicity in organs, such as the brain, the target organ. What is known about the mammalian metabolism of this compound (hint: bloodstream vs what is going on in organs) and about its organ distribution in mammalian model organisms (e.g., rats), humans and non-mammalian model organisms (e.g., zebrafish)? Identify some urgent questions that need to be addressed to better understand its metabolism and propose which techniques could be used to gain insight. (**part 3**)

12.2.9 Mercury-containing antibacterial

Thimerosal a mercury-containing compound has been used as an bactericidal agent in vaccines for more than 50 years and it is still used in the flu vaccine as well as other vaccines today. After describing its structure, summarize what is known about the stability of this compound at near physiological conditions (i.e., after its injection into the bloodstream). Where does this mercury-containing compound go after it is injected into the bloodstream? Briefly summarize results from studies in which mammalian organisms were used to gain insight (**part 1**). There is a controversial debate as to whether thimerosal is involved in the etiology of autism in children. Provide a succinct summary of the most important issues pertaining to this compound. Make sure to separately discuss all arguments that support either side of the debate using proper references. What advice would you give the prime minister if he consulted you to help him decide what Canada should do to deal with this widely used bactericidal agent? (**part 2**) Imagine that you are the director of a research lab and you were given $ 1,000,000 by the vaccine industry to conduct critical investigations as to find out if thimerosal is actually dangerous to humans, including children. If you plan to conduct animal experiments, you need to provide a clear rationale to justify the animal model that you propose to use. Clearly describe what experiments you would conduct and what you would measure with which instruments to gain much needed insight? (**part 3**).

12.2.10 Cadmium

The diet is the predominant source of human exposure to the highly toxic metal cadmium. Critically elaborate on what regulations are in place/not in place to protect humans from chronic cadmium exposure in North America and the European Union (discuss food regulations as well as regulations pertaining to consumer products). The detrimental effects of cadmium exposure in humans were brought to the attention of the world during the disaster that unfolded along the Jintsu river, Japan. Describe how this pollution epidemic unfolded and what daily doses people were exposed to. Which sub-population displayed the first symptoms and in what way were people adversely affected as cadmium exposure progressed? Belgium is another country in which humans have been exposed to cadmium. What is the source of cadmium pollution in Belgium? Are different age groups/ethnicities/men/ women at higher risk of chronic cadmium exposure? On a worldwide basis which food items are known to have comparatively high cadmium concentrations (tabulate the total Cd concentration of different food staples). Is the exposure of humans to cadmium going to decrease, remain constant, or increase in the near future? Summarize what is known about the mechanisms that are involved in ADME of cadmium (i.e., absorption, distribution, metabolism, excretion) in humans. What is the half-time of cadmium in humans and how is it excreted? What is known about the absorption of cadmium from the GI tract (i.e., how much is absorbed into the bloodstream)? How do other essential elements that are present in the diet affect the GI absorption of cadmium (hint: iron, zinc, selenium)? What are the toxicological target organs of cadmium? What is known about its mechanism of carcinogenicity?

Are chelating agents (oral or intravenous administration) available that can mobilize cadmium from human tissues? If not, why not?

12.2.11 Sodium nitrite

Nitrite is a widely used food additive in meats, but there is evidence that it is associated with adverse health effects in humans. Provide an overview of widely consumed food items that contain this chemical and why (why is it added) and elaborate whether there are alternatives for it (could it be replaced by some other chemical). Briefly describe what is known about its ADME profile (absorption, distribution, metabolism, and excretion). What percentage of an ingested sodium nitrite dose is absorbed from the GI tract into the bloodstream (summarize animal and/or human studies) and what is known about its potentially adverse biochemical reactions with components of the bloodstream (e.g., is it taken into red blood cells and if so what happens therein) and/or organs. How is sodium nitrite and or its metabolites predominantly excreted? Briefly describe what is known about its acute toxicity in mammals (is its LD50 value known?) and its mechanism of toxicity. What other adverse health effects have been described in mammals in the literature (e.g., is it teratogenic, immunotoxin?) and has any mechanistic insight been obtained as to how it does that? Is sodium nitrite a carcinogen, and, if so, how convincing is the information that it is a carcinogen in humans? Have carcinogenicity studies been conducted in animal studies and what information about it mechanism of toxicity have been unveiled? Briefly summarize the epidemiological evidence that sodium nitrite is a carcinogen in humans (consult Toxicol. Appl. Pharmacol. Vol. 199, 118–131, 2004 as a starting point).

12.2.12 Chromium (VI)

Hexavalent Cr can be present in ground and drinking water and featured prominently in the movie *Erin Brokovich* with Julia Roberts. Provide an overview of the prevalence of Cr^{VI} in drinking water in North America (i.e., what concentration ranges have been reported in the literature). Is there a maximum permissible concentration of Cr^{VI} in drinking water, and if so, what is it? While Cr^{VI} is an established human carcinogen, which carcinogen class does it belong to? What other adverse human health effects are associated with the chronic human exposure to Cr^{VI}? Are certain population groups particularly susceptible to Cr^{VI} (children vs adults, women vs men, Caucasian vs Inuit)? Briefly summarize what is known about the biochemistry of Cr^{VI} after it enters a human organism. Is it known where in mammalian cells it accumulates? Depict what we currently know about the mechanism of action by which Cr^{VI} causes cell damage and carcinogenesis in mammals.

12.2.13 Polycyclic aromatic hydrocarbons

Discuss PAHs in the aquatic environment and their effects on wildlife. The first part should be a review on the source and distribution of PAHs in aquatic ecosystems, and the uptake and elimination of PAHs within fish or aquatic macroinvertebrates. Focusing on either fish **or** invertebrates, discuss: (1) direct toxicological effects of PAHs, (2) indirect toxicological effects of PAH metabolites, (3) relative importance of (1) or (2) in acute vs. chronic exposures. The report should cover the molecular mechanisms of action of PAHs in affecting target tissues. You are expected to synthesize information and provide candidate biomarkers for risk assessment of PAHs in the aquatic environment.

12.2.14 Metals in the aquatic environment and their effects on wildlife

Review the sources and distribution of metals in aquatic ecosystems, and the uptake and elimination of metals within fish **or** aquatic macroinvertebrates. You may choose to focus on a single metal or group of metals. Discuss: (1) direct toxicological effects of metals, (2) indirect toxicological effects of metals, and (3) biological or chemical factors that may modify (1) and (2). The report should cover the molecular mechanisms of action of metals in affecting target tissues. You are expected to synthesize information and provide candidate biomarkers for risk assessment of metals in the aquatic environment.

12.2.15 Nanoparticle route of exposure

The vertebrate immune system plays a critical role in the sentinel protection of a host from foreign agents, including anthropogenic contaminants. Evidence that engineered metal nanoparticles (NP) can breach protective external barriers has generated concerns about their development in many industrial and commercial applications. Using evidence from recent literature, please (1)describe the NP physicochemical properties that make them a unique contaminant of concern, (2) identify the most-likely route of exposure in humans and fish and describe the mechanisms by which NPs could penetrate these barriers, and (3) discuss the various immunomodulatory effects NPs can have on immune tissue and cells and identify potential modes of toxicity. Finally, argue whether specific nano-based regulations should be created for metal NPs, or whether current regulations for heavy metals are sufficiently protective.

12.2.16 Are antibiotics pollutants?

In 2015, the UN put forth goals in their "Transforming Our World: The 2030 Agenda for Sustainable Development". Identify the specific goals that you think link to use of antimicrobials. Define how antimicrobials as toxins to the environment play a role in unsustainable world. Discuss the issue of dumping of antibiotics by industry, agriculture, and municipal wastewater sites as a toxic pollutant dump to the environment. Consider if you agree with the statement "antimicrobials are toxins in the environment". Are antibiotics a pollutant? Follow through your discussion with case studies to support your argument if such practices lead to antimicrobial resistance and damage to the environment.

12.2.17 Antibiotic resistance era

Antibiotics are the toxins we exploit for disease control. However, misuse has been linked to antibiotic resistance and multidrug resistance in bacteria (so-called superbugs). Such practices of overprescribing and particularly prevalent use in the agricultural industry have been considered to be of blame. The importance of using antibiotics to control agricultural pathogens not only to animals and fish but also to crops is thus considered critical to maintain world's food supply. As stated recently by Jake Yeston (*Science* editor), "Do you want to die quickly from antibiotic-resistant microbes, or do you want to die slowly because the pests and weeds kill your food supply?" Explore the idea of collateral resistance such as the use of metal-based antimicrobials in agriculture and possible link to antibiotic resistance. Discuss this link and define an argument if use policy should be changed.

12.2.18 Mine tailings question

Refer to the case study of Lake Karachay in Chapter 8. This is an example of radioactive metal release. Explore radiation poisoning on various life forms, and consider the entire food web and the consequences. Consider the issue of both the river and the release into marine environment.

Consider other case studies on gold mining or coal mining presented in this textbook. What are the issues, release distribution, types of toxins, and biochemical toxicological effects? Consider also acid-mine drainage. Why do they exist and what toxins to they carry?

Explore your local region for history of mine tailings failures. Study this site to obtain information described in the case studies. Consider the toxin, the effects, as well as policy issues.

12.2.19 Arsenic

The giant mine outside of Yellowknife in the Northwest Territories of Canada is the largest single source of arsenic pollution in the world. There is enough arsenic trioxide at this former mining site to kill every person on the planet multiple times. Provide a historical overview about what happened at this particular mine site and why arsenic trioxide is still remaining there, how it is stored safely? Summarize what efforts that are underway to ascertain that arsenic is not adversely affecting the groundwater resources in the area.

12.2.20 Air pollution

For the most part air pollution issues are measured in amount and size of particulate matter, ozone, and nitrous oxide compounds. But this does not tell the complete story. Rarely are noxious volatiles considered, some of which are toxic, and others considered a mere annoyance. Chapter 7 also brings to view that the PM is also a vehicle for toxic metal delivery. Chose a region of known air pollution problems (major metropolis, industry, agriculture, mine site, etc.). Note the challenges around monitoring air quality such as, climate, winds, microenvironments/microclimates, local natural and agricultural flora and fauna. Survey the toxins present; do you think all possible toxins are being reported (consider the surrounding industries)? What are the toxicity pathway(s) and mechanisms of the toxins you identified in humans? Are the effects acute or chronic? What are the social political economic drivers leading to this air pollution and whether it is being mitigated or not?

12.2.21 Toxin pollutant sampling, monitoring, toxicology

There are three major issues that are associated with monitoring toxins in environmental and biological systems. (1) Sampling. What does one sample, and how many, how much sample, model systems vs source contaminated life forms? (2) Analytical approach. What method, what level of resolution, separation technologies should be used to remove the natural background of metals and organics? (3) What toxicological parameters should be measured? Consider the recent toxic pollutant release noted in recent news and follow the toxin by considering these issues.

12.2.22 Research monitoring tools

12.2.22.1 X-ray spectroscopy

The Canadian Light Source (CLS) is a 200-million-dollar facility located in Saskatoon/ Saskatchewan in Canada. Among the many different research problems that can be tackled with this unique research tool, there is active research being conducted into the toxicology of metal(loid)s. Write an article for a newspaper (1,000 words maximum) for which you have to come up with an engaging title. In your article you should introduce readers to this facility and convey relevant information about this facility to a lay audience (avoid jargon!). You may want to first clearly state the main toxicological problems that the CLS can help to gain insight into the toxicology of metal(loid)s and provide relevant details about the facility itself (e.g., location, dimensions, number of beamlines, start of operation). Identify one visual that will be accompanying the article that will help the reader to better understand your article. Then briefly elaborate on what information can be obtained by analyzing environmental/biological samples for toxic metal(loid)s at the CLS. Provide an example of current research that is being conducted at this facility into the toxicology of metal (loid)s at a level that lay people can understand which you need to back up with proper references. After reading your article the reader should have a clear understanding about how research that is being conducted at the CLS contributes to societal needs.

12.2.22.2 Isotope tracing

Chapter 6 presents a compelling case study for the use of isotope signatures for monitoring. Think about what role isotope tracing can be used in defining the source of toxin and bioaccumulation pathways. Consider also if there should be a role in litigation and policy formulation by governing bodies and industry.

12.3 Specific questions to consider

Here we give some ideas for more direct questions related to specific biochemical processes of toxins. These can be expanded or limited in a variety of ways depending on the nature of a given course. Listing them here provides the learner a process to focus on key take-home lessons and ideas.

–   In many chapters the concept of transport of the toxin into the cell is discussed. Define the concepts of passive diffusion vs facilitated diffusion. Define different active transport systems as well as endocytosis and pinocytosis.

- Define and discuss the concepts of bioaccumulation, bioconcentration, and biomagnification.
- Define the process of excretion, from cells, organs, and the organism.
- How do bacteria become antibiotic resistant? Can you name cellular, biochemical, or molecular genetic mechanisms that underpin antibiotic resistance?
- Based on the fundamental biological and chemical principles presented in class on how prokaryotic cells withstand and/or succumb to toxic substances, choose a compound from EACH of the categories is listed below. Limit the answer of each category to <350 words. Welcome to use figures/cartoons to help answer your questions.
  a) Source of the toxin and/or reason the toxin exists/used.
  b) How toxic is it to bacteria, if not known give your opinion relative to similar compounds
  c) The expected biological target.
  d) A biochemical mechanism explaining how it is/may be toxic.
  e) A biochemical mechanism explaining how bacterial resistance/tolerance would be mediated against the compound, i.e., how does the bacteria respond to cope with the toxin and what would be a mechanism of resistance

Use the scientific literature to provide evidence for your answers and be sure to provide full citations. Categories
1. Organics: Trinitrotoluene (TNT), trichloroethylene (TCE), etc.
2. Metals: Chromate, silver, cadmium, etc.
3. Antibiotics/disinfectants/antiseptics: triclosan, benzalkonium chloride (BAC), paraquat, etc.

Some question ideas from thinking around why toxins are in the environment and policies.
1. What chemical forms of metals are most bioavailable and why? Explain how natural environmental factors can influence the bioavailability of metals from water and soil?
2. Explain why the uptake of metals and metalloids generally cannot occur via simple diffusion. Compare and contrast the mechanisms for epithelial uptake of essential and non-essential metals in organisms.
3. Describe the role of major proteins in the blood in transporting metals and metalloids from the site of uptake to tissues and organs where they are accumulated or stored.
4. What is the difference between biologically inactive metal (BIM) and biologically active metal (BAM) pools? Explain their importance in understanding the cellular toxicity of metals.

5. Explain the major mechanisms by which metals and metalloids cause adverse health effects in organisms at exposure levels that commonly occur in the environment.
6. What are four major steps for environmental risk assessment of metals? Explains how risks of metal exposure and toxicity in the environment are characterized? Compare and contrast the general approach for environmental risk assessment of metals between North America (Canada and USA) and EU?

Index

https://doi.org/10.1515/9783110626285-013

www.ingramcontent.com/pod-product-compliance
Lightning Source LLC
Chambersburg PA
CBHW080909220326
41598CB00034B/5522